PHOTODEGRADATION OF

WATER POLLUTANTS

Martin M. Halmann

Department of Environmental Sciences
and Energy Research
Weizmann Institute of Science
Rehovot, Israel

CRC Press
Boca Raton   New York   London   Tokyo

Library of Congress Cataloging-in-Publication Data

Halmann, Martin M.
   Photodegradation of water pollutants / Martin M. Halmann.
    p. cm.
   Includes bibliographical references and index.
   ISBN 0-8493-2459-9 (permanent paper)
   1. Water--Purification--Oxidation. I. Title.
   TD468.H35   1995
   628.1'68--dc20

                                                             95-352
                                                             CIP

No claim to original U.S. Government works
International Standard Book Number 0-8493-2459-9
Library of Congress Card Number 95-352
Printed in the United States of America 1 2 3 4 5 6 7 8 9 0
Printed on acid-free paper

Preface

Conventional biodegradation by activated sludge is very effective in the treatment of most domestic sewage waters. However, industrial effluents contain many obnoxious or toxic compounds and heavy metal ions that are resistant to bacterial biodegradation. Also, modern intensive agriculture uses large amounts of pesticides, herbicides and other agrochemicals, which accumulate in surface and groundwaters. Many of these compounds are recalcitrant to biotreatment. Such xenobiotic substances may, however, be treated by the Advance Oxidation Processes (AOPs), which include irradiation with UV or visible light, including sunlight, without or with additional oxidants such as oxygen, ozone, hydrogen peroxide and other peroxo compounds, and with homogeneous sensitizers such as dyes, or heterogeneous photocatalysts such as dispersed semiconductors.

The treatment options available are considered for the main groups of water pollutants: toxic inorganic ions (e.g., cyanides, heavy metals), hydrocarbon derivatives (e.g., oil spills, surfactants, pulp and paper wastes), halocarbons, organo-N, organo-P, and organo-S compounds. The naturally occurring 'self cleaning' of some pollutants in sunlit surface waters is discussed, as well as the economics of treatment of polluted groundwater and of industrial effluents. Several alternative non-photochemical approaches to water treatment are briefly evaluated.

This book should serve the needs of researchers and students interested in starting new projects in environmental chemistry. It will be useful as an auxiliary text in advanced undergraduate or introductory graduate courses in environmental science and engineering.

The Author

Professor Martin Mordehai Halmann, Ph.D., has been Professor Emeritus at the Weizmann Institute of Science, Rehovot, Israel, since 1990. He has been a member of the scientific staff there since 1949. He studied Chemistry at Hebrew University, Jerusalem, Israel, receiving his M.Sc. degree in 1949 and his Ph.D. in 1952.

Professor Halmann's main research interests (in chronological order) have been chemical reaction mechanisms, isotope effects on chemical reactions and on Franck-Condon factors, chemical effects of nuclear transformations, photochemistry of organic phosphorus compounds, photoelectrochemical and photocatalytic reduction of carbon dioxide using semiconductors, and photocatalytic oxidation of inorganic and organic compounds in water. His previous book, *Chemical Fixation of Carbon Dioxide*, was published by CRC Press in 1993.

Table of Contents

Chapter I. Introduction 1
  I.   Advanced Oxidation Processes 1
  II. Mechanisms: e_{aq}^{-}, $^1O_2$, ·OH and Peroxy Radicals 3
       A. Spin Trapping of Hydroxyl Radicals 3
       B. Molecular Probes for Reactive Transients 4
  III. Sensitized Excitation 5
  IV. Semiconductor-Promoted Photooxidation 5
       A. Primary Reaction Steps 5
       B. Kinetic Models 8
       C. Role of Oxygen 13
       D. Direct and Indirect Photodegradation 14
       E. Zeta Potentials and Surface Properties 15
       F. Catalyst Preparations 16
       G. Surface Density of OH Groups on TiO_2 Particles 18
       H. Quantum Yields and Turnover Numbers 19
  V.   Experimental Techniques 21
       A. Reactor Design 21
           1. Slurry Reactor 21
           2. TiO_2 as a Stationary Phase 22
           3. TiO_2 on Silica Gel 23
           4. TiO_2 on Cellulose and Polyester Membranes 24
           5. TiO_2 Anchored on Porous Alumina Ceramic Support 24
           6. Nanocrystalline TiO_2 25
           7. TiO_2 on Optical Fibers 25
           8. TiO_2 Aerogels 26
           9. Mass Transfer Limitations 26
           10. Falling Film Reactor 27
           11. Regeneration of Adsorbents 27
       B. Laser and Excimer Light Sources 28
       C. Light–Dark Cycling 29
       D. Concentrated Sunlight 29
       E. Photoelectrochemical Reactions 30
  References 31

Chapter II. Inorganic Ions and Molecules 41
  I.   Water and Hydrogen Peroxide 41
  II. Halide Ions 43
  III. Nitrogen Compounds 44
       A. Nitrate 44
       B. Nitrite 44
       C. Nitrite to Nitrate 45
       D. Nitrate to Nitrite, Hydroxylamine, and Ammonia 46

  E. Nitrite to Ammonia 47
  F. Ammonia to Nitrate 48
  G. Cyanides, Hexacyanoferrate(II) and -(III) Ions 48
  H. Thiocyanate 50
  I. Azide 50
 IV. Phosphorus Oxyanions 50
 V. Sulfur Oxyanions and Sulfide 51
 VI. Heavy Metal Ions 52
  A. Photooxidation of Fe(II) 52
  B. Photoreduction of Metal Salts 53
   1. Chromium 54
   2. Mercury, Lead, Manganese,
    Thallium, Cobalt, and Uranium 55
   3. Copper 56
   4. Mercury, Gold, Platinum, Silver, and Chromium 57
   5. Iron(III) 58
 VII. Organo-Metallic Compounds 58
  A. Mercurochrome 58
  B. Tributyltin Chloride 59
  C. Bis(tributylin) Oxide 59
  D. Phenylmercury Compounds 60
References 60

Chapter III. Hydrocarbon Derivatives 67
 I. Aliphatic Compounds 67
  A. Homogeneous Photolysis 67
   1. Phosphotungstates and Phosphomolybdates
    as Sensitizers 67
   2. Hg Salts as Sensitizers 67
   3. Benzophenones as Sensitizers 68
   4. Formic Acid with H_2O_2 68
   5. Acetic Acid with H_2O_2 69
   6. Formic Acid with Fe(III) 69
   7. 2-Propanol with H_2O_2 70
  B. Heterogeneous Photodegradation 70
   1. Carboxylic Acids 70
   2. Methanol and Ethanol 73
   3. Direct ESR Identification of Radical Intermediates 73
   4. 1-Propanol and Propanal 74
   5. Methyl Vinyl Ketone 75
   6. Polyvinyl Alcohol 75
 II. Aromatic and Other Cyclic Compounds 75
  A. Homogeneous Photolysis 75
   1. Toluene and Phenol — Direct Photolysis 75

2. Flash Photolysis of Phenol 75
3. Toluene and Phenol with H_2O_2 75
4. Photo Fenton Reaction 76
5. Dye Sensitization and QSAR Analysis of
   $^1O_2$ Reaction with Phenols 77
6. Nonylphenol 78
7. Phthalic and Maleic Anhydrides 78
8. α-Pinene 79
9. Hydroquinone 80
10. Ascorbic Acid Synthesis 81
B. Heterogeneous Photodegradation 81
1. Ethylbenzene 82
2. Polynuclear Aromatic Hydrocarbons 82
3. Benzoic Acid 83
4. Phenol 83
5. *o*-, *m*-, and *p*-Cresol 86
6. Phenoxyacetic Acid 87
7. Dimethoxybenzenes 88
8. Toluenes 88
9. Benzoquinone 89
10. Furfuryl Alcohol 89
11. Phthalic and Maleic Anhydrides 90
12. 2-Methylisoborneol 90
13. Cyclic Acetals 91
C. Oil Spills 91
III. Long-Chain Compounds 92
A. Surfactants 92
B. Lignin Sulfonates and Kraft Wastewater 95
C. Phthalate Esters 97
D. Polymers 98
IV. Photoinduced Nitrosation and Nitration 99
A. Dimethylamine 99
B. Aromatic Amino Acids 99
C. Phenol 100
D. 2-Phenylphenol 100
E. 2-Naphthol 101
F. Biphenyl 102
G. Pyrene 103
References 103

Chapter IV. Halocarbons 113
I. Aliphatic Halocarbons 114
A. Chlorocarbon Photodegradation — Relative Rates 114
B. Mixtures of Chlorocarbon Compounds 115

 C. Halomethanes 116
 D. Chlorinated Ethanes 117
 E. Chlorinated Ethenes (Ethylenes) 120
 F. Chloroacetic Acids 122
 G. Chloroalkyl Ethers 123
  1. 2-Chloroethyl Ether 123
  2. 1,2-Bis(2-chloroethoxy) Ethane 124
 H. Reductive Mechanism 125
 I. Halogenated Peroxy Radicals 126
 J. Bromocarbons 127
  1. 1,1- and 1,2 Dibromoethane 127
  2. 1,2-Dibromopropane 127
  3. 1,2-Dibromo-3-chloropropane (DBCP) 127
 K. Groundwater Remediation 128
 L. Fluorocarbons 129
  1. Fluoroalkenes 130
 M. Vacuum-UV Photolysis of Halocarbons 130
II. Aromatic and Other Cyclic Compounds 131
 A. Haloaromatics 131
  1. Reaction Rates with ·OH Radicals 131
  2. Semiconductor Photocatalysis 132
  3. Chlorobenzenes 132
  4. Polychlorobenzenes in Surfactant Micelle Solutions 135
  5. Photoinduced Electron Transfer by Organic Anions 136
  6. Chlorobenzoic Acids 136
  7. Haloaromatic Ethers 137
 B. Halogenophenols 138
  1. Homogeneous Photolysis 138
   a. UV/Ozone on 4-Chlorophenol 140
   b. Reaction Rates with ·OH Radicals 141
   c. Effect of H_2O_2 141
   d. Sensitization by Hydroquinone 141
   e. Dye-Sensitized Photooxidation of Chlorophenols 142
   f. Vacuum-UV Photolysis of 4-Chorophenol 143
   g. Fenton's Reagent 144
   h. Chlorophenoxy Herbicides 146
  2. Heterogeneous Photodegradation 147
   a. 1,2,4-Trichlorobenzene 147
   b. Chlorophenols 148
   c. 4-Chlorophenol 149
   d. Fluorophenols 152
   e. Hammett Correlation in p-Substituted Phenols 152
   f. 2,4-Dichlorophenoxyacetic Acid (2,4-D) 152
   g. Effects of Peroxydisulfate and Periodate 153

        h.   Di- and Tri-Chlorophenols 153
        i.   CdS as Photocatalyst 154
        j.   Pentachlorophenol... 155
        k.   Rose Bengal 156
   3.   TiO$_2$ on Photoelectrodes 157
  C. Polychlorinated Dioxins, Dibenzofurans, and Biphenyls 158
     1.   Homogenous Photolysis 159
     2.   Heterogeneous Photodegradation 164
  D. Halocarbon Pesticides 164
     1.   Permethrin and DDT 165
     2.   Chlordane 166
     3.   Mirex 166
References 167

Chapter V. Organic Nitrogen Compounds 181
  I. *s*-Triazines 182
    A. Homogeneous Photolysis 182
      1.   Propazine 183
      2.   Atrazine 184
      3.   Metazachlor 185
    B. Heterogeneous Photodegradation 185
      1.   Vacuum-UV Photolysis of Atrazine............................... 187
      2.   Cyanuric Acid 187
      3.   Solar Photodegradation of Triazine Herbicides
          in Ground Water 188
      4.   Bentazone 188
  II. Amines, Amides, and Carbamates 189
    A. Anilines 189
      1.   *p*-Phenylenediamines 189
    B. Aminopolycarboxylates 190
    C. Benzamide 192
    D. Isoxaben 193
    E. Carbetamide 193
    F. Aldicarb, Carbaryl, Carbofuran, and Baygon 194
    G. Urea and Uracil Derivatives 196
      1.   Urea 196
      2.   Monuron 196
      3.   Bromacil and Terbacil 197
      4.   Isoproturon 198
    H. Propachlor, Alachlor, and Pendimethalin 198
      1.   Propachlor 198
      2.   Alachlor and Pendimethalin 199
    I. Dequalinium Chloride 200
    J. Propyzamide 200

     K. Metalaxyl 201

     L. *p*-Aminophenol 201

     M. Aliphatic Diamines 202

     N. Rhodamine 6G 202

  III. Nitrobenzene and Nitrophenols 202

     A. Homogeneous Photolysis 202

        1. Nitrobenzene and Nitrotoluenes 202

        2. Nitrobenzyl Derivatives 203

        3. Nitrophenols 203

        4. Dye-Sensitized Photodegradation 205

     B. Heterogeneous Photodegradation 206

        1. Nitrobenzene 206

        2. Trinitrotoulene 207

        3. Nitrophenols 207

        4. 2,6-Dichloroindophenol 208

     IV. Bromoxynil and Chloroxynil 208

   V. Thymine 210

  VI. Triclopyr 211

 VII. Fenarimol 212

VIII. Flavins 213

  IX. Catecholamines 213

   X. Dyes 213

     A. Azo Dyes 213

     B. Tannery Dyes 216

     C. Dyes in Municipal Wastewater 216

     D. Methyl Violet 217

     E. Methyl Viologen 217

  XI. Polycyclic Aromatic Nitrogen Heterocycles 218

References 218

Chapter VI. Organic Phosphorus Compounds 227

   I. Homogeneous Photolysis 227

     A. Alkyl Phosphates 227

        1. Ethyl Dihydrogen Phosphate 227

        2. Trimethyl Phosphate 228

     B. Phosphonofluoridates 228

     C. Dimethyl Vinyl Phosphate and Dichlorvos 228

     D. Glycerophosphates 229

     E. Sugar Phosphates 229

     F. Vamidothion 230

     G. Ethylenediamine Tetra (methylenephosphonic Acid) 230

     H. Sodium Dodecyl Bis(oxyethylene) Phosphate 231

     I. Aromatic Phosphates 231

        1. Without Sensitizers 231

        a.   Parathion, Edifenfos, and Fenitrothion 231
        b.   Tolclofos-Methyl 233
        c.   Pyridoxal 5′-Phosphate 233
        d.   Chlorpyrifos and Fenamiphos 234
    J.   Dye-Sensitized Photooxidation 234
    K. Effect of Iron(III) Salts and H_2O_2 234
  II. Heterogeneous Photodegradation 235
    A. Trialkyl Phosphates 236
    B. Dichlorvos, Trichlorfon, and Tetrachlorvinphos 236
    C. O_3 and TiO_2 237
References 239

Chapter VII. Organic Sulfur Compounds 243
   I. Sulfonic Acids 243
    A. Benzenesulfonic Acid 243
    B. *p*-Toluenesulfonic Acid 243
    C. Metanilic Acid 244
    D. Anthraquinone Sulfonic Acid 244
    E. Thiolcarbamates 245
    F. Asulam 245
    G. Methylene Blue 246
    H. Thioacetamide 247
    I. Sethoxydim and Clethodim 247
    J. Benzo[b]thiophene and Dibenzothiophene 248
    K. 4-Thiomethyl-*N*-methylphenyl-carbamate 249
    L. Endosulfan 250
    M. Fenitrothion 250
References 250

Chapter VIII. Natural and Waste Waters 253
   I. Natural Transformations in Freshwater and Oceans 253
    A. Humic Substances and Singlet Oxygen 254
    B. Fulvic Acid Sensitization of Semiconductors 254
    C. Singlet Oxygen Generation by Photosensitized Soil 255
    D. Hydrated Electrons 255
    E   Superoxide Ion-Radicals (O_2^-) 255
    F. Nitrate-Induced Photodegradation 256
    G. Fe(III)/Fe(II) Reactions 257
    H. Hematite-Oxalate Photolysis 257
    I. Fog, Cloud, and Rain Waters 260
    J. Surface Waters 261
    K. Cr(VI) and Cr(III) 263
    L. Dissolved Gaseous Mercury 263
    M. Eutrophic Waters 263

N. Seawater 264
O. Anthraquinone Photosensitization 264
P. Photodegradation in Marine Surface Microlayers 265
II. Treatment of Polluted Groundwater 267
III. Treatment of Wastewater 268
A. Treatment of Landfill Leachates 268
IV. Photodynamic Sterilization 269
A. Dye Sensitization 269
B. Bactericidal Activity of TiO_2 269
V. Concentrated Sunlight 271
References 275

Chapter IX. Evaluation and Future Trends 283
I. Photodegradation Compared with Other Methods 283
A. Ozonation 283
B. Biodegradation 283
C. Radiolysis 284
D. Ultrasonics 284
E. Corona Discharge 285
F. Electrochemical Oxidation 285
G. Oxidation in Supercritical Water 286
II. Cost Estimates, Energetics, and Conclusions 286
A. Cost Estimates 286
B. Energy Requirement Evaluation 288
C. Future Trends 288
References 289

Index 293

*To my wife, Mirjam and my children,
Michal and Nahi, With love.*

Chapter 1

Introduction

Pollution is a scourge accompanying many human activities. The efforts required to limit and even reverse its obnoxious effects must be commensurate to its magnitude.

Water pollution is due both to microbial infection and to the presence of organic and inorganic waste materials, including harmful chemicals. A major breakthrough in medical disinfection was the introduction of chlorine as chlorinated lime by Holmes in Boston in 1835, and by Semmelweis in Vienna in 1847.[1] Chlorine, added in concentrations of 1 to 3 ppm, is now widely used to disinfect drinking water supplies and swimming pools. It has been a major factor in preventing the transmission of water-borne disease. However, chlorination is ineffective against most chemical pollutants, which include toxic heavy metals, anions such as cyanide, organic solvents such as hydrocarbons and chlorocarbons, agrochemicals such as pesticides and herbicides, and decomposition products such as the chlorinated dioxins. Also, chlorination of freshwater may cause the formation of trihalomethane compounds (THM), such as the carcinogenic chloroform, $CHCl_3$, and the lachrymatory and toxic bromoform, $CHBr_3$.

I. ADVANCED OXIDATION PROCESSES

Advanced oxidation processes (AOPs) for wastewater treatment include reactions with H_2O_2, with or without ultraviolet (UV) irradiation, ozonation, and O_3/UV treatment. While ozone has a high oxidation potential ($E° = +2.07$ V), ozone alone reacts very slowly with various compounds, such as with chlorinated alkanes, with chlorinated herbicides, such as 2,4,5-T (2,4,5-trichlorophenoxyacetic acid), and also with the triazine herbicides, such as atrazine. Improved treatment methods involve the combination of ozone with UV light, generating the ·OH radicals, which are even more reactive oxidants ($E° = +3.06$ V). However, in the case of wastewater from the pulp and paper industry, the combination O_3/UV was less effective than O_3 alone, and improved results were obtained by integration of ozonation and biotreatment. For removal of chloroalkanes and trichloroethylene, O_3/H_2O_2 was more effective than O_3 alone. A major problem preventing the widespread use of ozone is the high cost of treatment.[2] In contrast to the above AOPs in homogeneous solutions, in which the oxidants are consumed, heterogenous photocatalytic oxidations use near-UV and visible light as the energy source to overcome the required activation energy. In this case, water or dissolved oxygen usually provides the necessary oxidant. The photocatalysts, such as TiO_2, may be reused, or in a fixed-bed operation may be used for extended periods. Also, sunlight can be an inexpensive light source, particularly for well-insolated and arid regions.[3] Modern efficient UV/visible light sources, such as the novel excimer lamps (see Section V) may enable economic application of photodetoxification also in poorly insolated regions.

Photocatalytic reactions are by definition exoergic processes, in which the absorption of light energy results in acceleration of reactions. Many compounds and materials of our environment, such as all organic molecules and most components of living systems are unstable toward oxidation — but their oxidative degradation is delayed and hindered by steric constraints and by high activation energies. Photochemical processes are of major importance in the chemistry of gaseous molecules in the upper atmosphere, regulating the interaction between oxygen, ozone, nitrogen oxides, and nitrogen halides. In natural aquatic systems, in the earth's oceans, rivers, and lakes, photochemical reactions are often masked by biological and photobiological processes. However, recent evidence indicated that abiological photoassisted reactions may be important also in natural waters. Photocatalytic reactions in aqueous solutions that involve oxidation of solutes may be enhanced by the presence of molecular oxygen. Such photochemical reactions play an important role in the decontamination of organic pollutants in natural waters. An important advantage of photocatalytic processes is that they may be performed at low or ambient temperatures, and usually complete mineralization of the organic compounds may be achieved.[4] Basic concepts in photocatalysis and strategies for photochemical treatment of wastewater have been discussed in several reviews.[5–17]

In the present monograph, the primary photophysical process of photon absorption is only briefly reviewed. The main emphasis is on the secondary reactions, leading via chemical intermediates to stable products. Most of the compounds described are either toxic, carcinogenic, or otherwise harmful. However, since the requirements of treatment to drinking water quality impose essentially "complete" removal of all organic contaminants, including that of photolysis intermediates that by themselves are not toxic in small concentrations, such as formic and acetic acids, the photodestruction of such compounds is also described. Chapter I recounts the basic mechanisms of the various photoinduced reactions, as well as the main experimental techniques. The discussion in Chapters II through VII is organized by compound groups, illustrating for each group direct photolysis (usually by UV light), photosensitized degradation (usually with dye sensitizers), and heterogeneous photocatalysis (usually with TiO_2 or ZnO). Chapter VIII considers the natural photodegradation occurring in sea and freshwater, and the treatment of polluted groundwater and wastewater. The concluding Chapter IX evaluates the potential scope of photochemical methods in comparison with other technologies.

II. MECHANISMS: e_{aq}^-, $^1O_2$, ·OH, AND PEROXY RADICALS

Molecular oxygen is a poor oxidant because of the very low rates of reaction of ground-state triplet oxygen, $^3O_2$, with the singlet ground state of most molecules.[16]

Photoassisted oxidation in aqueous media usually involves several reactive transient species. The properties of these intermediates were revealed mainly by flash photolysis techniques, with detection by their optical absorption spectra, electron spin resonance, and sometimes by their electric conductivity. Further confirmation of the properties of these species was obtained by their creation and interactions following pulse radiolysis, using ionizing radiations.

The main reaction intermediates in photooxidation (and also in thermal, radiation-induced and electrochemical oxidation) are the hydrated electron (e_{aq}^-), singlet oxygen ($^1O_2$), the hydroxyl radical (·OH), and the superoxide radical anion (O_2^- or its conjugate acid HO_2). Rate constants for reactions of these and other inorganic radicals in aqueous solutions have been tabulated.[18] Of particular importance for the photodegradation of water pollutants is the highly reactive hydroxyl radical. One method of generating this radical is by UV illumination of aqueous solutions of hydrogen peroxide,

$$H_2O_2 + h\nu \rightarrow 2 \cdot OH \tag{1}$$

A. Spin Trapping of Hydroxyl Radicals

Due to its great reactivity, the concentration of the hydroxyl radical in illuminated solutions is usually very small. It can, however, be scavenged by a "spin trap," such as 5,5′-dimethyl-1-pyrroline *N*-oxide (DMPO), forming a stable radical, which can be identified and measured by electron spin resonance (EPR).[19,20]

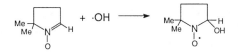

SCHEME I.1

Photolysis of hydrogen peroxide by ultraviolet light, the UV-H_2O_2 process, is useful for the treatment of low concentrations of organic contaminants in groundwater. An EPR technique with the spin trap DMPO was developed to measure the absolute concentration of the spin trap radical adduct DMPO-·OH formed by photolysis of H_2O_2, and to derive the initial rate of production of this adduct. The initial rate was found to be proportional to the light intensity and to the initial concentration of H_2O_2, but to be independent of the initial concentration of DMPO.[21]

The ESR spin-trapping technique, with DMPO as the spin trap, was applied to the determination of the quantum efficiency of hydroxyl radical production in aqueous semiconductor dispersions, without and with added hydrogen peroxide. The ESR spectrum revealed the DMPO spin adducts of both the hydroxyl and the hydroperoxyl (or superoxide ion) radicals. In water alone, the quantum efficiencies for $\cdot OH$ production under irradiation at 365 nm, with suspended anatase, rutile, WO_3, and ZnO, were 4.6, 3.2, 0.5, and 2.4%, respectively. In the presence of 1 mM H_2O_2, the quantum efficiencies of $\cdot OH$ production were considerably enhanced, reaching for anatase, rutile, and WO_3 the values of 14.8, 13.1, and 2.0%, respectively. In aqueous suspensions of TiO_2 anatase containing ethanol (20%), the DMPO adducts of C-centered radicals were observed, assigned as the $\cdot CH(CH_3)$ OH radicals, formed by reaction of hydroxyl radicals (or holes, h+) with the ethanol molecules.[22]

In environmental waters, the hydroxyl radicals may be scavenged by carbonate, bicarbonate, hydrogen phosphate, and phosphate ions. The rates of these reactions were determined by measuring the competition with the known rate of scavenging of the hydroxyl radical with the spin trap DMPO. Second-order rate constants thus derived for the reaction of $\cdot OH$ radicals with HCO_3^-, CO_3^{2-}, HPO_4^{2-}, and PO_4^{3-} were 5.7×10^6, 2.8×10^8, 5.9×10^5, and 7×10^6 M^{-1} s^{-1}, respectively.[23]

Hydroxyl radicals are produced also by the UV photolysis of ozone. The O_3/UV treatment of waste waters is one of the important advanced oxidation processes (AOPs). Hydroxyl radicals are formed directly by radiolysis of water, such as with β- and γ-rays from nuclear transformations, or with X-rays. The drawback of such radiolytic methods is in the elaborate and costly safety measures required.[16]

B. Molecular Probes for Reactive Transients

Light-induced reactions in solution involve very rapid formation of primary transient free radicals by photophysical processes, in the timescale of 10^{-12} to 10^{-8} s, followed by slower secondary reactions, mainly with oxygen molecules, at a timescale of about 1 μs. The rates of these secondary reactions, such as the formation of the peroxy ion radical from the hydrated electron, of organic peroxides from carbon-centered radicals, and of Br_2^- from Br^- and $\cdot OH$, are limited by the diffusion-controlled reactions of oxygen. These secondary reactions yield the relatively stable and isolable intermediates and final products of photochemical transformations. As a molecular probe for free radicals, the odd-electron stable nitric oxide (NO) has been applied. It reacts slowly with molecular oxygen in dilute solutions, but at practically diffusion-controlled rate with most radicals, including organic peroxides, carbon-centered radicals, and hydroxyl radicals, and can thus be used to determine the total radical flux in illuminated waters. Its rapid reaction with the superoxide radical-ion was used to measure the flux of this radical-ion by applying labeled ^{15}NO as a probe,

$$^{15}NO + O_2^- \rightarrow O^{15}NOO^- \rightarrow {}^{15}NO_3^- \tag{2}$$

Dialkyl nitroxides, which also are odd-electron stable radicals, fail to react with molecular oxygen, with the peroxy radical-ion, and with organic peroxides. They do react rapidly with carbon-centered radicals and with ·OH radicals. By comparing the rate of loss of nitric oxide with that of dialkyl nitroxide in illuminated seawater, it was possible to derive the flux of O_2^- relative to the total radical flux in surface seawater.[24]

III. SENSITIZED EXCITATION

In photosensitized reactions electronically excited molecules, which had undergone intersystem crossing (ISC) from a primary very short-lived singlet state to a longer lived triplet state, transfer their energy to other molecules. Such reactions may be either of Type I, in which this triplet-state photosensitizer directly oxidizes the substrate, or they may be of Type II, in which the photosensitizer S transfers energy to molecular oxygen, which thus becomes excited to its singlet state $(^1\Delta_g)$:[25]

$$^1S_0 + h\nu \rightarrow {}^1S_1 \tag{3}$$

$$^1S_1 \rightarrow {}^3S_1 \ (ISC) \tag{4}$$

$$^3S_1 + {}^3O_2 \rightarrow {}^1S_0 + {}^1O_2\left(^1\Delta_g\right) \tag{5}$$

See also Chapter VIII, Section I, on the transformations mediated in natural water by the primary reactive transients, e_{aq}^-, O_2^-, and ·OH radicals.

IV. SEMICONDUCTOR-PROMOTED PHOTOOXIDATION

Titanium dioxide has a band gap of 3.2 eV, equivalent to about 400 nm. Therefore TiO_2 is photoexcited by near-UV illumination, in the 300 to 400 nm range, which is part of the solar spectrum transmitted by the atmosphere. Gerischer[26] developed a model relating the quantum yields of photocatalytic reactions on semiconductor particles with the rate constants, light intensity dependence, and particle sizes. The driving force for the catalysis of photochemical reactions by photoexcited semiconductor particles is the excess free energy in the electron–hole pairs thus produced.

A. Primary Reaction Steps

Excitation of semiconductors with light of wavelengths shorter than the band gap results in formation of electron–hole pairs. Thus, with titanium dioxide,

$$TiO_2 + h\nu \rightarrow e_{cb}^- + h_{vb}^+ \tag{6}$$

Howe and Grätzel[27] showed that with colloidal TiO_2 particles in oxygen-free aqueous solution, the conduction-band electrons (e_{cb}^-) within a few picoseconds become trapped at the particle surface, possibly with the formation of Ti^{3+} centers — which were detected by EPR spectroscopy. These electrons and holes may either recombine, releasing heat, or they may migrate to the surface of the semiconductor, where they can undergo redox reactions with molecules and ions on and near the surface. In colloidal TiO_2 the process of electron–hole recombination was shown by Rothenberger et al.[28] to occur within a few hundred picoseconds of their formation.

The positively charged holes reaching the surface react within a few picoseconds with water or hydroxide ions to produce hydroxyl radicals, which are the primary oxidizing species due to their highly electrophilic character.[6,15]

$$h^+ + H_2O \rightarrow H^+ + \cdot OH \tag{7}$$

$$h^+ + OH^- \rightarrow \cdot OH \tag{8}$$

Hydroxyl radicals adsorbed on the surface of semiconductors like TiO_2 were proposed not to be experimentally distinguishable from surface trapped holes. It was thus suggested to equate photogenerated holes on metal oxide powders as chemically equivalent to surface-bound hydroxyl radicals. Hydroxyl radicals reacted with 13-nm colloidal TiO_2 particles at near diffusion-controlled rate. The $\cdot OH$ trapped on the TiO_2 surface had a broad absorption band with its maximum at 350 nm.[29,30]

The hydroxyl radical is a most reactive oxidizing agent, which reacts readily with surface adsorbed organic molecules, either by electron or hydrogen atom abstraction, forming organic radical cations, or by addition reactions to unsaturated bonds.[31]

In an effort to differentiate between the action of hydroxyl radicals and holes in semiconductor photocatalysis, Goldstein et al.[32] compared the distribution of the hydroxylation products of phenol in aqueous solutions, as obtained by three different methods: (1) radiolysis with γ-rays from a ^{137}Cs source, in which $\cdot OH$ radicals are known to be involved, (2) chemical oxidation with $\cdot SO_4^-$, obtained by reduction of persulfate in deaerated solution,[30]

$$Fe^{2+} + S_2O_8^{2-} \rightarrow Fe^{3+} + SO_4^{2-} + \cdot SO_4^- \tag{9}$$

in which oxidation operates by direct electron transfer, and (3) photolysis in the presence of colloidal TiO_2. In all experiments, Cu^{2+} ions were present as oxidants of the hydroxylation adducts, to freeze the isomer distribution. In the

γ-radiolysis, N_2O was present in addition to Cu^{2+}, so that hydrated electrons were trapped by the two very rapid reactions,

$$e_{cb}^- + N_2O + H_2O \rightarrow \cdot OH + N_2 + OH^- \tag{10}$$

$$Cu^{2+} + e_{cb}^- \rightarrow Cu^+ \tag{11}$$

With initially 10 mM phenol, the main products were hydroquinone and catechol, with only minor yields of resorcinol and benzoquinone. By the chemical oxidation with the sulfate radical-anion, the products were also hydroquinone and catechol, but in much lower yields and in a different ratio than that from the γ-radiolysis. The photolysis of phenol in colloidal solutions of TiO_2 also produced mainly catechol and hydroquinone, and only traces of resorcinol. The general mechanism of aromatic orientation suggests that the electrophilic reagent $\cdot OH$ will be directed in phenol to the *ortho* and *para* positions at equal rates, and since there are two free *ortho* positions in phenol, the predicted isomer ratio for the attack by the hydroxyl radical should be *ortho:para* = 2:1. At low initial concentrations of phenol (1 to 30 mM), the γ-radiolytic and the photochemical hydroxylations were similar, while at high concentrations of phenol (>30 mM), the photochemical system was more similar to that with $\cdot SO_4^-$ as the oxidant. Presumably, at the lower phenol concentrations, the photolytic hydroxylation involved $\cdot OH$ radicals, while at higher phenol concentrations, direct hole oxidation of surface-adsorbed phenol may be the predominant reaction.[32]

In the presence of molecular oxygen, the negatively charged electrons are preferentially trapped by oxygen:

$$O_2 + e^- \rightarrow O_2^- \tag{12}$$

$$O_2 + 2H^+ + 2e^- \rightarrow H_2O_2 \tag{13}$$

$$H_2O_2 + e^- \rightarrow \cdot OH + OH^- \tag{14}$$

Photocatalytic oxidation of organic compounds on TiO_2 in aqueous dispersions is considerably enhanced in the presence of low concentrations of hydrogen peroxide. Often an optimal concentration of hydrogen peroxide was observed, in the range of 10^{-4} to 10^{-2} M H_2O_2, which depended on the temperature of calcination of the TiO_2. The favorable effect of low concentrations of hydrogen peroxide was ascribed to its ability to accept conduction band electrons, thus assisting in the charge separation, while forming the strongly oxidizing hydroxyl radicals, as in Equation 14, in addition to reaction with O_2^- radical anions:[33,34]

$$O_2^- + H_2O_2 \rightarrow \cdot OH + OH^- + O_2 \qquad (15)$$

Excess of hydrogen peroxide results, however, in the formation of peroxo compounds on the surface of the TiO_2 particles, such as of $Ti(O)_2(OH)_2$ and $Ti(OOH)(OH)_3$, which inhibit the photocatalysis by TiO_2.[35,36]

See also Chapter II, Section I, on the formation and decomposition of hydrogen peroxide.

Heterogeneous reactions at catalytic surfaces of particles may be considered as occurring in microelectrolysis circuits, with cathodic reduction and anodic oxidation of adsorbed molecules on the surface of the particles, while electric current is shunted through the interior of the particles.

In the absence of molecular oxygen, other reactions by electrons are possible, such as dissociative electron capture by halocarbons,

$$RX + e^- \rightarrow R\cdot + X^- \qquad (16)$$

resulting in stable chloride or bromide anions and highly reactive alkyl radicals (see Chapter III, Section I).

With decreasing particle size, dispersed semiconductor materials reach the dimensions and properties of colloidal particles. Some of these atomic and molecular clusters may play an important role in the photodegradation of organic compounds in natural aquatic systems.[35,37]

Photochemical and other reactions of particles of hydrous oxides are affected not only by the electric double layer on their surface, but also by specific chemical reactions. Such reactions involve the surface hydroxyl and other functional groups, which may interact coordinatively by ligand exchange with anions and weak acids.[38,39]

B. Kinetic Models

In heterogeneous media, the dependence of the rate of oxidation on the concentration of the solutes often follows the Langmuir–Hinshelwood expression, with the initial rate R being[40–43]

$$R = kK(S)/[1 + K(S)] \qquad (17)$$

where (S) is the initial concentration of the substrate, K is taken to represent the Langmuir adsorption constant, and k is a "reactivity constant," providing a measure of the reactivity of the surface of the catalyst with the substrate. At low substrate concentrations, the rate R is thus proportional to the substrate concentration, with kK being the observed first-order rate constant. At higher substrate concentrations, at which most of the available surface sites are saturated with solutes, R will approach the limiting value of k. This expression is conveniently used as a plot of reciprocal rate vs. reciprocal initial substrate concentration, resulting in the linear dependence,

$$1/R = (1/k) + (1/kK)[1/(S)] \qquad (18)$$

Turchi and Ollis[42] suggested four possible general mechanisms for the photodegradation of organic molecules in illuminated aqueous TiO_2 slurries, assuming that hydroxyl radicals ·OH are the primary oxidants. These reactions may occur:

1. While both the ·OH radical and the organic molecule are adsorbed on the TiO_2 surface.
2. Between an adsorbed organic molecule and a nonbound ·OH radical (free in solution).
3. Between an adsorbed ·OH radical and a free organic molecule hitting the catalyst surface.
4. Between a free ·OH radical and a free organic molecule in solution.

Kinetic expressions were developed for these four types of reactions, which all were consistent with the kinetic form of the preceding Langmuir–Hinshelwood model. A surprising observation was that in a given photoreactor design, the same illumination conditions, and with a given TiO_2 photocatalyst at a given concentration, in oxygenated solutions, the disappearance rates of various very different organic compounds were quite similar. This is in contrast to direct homogeneous photolysis, or to thermochemical reactions, in which reactions rates almost always very strongly depend on the structure of the reactants. In order to explain the insensitivity of the photocatalyzed reaction to the chemical structure, Turchi and Ollis[42] proposed that the rate-limiting step may be the formation of the hydroxyl radical on the TiO_2 surface.

Domestic and industrial wastewater usually contains not only organic contaminants, but also considerable concentrations of common inorganic ions, such as chloride, carbonate, sulfate, nitrate, and phosphate. Photocatalytic oxidation of organic compounds with UV-illuminated aqueous slurries of TiO_2 was found to be appreciably inhibited by the presence of chloride ions. Pruden and Ollis[43] and Ollis[44] observed that for various halocarbon compounds studied, the reciprocal rate of photooxidation was proportional to the chloride or bromide ion concentration, indicating competitive adsorption of the halide ions on the TiO_2 surface. Abdullah et al.[45] found that sulfate and phosphate ions also inhibited the photocatalytic oxidation of organic solutes, and may reduce the rate by up to 70%. Carbonate, nitrate, and perchlorate ions had little effect on the rates of photooxidation. The inhibitory effects of chloride, sulfate, and phosphate could be accounted for by a Langmuir–Hinshelwood type kinetic model, in which inorganic ions compete with organic solutes for oxidizing sites on the TiO_2 surface. Regeneration of photocatalysts with adsorbed sulfate or phosphate was possible by rinsing with 0.1 M NaHCO$_3$. For removal of adsorbed chloride ions, even rinsing with water was sufficient.[45]

In a careful study of the TiO$_2$-sensitized mineralization of 4-chlorophenol, Mills and Morris[46] showed that a variant of the Langmuir–Hinshelwood equation more accurately represented the rate of release of carbon dioxide, R_{CO_2}, as a function of the concentrations of oxygen [O$_2$] and of the 4-chlorophenol [CP]:

$$R_{CO_2} = \gamma K_{O_2}[O_2](I_a)^\theta K_{CP}[CP] \Big/ \left\{ \left(1 + K_{O_2}[O_2]\right)\left(1 + K_{CP}\right)[CP]\right\} \quad (19)$$

in which $K_{O_2} = 0.044 \pm 0.005[O_2]^{-1}$ is the Langmuir adsorption coefficient for oxygen on TiO$_2$, γ is a proportionality constant, and $K_{CP} = (29 \pm 3) \times 10^3$ dm^3 mol^{-1} is a reactivity constant of the substrate. Oxygen appeared to be noncompetitively adsorbed on titanium oxide, and only on Ti(III) sites, and therefore the Langmuir adsorption coefficient was adequate to represent the oxygen reactivity in the preceding equation. On the other hand, this value of K_{CP} was much higher than the observed value of the "dark" Langmuir adsorption coefficient for 4-chlorophenol, $K_{L(CP)} = 1.3 \times 10^2$ dm^3 mol^{-1}. Similar results of much higher reactivity constants than Langmuir adsorption coefficients had been observed by Cunningham and Al-Sayyed[47] for three substituted benzoic acids. Adsorption isotherm data for salicylic acid, 4-aminobenzoic acid, and 3-chloro-4-hydroxybenzoic acid onto TiO$_2$ particles in dilute aqueous suspensions indicated competitive adsorption–desorption equilibria by solute and solvent molecules. Each adsorbed solute molecule coexisted with about 40 adsorbed H$_2$O molecules. Considerable discrepancies occurred between the concentration dependences of "dark" adsorption and of photodegradation rates for these benzoic acids. Explanations involve rapid release of ·OH radicals from the TiO$_2$/solution interface and subsequent oxidation of solute molecules, or a contribution of photoadsorption — in addition to the "dark" adsorption measured in the adsorption isotherms.[47]

The observed lack of correlation between parameters obtained from "dark" equilibrium adsorption studies of organic pollutants of TiO$_2$ and the parameters deduced from the pollutant concentration dependence of the photocatalytic degradation rates prompted the hypothesis that the simple Langmuir–Hinshelwood model is inadequate for low initial concentrations (5 to 200 ppm). Instead of adsorbed substrates reacting in the monolayer with the oxidizing species (·OH/h$^+$), it was proposed that reaction took place in solution multilayers, either by inward movement of pollutant molecules toward the oxidizing species in the surface solution monolayer, or by outward tunneling of the oxidizing species toward the pollutant molecules in the multilayer.[48]

The dependence of the initial rates of the TiO$_2$-photocatalyzed decomposition of monochlorophenols on the concentration in solution of these chlorophenols (after prior dark equilibration on the catalyst) was studied at different light intensities. At a high light flux (2×10^{18} min^{-1}; $\lambda = 365$ nm), the initial rate of photodecomposition of 4-chlorophenol in an aqueous slurry of TiO$_2$ (2 g/L) was independent of its initial concentration. On the other hand, at a low light flux (1.7×10^{16} min^{-1}), the dependence of the initial rate increased with the initial concentration of the chlorophenol, reaching saturation in a form analogous to the Langmuir–Hinshelwood adsorption isotherm. The initial rates of photodecomposition of 4-chlorophenol at high photon flux increased when changing from aerated to oxygen-saturated mixtures, and even more (up to 350%) when adding AgNO$_3$ (500 ppm) to the aerated suspension before illumination. The explanation for the rate-enhancing effect of these electron acceptors is that Ag$^+$ and O$_2$ efficiently trap photogenerated electrons at the TiO$_2$-solution interface, thus decreasing electron–hole recombination, and re-leasing more holes to the photooxidation of the organic substrates. At the high light flux, the transfer of electrons may therefore be the rate-determining step. At moderate and low light intensity, the rate-determining step seemed to be the interaction of adsorbed chlorophenol molecules with ·OH (or h$^+$) species photogenerated on the TiO$_2$ surface.[49]

D'Oliveira et al.[50] also concluded from a study of the TiO$_2$-promoted photodegradation of di- and trichlorophenols that the Langmuir–Hinshelwood kinetic model (based upon measurements of adsorption of these chlorophenols from dilute aqueous solutions on the TiO$_2$ in the dark) is an inadequate representation of the observed kinetics. A much better correlation of the apparent first-order rate constants k$_{app}$ (h^{-1}) of disappearance of the various isomeric chlorophenols was obtained with their Hammett constants σ and their 1-octanol/water partition coefficients K$_{ow}$:

$$k_{app} = -10\sigma + 5.2 \log K_{ow} - 7.5 \tag{20}$$

This model relates the degradability of a given compound to its molecular structure.[50]

Davis and Huang[51] developed a model of the photocatalytic oxidation of organic compounds in aqueous suspensions of illuminated semiconductors, using the steady state approximation. The rate of disappearance of the amount M of organic substrate in the bulk solution (moles of substrate) was represented by

$$-dM/dt = k_o[\cdot OH]M_s\left\{(V + KM)^2 \Big/ \left[(V + KM)^2 + C_m wKV\right]\right\} \tag{21}$$

where [·OH] and M$_s$ are the concentration of surface-adsorbed hydroxyl radicals and amount of adsorbed organic substrate, respectively, V is the volume of the reactor, w is the weight of the photocatalyst, and C$_m$ is the monolayer coverage for organic adsorption. Langmuir isotherms were assumed both for the surface organic concentration and for the adsorbed oxygen concentration,

$$M_s = C_m wKM/(V + M) \qquad (22)$$

with K being the adsorption constant for either the organic substrate or oxygen. The oxidation rate depended on the light intensity, its attenuation by absorption in the semiconductor, losses due to electron–hole recombinations, and the rate of formation of hydroxyl radicals. The values of K, C_m, and M_s were all a function of pH. The model was fitted to the data of Okamoto et al.[52,53] on the TiO_2-photocatalyzed oxidation of phenol.[51]

For the case of the photodegradation of 3-chlorosalicylic acid, Zeltner et al.,[54] developed a rate equation based on the Langmuir–Hinshelwood/Hougen–Watson model:[55]

$$Rate = kK_a K_b C_a C_{ox} / \left[\left(1 + K_a C_a \right)\left(1 + K_b C_{ox} \right) \right] \qquad (23)$$

where k, K_a, and K_b are constants, and C_a and C_{ox} are the concentrations of 3-chlorosalicylic acid and dissolved oxygen, respectively.

Cunningham and Srijaranai[56] applied a deuterium oxide solvent isotope effect to support the hypothesis that photogeneration of hydroxyl radicals from water molecules is the rate-determining process (RDP) in the TiO_2- and ZnO-photocatalyzed oxidation of secondary alcohols to ketones in oxygenated solutions. Using illumination through a 365-nm narrow bandpass filter (to prevent the direct excitation of the carbonyl products by their n–π* transition at about 280 nm), the rates of the photocatalyzed oxidation were compared for the normal and deuterated isopropanol in water and in D_2O, respectively:

$$\left(CH_3\right)_2 CHOH-H_2O-TiO_2-O_2 \quad \text{and} \quad \left(CD_3\right)_2 CDOD-D_2O-TiO_2-O_2$$

The solvent isotope effect on initial rates in these two reactions was k_{H_2O}/k_{D_2O} = 2.8 ± 0.1. With ZnO as photocatalyst for the same reactions, the solvent isotope effect for isopropanol oxidation was even larger, k_{H_2O}/k_{D_2O} = 4.3. Similarly, with cyclobutanol as substrate and TiO_2 as photocatalyst, k_{H_2O}/k_{D_2O} = 2.5 ± 0.2. In these semiconductor photocatalyzed heterogeneous oxidations, the RDP was proposed to be the *formation* of the hydroxyl radical. On the other hand, in the homogeneous oxidation of isopropanol and of deutero-isopropanol, photosensitized by the uranyl ion, (UO_2^{2+}), there was no solvent isotope effect, and k_{H_2O}/k_{D_2O} = 1.0 ± 0.1. For the UO_2^{2+}-photosensitized oxidation of

$$\left(CH_3\right)_2 CHOH - H_2O - UO_2^{2+} \quad \text{and} \quad \left(CD_3\right)_2 CDOD - H_2O - UO_2^{2+}$$

the relative rates were $k_{H_2O}/k_{D_2O} = 1.89 \pm 0.1$, indicating that breakages of C-H and C-D bonds were the rate-determining steps.[56] Further evidence for hydroxyl radicals as intermediates in photocatalytic reactions was obtained by the "spin-trapping" technique and electron spin resonance detection (see Section II.).

C. Role of Oxygen

Gerischer and Heller[57,58] and Gerischer[59] analyzed the kinetics of the photooxidation under sunlight of organic molecules, and the concurrent reduction of molecular oxygen, in aerated seawater at TiO_2 particles of different sizes. Photoexcitation of semiconductors results in the formation of electron–hole pairs. Since hole transfer to adsorbed organic molecules or to water is very fast, the two-electron transfer to O_2 may become rate-limiting. In n-type semiconductors, the more reactive species are the holes, which carry the major part of the energy of the light quantum. Either the holes may react by recombination with electrons (thus lowering the quantum efficiency), or they may react with organic molecules RH_2,

$$h^+ + RH_2 \rightarrow \cdot RH + H^+(aq) \tag{24}$$

as well as with surface-adsorbed OH^- anions, forming hydroxyl radicals,

$$h^+ + OH^-(ads) \rightarrow \cdot OH \tag{25}$$

Electrons usually react more slowly than holes.

The known variability in the photocatalytic efficiency of different preparations of TiO_2 may be due to differences in the distribution of surface traps, on which electrons are immobilized, and thus are available for transfer to oxygen.

Since, as noted earlier, holes at the particle interface usually react faster than electrons, the particles under illumination contain an excess of electrons. Removal of this excess of electrons is necessary to complete the oxidation reaction, by preventing the recombination of electrons with holes. The most easily available and economic electron acceptor is molecular oxygen. Thus in the presence of air or oxygen, the predominant reaction of electrons is that with O_2 as electron acceptor,

$$e^- + O_2 \rightarrow O_2^- \tag{12}$$

Oxygen diffusion was shown not to be the rate-limiting reaction in the case of illumination of n-TiO_2 particles in water under ~1 sun. For TiO_2 particles larger than 1 mm, the rate of electron transfer to adsorbed O_2 may be the limiting factor, resulting in losses in quantum efficiency. The catalytic effect of noble metal catalysts on TiO_2 particles could possibly be due to their facilitating the electron transfer from the particles to molecular oxygen. The rate-controlling process in oxygen reduction on illuminated semiconductor particles has been proposed to be either (1) with electrons trapped near the surface of the particles, which may transfer to surface-adsorbed oxygen molecules, or (2) with freely moving electrons in the bulk of the semiconductor particles, which react with O_2 molecules when reaching the surface. Gerischer and Heller[57,58] proposed that case 1 appeared more realistic, and that therefore highly efficient photo-oxidation of organic molecules requires fast O_2 reduction, which may be achieved by incorporating catalytic sites on the TiO_2 surface. The required electron traps may be oxygen vacancies, which appear in the X-ray photoelectron spectroscopy (XPS) measurements as Ti^{3+} ions, as was demonstrated by Henrich et al.[60]

The maximal rate of electron uptake by oxygen was shown to depend on the particle size and the oxygen concentration in solution. Gerischer therefore suggested that high yields in the photooxidation of organic molecules may be achieved only with very small size TiO_2 particles as photocatalysts.[59]

In case oxygen is the electron acceptor in photocatalytic reactions on semiconductor particles, and in the absence of the catalytic activation of their surface by metal atoms or clusters, the rate-limiting step was thus electron transfer to oxygen. In addition, the reactions with oxygen are also required to stabilize the primary organic radicals formed from the reaction of holes with organic molecules. Experiments by Zeltner et al.[54] on the photodegradation of the 3-chlorosalicylate anion at pH 6 in aqueous suspensions of TiO_2 indicated that the rate of reaction increased with the oxygen concentration. The conclusion that electron transfer to oxygen on semiconductor surfaces is a relatively slow process has been supported by data on the heterogeneous rate of O_2 reduction on colloidal TiO_2, $k = 10^{-7}$ cm^{-1} s^{-1}.[61,62]

D. Direct and Indirect Photodegradation

Photodegradation in the presence of suspended semiconductor particles may be either by a "direct" process, by organic molecules adsorbed on the surface of the particles which interact with holes and hydroxyl radicals on the surface — or they may be "indirect," by interaction of the organic molecules with hydroxyl radicals in the bulk of the solution. Zeltner et al.[54] proposed to distinguish between the direct and indirect pathways by the selective action of added methanol. Methanol scavenges both holes and hydroxyl radicals. In the case of the photodegradation of 3-chlorosalicylic acid in aqueous TiO_2 at pH 4, the rates of reaction were almost equal in the presence and absence of methanol. This indicated that degradation occurred mainly by direct charge transfer between the adsorbed 3-chlorosalicylate anion and the TiO_2. On the other hand, at pH 7, addition of methanol considerably decreased the rate of degradation, which suggested that the reaction occurred mainly in the bulk of the solution, where hydroxyl radicals were effectively scavenged by methanol molecules.[63,64]

E. Zeta Potentials and Surface Properties

TiO_2 particles in aqueous dispersions bind both molecular and dissociated water to their surface. Changes in the pH of the dispersion lead to variations in the surface charge of the particles:[65]

$$\left[OH^-\right]_{ads} + H^+ = \left[H_2O\right]_{ads} \qquad (26)$$

$$\left[H_2O\right]_{ads} + H^+ = \left[H_3O^+\right]_{ads} \qquad (27)$$

The electric charges of the particles are represented by their ζ (zeta) potentials. These potentials may be measured in electrophoretic cells containing the TiO_2 suspension. In these the motion of the particles is observed either with a telescope or a video system, determining the velocity of the particles in an electric field.[64] In aqueous dispersions of TiO_2, the ζ potential was found to decrease linearly with the pH of the medium. The isoelectric point (pI), or point of zero charge, represents both the pH at which an immersed solid oxide electrode would have zero net charge, and the pH resulting in electrically equivalent concentrations of positive and negative complexes. An isoelectric point of about pI = 6.8 was found for aqueous dispersions of TiO_2 in the absence of additional solutes. The ζ potentials of TiO_2 were positive in acidic (pH < 6.8) dispersions, and were negative in alkaline (pH > 6.8) dispersions — indicating positive or negative charges on the TiO_2 particles. Illumination of the dispersions caused acidification and formation of positive charges on the TiO_2 particle surfaces.[65]

An indication that most of the chemical effects of TiO_2-promoted photodegradation of organic compounds occur at or near the photocatalyst surface come from considering the influence of the pH of the medium, and of the effects of inorganic ions. The point of zero charge of TiO_2 Degussa P25 is at pH 6.3. Therefore, at pH > 6.3, the TiO_2 surface is negatively charged and should repel negative ions, which should not interfere in the action of hydroxyl radicals. In the photodegradation of 3-chlorophenol at pH 8, the presence of concentrations as large as 0.1 M NaCl, Na_2SO_4, $NaNO_3$, or $NaHCO_3$ was found not to have any effect on the initial rate of disappearance. At pH < 6.3, the positive charge on the TiO_2 surface attracts anions, which may considerably interfere with the hydroxylation of the organic substrates. Chloride, sulfate, and phosphate ions inhibited photodegradation, while nitrate and perchlorate had no significant effect.[66] This effect could be correlated with the order of adsorption of these ion by binding to the inner coordination sphere of TiO_2:[67]

$$H_2PO_4^- > HSO_4^- > Cl^- > ClO_4^-$$

The insensitivity of TiO_2-promoted photodegradation, at basic pH, to high concentrations of common inorganic ions is an important advantage vs. the other advanced oxidation processes, using H_2O_2–O_3, H_2O_2–UV light, or O_3–UV light, which are strongly inhibited by the presence of bicarbonate ions. In these processes, which occur in homogeneous solutions, the HCO_3^- ions effectively scavenge the ·OH radicals. In the heterogeneous TiO_2–UV process, the action of ·OH radicals occurs mainly at the particle surface and not in the solution.[66]

F. Catalyst Preparations

The photocatalytic activity of TiO_2 for the degradation of organic compounds increased with the anatase content of the catalyst and also with the crystallite size of TiO_2 per unit surface area of the catalyst. This was observed by measuring the photooxidation of trichloroethylene, dichloroacetic acid, and phenol over 12 commercial samples of TiO_2 with different anatase/rutile ratios and crystallite sizes. The more efficient photocatalytic action with the larger crystallites was explained by the longer migration distances available for the photoproduced electrons and holes, thus decreasing the probability of electron–hole recombination. The lower degradation rates with rutile vs. anatase was attributed to the more positive conduction band of rutile, which retarded the reduction of oxygen. Platinization of TiO_2 did not affect the rates of photocatalyzed degradation of trichloroethylene and dichloroacetic acid, but enhanced the rate of degradation of phenol. Also, platinization had more effect with rutile than with anatase.[68]

The photocatalytic activity of preparations of TiO_2 depended strongly on the temperature used to anneal the photocatalyst. Heat treatment above 600°C resulted in a dramatic decrease in the photocatalytic activity toward the mineralization of 4-chlorophenol. This decrease in photoactivity was shown to be correlated most closely with the decrease in surface area of the TiO_2 particles, and not with the anatase to rutile transformation, which occurred only above 700°C. Annealing of TiO_2 above 600°C caused considerable sintering of the particles, and thus decreased the surface area.[69] In a comparative study of the photodegradation of 4-chlorophenol mediated by oxygen-saturated aqueous suspensions of various preparations of TiO_2, highest rates of disappearance of the pollutant were achieved with the commercial products Tioxide A-HR, followed by Aldrich anatase and by Degussa P25. However, for the photodestruction of the intermediates 4-chlorocatechol and hydroquinone the most active catalyst was Aldrich anatase. No simple correlation was observed between the photoactivity and the surface area of the catalysts.[70]

While the anatase modification of TiO_2 in aqueous dispersions is usually photoactive for the degradation of organic compounds, the rutile modification is usually inactive. A further explanation for the lower photoactivity of rutile is based on the drastic irreversible dehydroxylation on the surface of the TiO_2 that occurs during the heat treatment (above 700°C) in the transformation of anatase to rutile. The surface hydroxyl groups are required to trap holes h^+ and produce ·OH radicals, and to adsorb both O_2 (to trap electrons) and organic molecules. In the photooxidation of phenol, rutile was active as a photocatalyst when Ag^+ ions were also present, presumably because these ions served to trap electrons:

$$Ag^+_{ads} + e^- \rightarrow Ag_{ads} \qquad (28)$$

The addition of silver ions enabled the photooxidation of phenol on rutile to occur at a rate similar to that on anatase. The presence of both oxygen and hydrogen peroxide are often beneficial for the photocatalytic effect of TiO_2. It was proposed that at least two types of sites exist on the surface of TiO_2, one of which serves to adsorb the organic substrate, while the other adsorbs both O_2 and H_2O_2 in competition.[71,72]

In a study of the effect of "calcination" temperature of TiO_2 prior to its use as photocatalyst for the degradation of various organic compounds, it was observed that for water-insoluble compounds like trichloroethylene, tetrachloroethylene, and benzene (group I), the rate of degradation depended strongly on the temperature of calcination. The rate was maximal with TiO_2 that had been annealed for 3 h in air at about 500°C. For water-soluble compounds, like chloroacetic acid, bromoacetic acid, and benzoic acid (group II), the rates of photodegradation were unaffected by the temperature of calcination of TiO_2 in the range of 200 to 500°C, but decreased with higher calcination temperatures. The difference in behavior between the two groups of compounds was proposed to be due to the different absorbabilities to the OH groups on the surface of TiO_2. At calcination temperatures below 600°C, the predominant crystal form was anatase, which was transformed to rutile at higher temperatures of calcination. For the photodegradation of water-insoluble compounds like trichloroethylene, it was shown that the logarithm of the rate of degradation was linearly related to the crystal size of anatase. With the water-soluble and strongly adsorbed compounds, holes were efficiently scavenged by the organic molecules, while electrons accumulated in the conduction band. The scavenging of electrons (e.g., by oxygen) thus became the rate-determining process for group II compounds, and therefore the rate of photodegradation was less dependent on the crystal size of the anatase.[34]

There have been many attempts to enhance the photoactivity of TiO_2 by doping with various heavy metal ions. With Fe^{3+} ions, there was no influence on the photodegradation of 4-nitrophenol in aqueous dispersions as long as the concentration of the dopant was below 0.1 mol%. At higher contents of iron, the photoactivity decreased considerably, presumably because of a decrease in the density of surface active centers.[73] In experiments on the photocatalytic oxidation of oxalic acid in the presence of lattice-doped TiO_2 colloids, doping with Cr^{3+}, Fe^{3+}, and V^{5+} resulted in decreased degradation.[74]

Several catalysts were tested for their photoactivity, using trichloroethylene (TCE), toluene, methyl ethyl ketone (MEK), salicylic acid, and 2,4-dichlorophenol as model water pollutants. For the photodestruction of TCE, both $SrTiO_3$ and 1.5% NiO–$SrTiO_3$ were completely inactive. Several TiO_2 preparations (each 1 g/L) were compared for the photodegradation of TCE (initially 10 mg/L), both under artificial illumination (450-W high-pressure Hg lamp) and under sunlight. The times for 90% destruction of TCE under artificial light by Aldrich TiO_2, Degussa P25, and 1% Pt/Aldrich TiO_2 were 48, 38, and 10 min, respectively. Under sunlight, Aldrich TiO_2 and 1% Pt/Aldrich TiO_2 required only 12.2 and 2.5 min for 90% destruction. The 1% Pt/Aldrich TiO_2 was also the best catalyst for the photodestruction of toluene and MEK. On the other hand, for the photodegradation of salicylic acid and 2,4-dichlorophenol, the best catalyst was a 1% Pt TiO_2, with the TiO_2 prepared by a sol-gel process.[3]

In order to clarify the observed deactivation of TiO_2 catalysts during the photodegradation of TCE in aqueous solutions, the surface properties of the catalyst were compared before and after deactivation. X-ray photoelectron spectroscopy (XPS) indicated that there was no significant difference between fresh and deactivated catalysts. Although chloride ions in solution had been shown to inhibit the aqueous phase photodestruction of TCE, no chlorine was detected in the XPS spectra of deactivated TiO_2. Temperature-programmed desorption (TPD) of the deactivated catalyst revealed the desorption of water, carbon monoxide, and carbon dioxide. Presumably, strongly absorbed groups such as carbonate had accumulated on the catalyst surface, and were dissociated to CO and CO_2 by high-temperature decomposition.[75]

G. Surface Density of OH Groups on TiO_2 Particles

The surface density of OH groups on TiO_2 particles suspended in water was determined by measurements of the adsorption of the cations Cu^{2+}, Cd^{2+}, and Zn^{2+} on TiO_2 in aqueous suspensions. The adsorption behavior of the cations could be described by a modified Langmuir–Hinshelwood model. The adsorption of these cations on TiO_2 resulted in the release of H^+ ions (and decrease in the pH of the solution), due to the dynamic equilibrium,

$$\equiv Ti - OH + M^{2+} \leftrightarrow \equiv Ti - OM^+ + H^+ \qquad (29)$$

In the modified model, both the rate of adsorption of the cations M^{2+} on the TiO_2 and the rate of desorption by H^+ ions were taken into account. At steady state, the ratio of the concentrations M^{2+}/H^+ in solution was proportional to the reciprocal of the number of adsorbed cations. From the slopes and intercepts of the resulting straight lines, the adsorption ability of the cations and the total number of OH groups on the TiO_2 could be derived. The relative adsorption abilities of the cations were in the order $Cu^{2+} \gg Cd^{2+} > Zn^{2+}$. The observed intercepts were practically identical, resulting in a surface density of 1.4 ± 0.2 OH groups/nm^2.[76]

H. Quantum Yields and Turnover Numbers

In homogeneous-phase photochemistry, a basic parameter for comparing the efficiency of different photochemical processes and under various conditions is the quantum yield ϕ, which may be defined as the rate at which reactant molecules disappear or product molecules are formed, divided by the number of photons absorbed per unit time. In heterogeneous photocatalysis, the species absorbing light is the photocatalyst, such as TiO_2 or ZnO, which may be dispersed as a slurry in an aqueous (or other fluid) medium, or may be coated on a solid support in contact with the reactants. With such dispersed or coated catalysts, a large fraction of the incident light is either reflected or scattered. There does not usually exist any possibility to determine experimentally the amount of light absorbed by the photocatalyst. In order to bypass the difficulty of determining quantum yields in heterogeneous photocatalysis, Serpone et al.[77] suggested the use of photonic efficiency, defined as the number of reactant molecules transformed (or product molecules formed) divided by the number of photons, at a given wavelength, incident inside the front window of the cell. This parameter still depends on the reactor geometry and light source. For comparing different reactions by a parameter independent of reactor geometry and light source, the relative photonic efficiency was introduced, defined as the photonic efficiency relative to a standard process. As a convenient standard, the TiO_2 (Degussa P25) photocatalyzed reduction of Au(III) to metallic Au in 10% HCl was proposed. At initial Au(III) concentrations of 0.25 to 2.3 mM and pH 3, the reduction followed zero-order kinetics, with an apparent rate constant $k_{obs} = 3.8$ μM min^{-1}. Another parameter often reported is the "Apparent quantum yield," defined as the number of reacted molecules divided by the number of photons entering the reactor in a given time.[78]

Experimental determination of the fractions of incident light on a semiconductor dispersion that were absorbed, reflected, and transmitted enabled direct determination of the quantum yield. This was applied to the photodegradation of phenol in aqueous dispersions of TiO_2. The neglect of scattering was proposed to be applicable if the particle dimensions are much larger than the wavelength of the incident radiation, so that the laws of geometrical optics are valid.[79,80]

In homogeneous thermal or photochemical catalysis, another very useful measure of the efficiency of catalysis is the turnover number (TN), defined as the number of molecules undergoing reaction per molecule of catalyst. In a catalytic reaction, TN > 1. In thermal or photochemical heterogeneous catalysis, a primary step is adsorption of reactant molecules on specific active sites on the surface of the catalyst. In heterogeneous photocatalytic reactions, these active sites are due to charge carriers, electrons and holes, which are mobile or trapped on surface states of the semiconductor. Photoreactions on illuminated semiconductors may be considered to be catalytic if the photocatalytic turnover numbers (PTN), defined as

Molecules reacted/[(time) (active site) (photon absorbed)]

are substantially larger than 1, say 100 or more. If these turnover numbers are lower than 1, it is very probable that the chemical reactions observed are due to conversion of the semiconductor lattice, such as by corrosion and dissolution.[81]

However, there exists a major difficulty in determining the number of active sites.[77] Since the number of active sites may be assumed to be proportional to the surface area, which can be experimentally determined by the Brunauer–Emmett–Teller (BET) method, a more convenient definition of the turnover number uses the number of reacted molecules divided by the surface area and the unit time, with the dimension (mol/s). By measuring both the rates of reaction and the rates of photon absorption, it may be possible to derive values for the quantum yield.[80,82]

In an effort to overcome the difficulty of quantum yield determinations in heterogeneous photocatalysis, Lepore et al.[83,84] prepared aqueous suspensions of polycrystalline TiO_2 that were optically transparent in the visible spectrum.[83,84] These were prepared by 2 or 3 times sequential centrifugations of aqueous slurries of TiO_2 (0.2%). This material (Degussa P25) had an anatase/rutile ratio of 76/24, a surface area (determined by the dynamic BET method) of 55.4 m^2 g^{-1}, and consisted of 50 to 200 nm aggregates of microcrystalline TiO_2 particles, which themselves were 20 to 35 nm in diameter. Centrifugation precipitated and removed most of the larger aggregates, which mainly contributed to light scattering. The aggregates remaining in the supernatant had an average measured radius (by dynamic light scattering) of 62 nm, with a broad size distribution. The resulting solution was considered optically transparent and had an onset of absorption at about 360 nm, a shoulder at about 315 nm, and a maximum at 255 nm. The quantum yield of the photooxidation at 313 nm of I^- in these TiO_2 solutions was determined to be 0.02. With propanal, the quantum yield of photooxidation catalyzed by these TiO_2 solutions was found to increase with the initial propanal concentration, suggesting dependence on Langmuir–Hinshelwood type adsorption. A limiting value for the quantum yield was 1.0.[83,84]

A simpler approach to the determination of quantum yields in heterogeneous systems was taken by Valladares and Bolton[79] and was applied to the TiO_2-photocatalyzed bleaching of methylene blue. The reaction cell was surrounded by a tightly fitting aluminum jacket that had only a small aperture to let the light penetrate into the cell. Thus, almost all the light scattered and reflected from the TiO_2 particles was reflected back into the reaction mixture. Optimal conditions were found with an aperture area of 180 mm², methylene blue initially 27 μM, TiO_2 80 mg/L, illumination at 320 nm, and a light quanta dose of 4.3×10^{-7} einsteins. The maximal quantum yield obtained under these conditions, $\phi = 0.056 \pm 0.002$, was proposed to be the accurate value.[85]

Since the rate of photodegradation of a given substrate depends on a large variety of factors (such as the concentrations of the substrate, photocatalysts and sensitizers, light intensity, and spectral distribution, as well as light absorption coefficients, oxygen concentration, temperature, pH, etc.), Mills et al.[86] proposed referring rate measurements under carefully defined conditions to those of a standard test system, using the TiO_2-promoted photomineralization of 4-chlorophenol under oxygen as such a system.

V. EXPERIMENTAL TECHNIQUES

A major problem in the practical application of photocatalytic reactions to the degradation of environmental pollutants is the design of reactor systems, which will achieve a considerable reactant conversion rate at a high throughput. Several basic reactor systems have been tested by Anderson et al.[54,87–90] The modeling and design of experimental reactors for heterogeneous photocatalytic reactors and their upscaling to industrial processes have been reviewed.[14,91–93]

A. Reactor Design

A convenient laboratory reactor system, which has been used in many photochemical studies, is the annular photoreactor (see Figure I.1), which includes an inner light source, such as a high-pressure mercury lamp, surrounded by a concentric tube of Pyrex® glass or quartz containing either a coolant water flow or directly the reactant solution. Often a fritted glass inlet tube at the bottom of the reactant solution enables introduction of gases such as air, oxygen, argon, or nitrogen, bubbling through the reaction mixture and mixing the components.

1. Slurry Reactor

The simplest reactor configuration uses an aqueous dispersion or slurry of the photocatalyst. The preferable catalyst, due to its high activity, insolubility, resistance to corrosion, nontoxicity, and relatively moderate cost, has been titanium dioxide. Many experiments have been performed using the product TiO_2 P25 (Degussa Corporation), which is produced by high-temperature (>1200°C) flame hydrolysis of $TiCl_4$ in the presence of O_2 and H_2. The product, after steam treatment to remove HCl, is nonporous, with cubic particles having an average primary particle size of 20 nm, a surface area of 35 to 65 m²/g, and

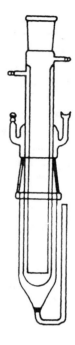

FIGURE I.1.   Annular photoreactor for illumination with medium- or high-pressure Hg lamp. Reactant mixture is in outer compartment.

an anatase to rutile ratio of 70:30.[94] Good photocatalytic activity for the photodegradation of organic compounds was also achieved with TiO_2 photocatalysts prepared by the hydrolysis of titanium tetra-isopropoxide, which yielded particles with crystallite sizes in the range of 20 to 30 nm.[95,96]

2. TiO_2 as a Stationary Phase

For heterogeneous photocatalysis with semiconductors such as TiO_2, either the photocatalyst may be dispersed in suspension, or it may be coated on the inside walls of the reactant chamber.[97] The photodegradation of organic pollutants in illuminated aqueous dispersions of TiO_2, while very effective, has the drawback of the difficult recovery of the photocatalyst, such as by filtration, and thus usually is performed in a closed system.[63] Serpone et al.[98,99] and Matthews[100] introduced open-flow continuous operation by coating the TiO_2 onto glass surfaces. Thus, the photocatalyst became a stationary phase. Coating of the photocatalyst was made on the inside wall of a borosilicate glass tube (4 mm i.d.), which for illumination was either coiled around a blacklight fluorescent tube, or mounted at the focus of a solar parabolic trough. The aqueous solution of the pollutants was then pumped through the tube. The open-flow systems were tested using 2- and 4-chlorophenol, benzoic acid, 2-naphthol, naphthalene, and fluorescein as substrates, both in single-pass and recirculation mode. The oxidation rates were usually lower with the immobilized catalysts than with the free suspensions.[12,100–102]

Since the photocatalytic oxidation is quite nonselective with respect to the chemical nature of the organic substrate, this process with TiO_2 supported on glass has been developed as a convenient and fast method for total organic carbon (TOC) analysis in water.[103] Glass fiber cloth coated with TiO_2 has also been applied to the photocatalytic degradation of chloroform (see Chapter IV, Section I).[104]

These reactor systems are convenient if the light source is of relatively low intensity (e.g., less than 150 W). With more intense light sources (e.g., 500-W high-pressure Hg lamps), this arrangement does not provide adequate cooling. Another drawback of this design is that the TiO_2 coating is quite thin (about 1 μm), resulting in low overall conversion efficiency.[54]

Improved TiO_2 thin films for photocatalytic degradation of water pollutants were obtained by repeatedly dip-coating on a solid substrate a solution prepared from titanium tetraisopropoxide, ethanol, diethanolamine, water, and polyethylene glycol. With careful repeated heat treatment, the TiO_2 crystallized in the anatase phase. This catalyst provided efficient photodegradation of acetic acid in aqueous solution.[105] In a similar study, titanium tetraisopropoxide was hydrolyzed on Pyrex® glass tubes and heat-treated at 550°C. This catalyst enabled rapid photodecomposition of tetrachloroethylene (100 ppm) in aqueous solutions containing Fe^{3+} ions under illumination with a 300-W xenon lamp.[106]

3. TiO_2 on Silica Gel

An alternative approach was to attach TiO_2 to silica gel, by mixing TiO_2 (0.15 g) with silica gel (35 to 70 mesh, 60 g) in water (200 ml), sonicating, and rotary evaporating under vacuum at 95°C. The TiO_2-coated silica gel was packed into a borosilicate glass tube, which was illuminated while the reactant solution was pumped through it.[107]

In a systematic optimization of fixed-bed photocatalysts, 25% platinized TiO_2 (1 wt% Pt) supported on silica gel (50 to 60 mesh size), and calcined at 600°C in air for 24 h, was packed in a continuous-flow reactor system. This photocatalyst was found to give highest rates of degradation and mineralization of trichloroethylene under sunlight. Each reactor was constructed by connecting in series four 25-cm-long plastic tubes ($1/4$ in. i.d., made of plastic material with 80% transmittance in the 300 to 400 nm range). By connecting in series up to 16 reactors, the total illuminated length could reach up to 16 m. The reactor system was mounted in front of a flat UV-reflecting metal mirror. With the silica gel supported Pt/TiO_2, remarkably rapid destruction of trichloroethylene was achieved even on a rainy late afternoon, within about 2.5 min. Under such low-light conditions, the apparent quantum yield (moles of compound degraded per incident einsteins) for the 300 to 388 nm range was 40%. The effect of hydraulic loading (flow rate) was tested in order to determine if the degradation rate may be mass transfer limited. At the hydraulic loading of

19 m/h used in most experiments, there was no evidence of mass transfer limitation. With this fixed-bed catalyst system, the photodegradation rate was about four times higher than with a slurry of Pt/TiO_2 (0.25 g/L) tested in the same flow-through reactor.[108]

4. TiO_2 on Cellulose and Polyester Membranes

TiO_2 was also immobilized using polymeric membranes (from an epoxy-diacrylate copolymer as prepolymer) photografted on the surface of microporous cellulose membranes (of initial pore size 28 to 35 μm). The composite material had an apparent surface of 12 to 16 mg cm^{-2}, and contained 25 to 35% TiO_2. These membranes were tested for the photodegradation of aerated aqueous solutions of chloroacetic acid, trichloroethylene, and tetrachloroethylene. The initial rates of degradation as a function of the initial concentrations could be fitted to the Langmuir–Hinshelwood equation. The results compared favorably with those observed in aqueous suspensions of TiO_2.[109]

Photocatalytic membranes (PHOTOPERM™ CPP/313 membranes, produced industrially by Chimia Prodotti e Processi, Milan, Italy) contained 30% immobilized TiO_2, which had been photografted onto a polyester support, resulting in an overall loading of 930 ± 45 g m^{-2}. The efficiency of these photocatalytic membranes for the degradation of chloroaliphatic compounds such as dichloromethane, trichloroethylene, and mono-, di-, and trichloroacetic acids was considerably enhanced by the presence of hydrogen peroxide, and by incorporating certain promoting photocatalysts, such as triethyl vanadate, or a mixture of tri(*tert*-butyl)- and tri(*iso*-propyl) vanadate, or Fe(III) potassium oxalate. Both in laboratory-scale experiments and in pilot-plant reactors, the dependence of reaction rates on the initial substrate concentration could be fitted to the Langmuir–Hinshelwood equation.[110]

5. TiO_2 Anchored on Porous Alumina Ceramic Support

An effective support for a TiO_2-based photocatalyst was prepared by coating TiO_2 on a porous alumina–silica ceramic. This ceramic was prepared in the shape of a coaxial cylinder (for inserting a UV lamp inside the cylinder) by spraying a slurry of Al_2O_3 (70 wt%) and SiO_2 (30 wt%) on polyurethane foam, and heating for 30 min at 400°C, and for 1 h at 1000°C. Such ceramic cylinders had macropores of 1 to 3 mm diameter (for deep penetration of light) and micropores of about several micrometers diameter (for adequate anchorage of TiO_2 particles). The TiO_2 coating was obtained by dipping the ceramic cylinders into a solution of titanium tetraisopropoxide in ethanol, and adding a controlled amount of water for hydrolysis. The cylinders were then dried in air at 100°C for 1 h, and calcined at 700°C for another hour. Optimal homogeneous coverage of the ceramic by TiO_2 was obtained when the alkoxide solution used had a pH of 8, which is between the isoelectric points of TiO_2 and Al_2O_3. Also, best photocatalytic activity for the oxidation of ethanol was achieved with a ceramic coated by a water to alkoxide ratio of 100:1. TiO_2 powders prepared under similar conditions had been shown to consist of particles of 28-nm diameter, and to have in aqueous slurries excellent activity for the photooxidation of ethanol. Such a particle size may correspond to the distance through which the electrons and holes photogenerated in the particles can diffuse to the surface without recombination, and to be available for redox reactions with the solutes.[111]

6. Nanocrystalline TiO_2

Nanocrystalline TiO_2, with particle sizes of 7 nm for the anatase structure and 12 nm for the rutile structure, were prepared by a liquid-phase synthesis. These catalysts were almost twice as effective for the photodegradation of tetrachloroethylene than the commercial anatase-rutile material (Degussa P25), which has an average particle size of 30 nm.[112]

Very high photocatalytic activity was also achieved with a glass-supported pure anatase film, obtained from a TiO_2 sol prepared by thermal hydrolysis of titanyl sulfate.[113] The sol was aged in an autoclave at 180°C for 15 min, was coated on glass plates, and was sintered in air at 450°C for 30 min. The film thus formed had a thickness of about 10 μm and contained crystalline TiO_2 particles with an average diameter of about 50 nm. The photoactivity of this film on the degradation of gaseous acetaldehyde (initially 3000 ppm) under illumination with black fluorescent light (λ = 300 to 400 nm; 1.8 mW/cm²) resulted in a quantum yield of 28.9%, while the quantum yield using TiO_2 powder (Degussa P25) was 19.3%.[114]

7. TiO_2 on Optical Fibers

In an ingenious experimental design, a fiber-optic bundle (which had been stripped of its polymer cladding) was coated with a titania sol and fired. This bundle of about 200 silica optical fibers, each 110 μm in diameter and with 0.5 μm of the TiO_2 coating, immersed in a tube of the flowing contaminated aqueous solution, was illuminated from its end through a coupler. This reactor system seemed to provide efficient illumination. However, the extreme fragility of the bare fibers (without their polymer cladding) was a major problem.[54] In a further development, fused-silica optical fibers (140 pieces, which had been doubly coated with TiO_2 by dipping in a methanolic solution of tetrapropyl orthotitanate, heating in a humid atmosphere, and calcining at 600°C for 3 h) were packed into a glass tube, and were illuminated from one end of this tube with a collimated beam from a 400-W high-pressure Hg lamp. The light entering the optical fibers underwent multiple reflections, and at each contact with the interface lost part of its intensity by reaching the TiO_2 surface — thus generating electron and hole pairs. This reactor was applied to the photodegradation of 4-chlorophenol in oxygenated aqueous solutions, which were pumped through the space between the coated fibers. The rate of degradation increased with temperature, in the range of 12 to 57°C, and the activation energy derived by the Arrhenius equation for the initial attack of ·OH radicals on 4-chlorophenol was 20.6 kJ mol^{-1}. A comparison of the quantum yields of destruction of 4-chlorophenol by light in the range of 320 to 380 nm with this coated optical fiber reactor vs. a suspension of TiO_2 (Degussa P-25) gave $\Phi = 2 \times 10^{-4}$ and 5×10^{-5}, respectively. The destruction of total organic carbon (TOC) was 2.8 times faster with the optical fiber reactor than with suspended TiO_2.[115]

8. TiO$_2$ Aerogels

Highly active TiO_2 photocatalysts were obtained as aerogels or cryogels by a sol-gel method. These materials were monolytic structures, with 85% porosity, with a very high surface area, and translucency to visible light. Due to their low density, they floated on water, and thus should be attractive for the solar photooxidation of oil spills. The sols were prepared by careful mixing of titanium isopropoxide, isopropanol, water, and nitric acid (molar ratio 1:20:3:0.08). After gelation of the sols, the solvents were removed either by freeze-drying, forming cryogels, or by supercritical-point drying, forming aerogels. In the latter method, the solvent was replaced by liquid CO_2, which was then removed slowly by raising the material above the CO_2 critical point (above 35°C and 1200 psi). The BET (Brunauer–Emmett–Teller) surface areas measured for the TiO_2 aerogel, TiO_2 cryogel, and the commercial powdered TiO_2 (Degussa P25) were 607, 237, and 54 m^2/g, respectively. The adsorptive capacities and the photocatalytic activities of the above materials were compared for aqueous solutions of salicylic acid. The "dark" equilibrium adsorption values of salicylic acid amounted to 373, 273, and 41 µmol/g. Minimal values of the quantum efficiencies (number of molecules of salicylic acid disappearing per number of incident photons in the 300 to 400 nm range) with these materials were 0.66, 0.53, and 0.07, respectively. The photocatalytic activity of TiO_2 aerogels toward the decomposition of salicylic acid in aqueous solutions was thus considerably superior to that of the powdered TiO_2 particles. The higher activity of the aerogels correlated with their larger surface area. One of the advantages of these aerogels was that they could be prepared with a density smaller than that of water, enabling them to float on water. This facilitates light absorption and recovery after treatment of wastewaters or of oil spills. These excellent results were obtained with aerogels and cryogels that had not undergone high-temperature annealing. Such additional treatment (1 h at 400 to 500°C) decreased the BET surface area and markedly deteriorated both the adsorption capacity and the photoactivity for salicylic acid.[116,117]

9. Mass Transfer Limitations

Turchi and Ollis[118] estimated that in heterogeneous photocatalysis with slurries of TiO_2 (Degussa P25), having 30 nm average particle sizes, at 0.1 wt%

concentration, the maximum diffusion distance of the reactant to the photocatalyst is half the mean particle spacing, or 0.5 μm. On the other hand, in a coated tubular reactor of 2 mm radius, the maximal transport distance is 2 mm, or 2000 μm. Thus, in coated-tube reactors (and similarly with coated glass-bead-packed columns), mass transfer rates may limit the overall photocatalytic conversion rates. Evidence for diffusionally controlled kinetics appeared in the observation of Matthews[99] that the degradation rates of several aromatic compounds in the coiled-glass TiO_2-coated tubular reactor increased with the flow rate. Turchi and Ollis[118] could account quantitatively for this flow-rate effect by calculating the mass-transfer kinetics in addition to the chemical kinetics.

A careful study was performed on the importance of mass transfer limitations on the photodegradation of phenol. Oxygen-saturated aqueous solutions of phenol (pH 3) were pumped through an annular Pyrex® glass continuous photoreactor containing TiO_2 (anatase form) shaped as spherical beads of 1.2 mm diameter and illuminated with a 500-W high-pressure Hg lamp. From the effects of flow rates and initial phenol concentrations on the rates of conversion of phenol, and assuming plug flow conditions, a model was developed that accounted for the experimental data. The rates of conversion were mainly determined by diffusional processes. At low initial phenol concentrations, the mass transfer limitation depended on the diffusion of phenol. At high initial phenol concentrations, the mass transfer limitation was that due to oxygen diffusion.[119]

10. Falling Film Reactor

A novel reactor design, providing considerably improved mass transfer both at the solution/photocatalyst interface and for oxygen uptake at the gas/liquid interface, is the falling film reactor. Using slurries of TiO_2 (0.5 g/L; Degussa P25 or Tioxide) flowing down at the rate of up to 7.5 L/min over the inner surface of a glass tube surrounding a 2.5-kW medium pressure Hg lamp, essentially complete destruction of 4-chlorophenol (initially 0.25 mM) was achieved within about 20 min. In spite of the very high light intensity, no external oxygen sparging was required, and the falling film design seemed to provide higher destruction rates than conventional tubular or stirred slurry reactors.[120] A further development of the falling film reactor may be by immobilizing the photocatalyst, such as by electrochemical oxidation of Ti metal.[121]

11. Regeneration of Adsorbents

An interesting application of photocatalytic oxidation is the regeneration of spent adsorbents, such as granular activated carbon (GAC). The usual methods of disposal by incineration or landfills are undesirable as they cause air or water pollution. For the photocatalytic regeneration, commercial adsorbents (Filtrasorb

400 carbon and Ambersorb 527F carbonaceous resin) loaded with chloroform or trichloroethylene were dispersed in TiO_2 slurry reactors and were illuminated with blacklight UV or high-pressure Hg lamps, or with sunlight. Adsorption recoveries of >90% were achieved by photooxidation of the halocarbons pollutants. The regeneration time could be decreased by impregnating the adsorbent with platinized TiO_2.[122]

B. Laser and Excimer Light Sources

Laser flash photolysis has often been applied to the detection and identification of primary transient species formed in photochemical reactions. Laser-induced photodegradation was also tested as a method for the destruction of toxic water pollutants. Laser irradiation of polychlorobiphenyl and triazine herbicides and of polycyclic aromatic hydrocarbons led in most cases to nontoxic products.[123]

The acronym "excimer" (*exci*ted di*mer*s, tri*mer*s) describes complexes of molecules having weakly bound excited states — for which there does not normally exist a stable ground state. Useful excimers include Xe_2^*, $XeCl^*$, KrF^*, and Ar_2F^*. Excimers may be produced in a dielectric barrier discharge. They have a lifetime of the order of 100 ns, and decay with intense UV radiation. Excimers of Xe_2^*, KrF^*, and $XeCl^*$ emit radiation at 172, 248, and 308 nm, respectively.[124]

The novel Xe_2^* light source (developed by ABB, Baden, Switzerland) emits an intense light beam peaking at 172 nm, with a half-width of 12 to 14 nm. The Xe_2^* source has no other radiation emission between 150 and 400 nm, besides the 172 nm peak. The efficiency of UV output/electrical input with the Xe_2^* source is very high, 5 to 9%, because the ground-state Xe does not absorb in this spectral region. The excitation of the xenon gas is attained by a high-frequency and high-voltage power supply (220 to 230 kHz and 150 W) applied outside two concentric Suprasil quartz tubes, which serve as a dielectric barrier.[124,125] In an application to the degradation of 4-chlorophenol, the reaction solution was circulated through an annular flow reactor, while the lamp was placed in the axis of the reactor (see Chapter IV for details).[126]

An improvement in the development of such light sources as an advanced oxidation procedure for wastewater treatment was achieved by double irradiation with both UV and vaccum-UV (VUV) light. The wastewater was injected into a module illuminated in series by a $KrCl^*$ excimer ($\lambda = 222$ nm) and by an Xe_2^* excimer ($\lambda = 172$ nm). The first light source was designed to initiate direct photolysis of organic water pollutants, and the second excimer caused efficient water cleavage, creating ·OH radicals. This combination provided improved mineralization.[127]

In order to compare the cost effectiveness of different advanced oxidation processes (AOPs) for wastewater treatment, the volume-corrected efficiency of electrical energy consumed to achieve the half-time ($t_{1/2}$) of pollutant degradation of TOC elimination was determined. This was applied to a comparison of the H_2O_2/UV process using medium-pressure Hg lamps with the VUV process using 172 nm or 172 nm + 222 nm excimer lamps. The pollutants tested included aqueous solutions of rhodamine B, formaldehyde containing industrial wastewater, various volatile chlorohalocarbons, and the dyes naphthol blue black B, orange G, and orange I. From these experiments it was concluded that the VUV-excimer irradiation process has a high potential for efficient purification of industrial waste, if the energy levels of the plasma generator and the VUV-excimer source are carefully optimized.[128]

Excimer lamps for the far-UV photolysis of polluted water seem attractive for the treatment of pollutants that are recalcitrants to other advanced oxidation processes. However, the excimer lamps now available do not have the power required for large-scale installations.[129] Also, their cost is yet prohibitive for many laboratories.

C. Light–Dark Cycling

A remarkable increase in the photoefficiency of aqueous heterogeneous photocatalysis was attained by using a regime of light–dark cycling. The method was performed in an open-channel flow reactor, in which sections of the reactor channel were alternatively in the dark or exposed to illumination from a blacklight UV source (average wavelength 365 nm). The reaction mixture was an aqueous solution of sodium formate in a dispersion of TiO_2 (Degussa P25). The reaction was the oxidation of formate ion to carbon dioxide,

$$2HCOO^- + O_2 + h\nu \rightarrow 2CO_2 + 2OH^- \tag{30}$$

Optimal efficiency for formate degradation was achieved with an illumination period of 0.072 s and a dark recovery period of 1.45 s. In such a regime, the photoefficiency (defined as the ratio of the formate reaction rate to the incident photon flux) was 500 times larger than with continuous illumination. The mechanism of this effect is yet unclear. Presumably, the very short illumination periods are sufficient to create a sufficient number of electron–hole pairs, while the longer dark periods facilitate the product-forming reactions.[130]

D. Concentrated Sunlight

Sunlight is an attractive energy source for photochemical reactions. However, because of the relatively low intensity of sunlight, its direct large-scale application would require either solar concentrators or large areas of shallow ponds. Since the rates and yields of photochemical reactions usually increase with rising light intensity, some effort has been made to apply concentrated sunlight for the photoassisted degradation of water pollutants. These experiments are described in detail in Chapter VIII, Section IV.

For photocatalyzed reactions with aqueous slurries of TiO_2, the dependency of reactions rates on light intensity is usually linear at low light intensity, and changes over to a square-root dependency with increasing light intensity.[42] This decreasing photoactivity at higher light intensities could be due to increased electron–hole recombination, thus competing with the formation of hydroxyl radicals. An alternative explanation could be that of mass transfer limitation at high light intensities. Kawaguchi[131,132] examined the combined effects of the variation of light intensity and of the TiO_2 photocatalyst concentration on several photodegradation reactions. At high illumination intensity, the rate constants showed a transition from first-order to square-root dependency on the photocatalyst concentration, while at low illumination intensity, the rate constants had a transition from first-order to zero-order dependency on photocatalyst concentration.[131,132]

E. Photoelectrochemical Reactions

Enhanced photocatalytic oxidation of organic compounds was achieved by illuminating electrodes of conductive glass coated with TiO_2 and maintained at an anodic bias. Such photoelectrochemical oxidation was applied to the mineralization of formic acid.[54,133] High-efficiency porous TiO_2 thin-film electrodes were prepared by coating several layers of a TiO_2 sol (obtained by refluxing a mixture of titanium tetraisopropoxide, water, and HNO_3, 15:150:1 v/v) for 3 d at 80°C on SnO_2-covered glass, drying, and annealing at 400°C for 2 h — resulting in a layer of anatase, with a surface area of 124 m^2 g^{-1}, and a pore radius of 18.2 Å. The TiO_2 particles in the sol had a diameter of 60 nm. The rate of photoelectrochemical oxidation of formic acid, in an oxygen-purged solution, under a bias of 0.3 V (vs. SCE), with a Pt counterelectrode, increased about linearly with the number of TiO_2 coatings on the electrode, but seemed to reach a plateau above six layers. The photooxidation of formic acid by the photoelectrochemical reaction, with the 0.3 V bias, was about 35% times faster than the photocatalytic oxidation in the same cell (without an external bias). Even a bias potential of 0 V (vs. SCE) on the TiO_2 electrode was adequate for rapid photodegradation of formic acid. In the absence of oxygen, the rate of oxidation decreased only by about 20%. Presumably, in the anoxic conditions, the cathodic reaction was reduction of water to hydrogen. The presence of KCl, $NaNO_3$, $NaClO_4$, or Na_2SO_4, in 1 M concentrations, only moderately affected the photoelectrochemical oxidation of HCOOH. In the photocatalytic reaction, these ions, or the absence of oxygen, caused marked decreases in the rates of oxidation. The advantages of such photoelectrochemical cells are that the catalytic surface is bound to a solid support, permitting the operation in a flow system, and the successful oxidation even in oxygen-free conditions, and in the presence of high salt concentrations (see also Chapter III, Section I).[133]

See Chapter IV, Section II.B, on the photoelectrochemical oxidation of 4-chlorophenol.

REFERENCES

1. Davis, B. D., Dulbecco, R., Eisen, H. N., Ginsberg, H. S., and Wood, W. B., *Microbiology;* Harper & Row: New York, 1967, p. 346.
2. Masten, S. J. and Davies, S. H. R., The use of ozonation to degrade organic contaminants in wastewater, *Environ. Sci. Technol.,* 28, 180A–185A, 1994.
3. Suri, R. P. S., Liu, J., Hand, D. W., Crittenden, J. C., Perram, D. L., and Mullins, M. E., Heterogeneous photocatalytic oxidation of hazardous organic contaminants in water, *Water Environ. Res.,* 65, 665–673, 1993.
4. Pelizzetti, E., Photocatalytic processes for destruction of organic water contaminants, in: *Chemical Reactor Technology for Environmentally Safe Reactors and Products,* De Lasa, H. I. et al. (Eds.), Kluwer Academic Publishers, 1993, pp. 577–608.
5. Bard, A. J., Photoelectrochemistry and heterogeneous photocatalysis at semiconductors, *J. Photochem.,* 10, 59–75, 1979.
6. Schiavello, M., Basic concepts in photocatalysis, *NATO ASI Ser., Ser. C.,* 237, 351–360, 1988.
7. Schiavello, M., Some working principles of heterogeneous photocatalysis by semiconductors, *Electrochim. Acta,* 38, 11–14, 1993.
8. Pelizzetti, E., Minero, C., Pramauro, E., and Vincenti, M., Recent issues on environmental detoxification by solar photocatalysis, in: *Photochemical and Photoelectrochemical Conversion and Storage of Solar Energy,* Tian, Z. W. and Cao, Y. (Eds.), International Academic Publishers, 1992, pp. 217–233.
9. Larson, R. A., Marley, K. A., and Schlauch, M. B., Strategies for photochemical treatment of wastewater, *ACS Symp. Ser.* (Emerging Technol. Hazard. Waste Management), 468, 66–82, 1991.
10. Fox, M. A., Photocatalysis: decontamination with sunlight, *ChemTech,* 22, 680–685, 1992.
11. Mills, A., Davies, R. H., and Worsley, D., Water purification by semiconductor photocatalysis, *Chem. Soc. Rev.,* 22, 417–425, 1993.
12. Matthews, R. W., Photocatalytic oxidation of organic contaminants in water. An aid to environmental preservation, *Pure Appl. Chem.,* 64, 1285–1290, 1992.
13. Matthews, R. W., Photocatalysis in water purification: Possibilities, problems and prospects, in: *Photocatalytic Purification and Treatment of Water and Air,* Ollis, D. F. and Al-Ekabi, H. (Eds.), Elsevier Science Publishers, Amsterdam, 1993, pp. 121–138.
14. Legrini, O. Oliveros, E., and Braun, A. M., Photochemical processes for water treatment, *Chem. Rev.,* 93, 671–698, 1993.
15. Pichat, P. and Fox, M. A., Photocatalysis on semiconductors, in: *Photoinduced Electron Transfer,* Fox, M. A. and Chanon, M. (Eds.), Elsevier, Amsterdam, 1988, pp. 241–302.
16. Bahnemann, D. W., Bockelmann, D., and Goslich, R., Mechanistic studies of water detoxification in illuminated TiO_2 suspensions, *Sol. Energy Mater.,* 24, 564–583, 1991.
17. Serpone, N., A decade of heterogeneous photocatalysis in our laboratory: pure and applied studies in energy production and environmental detoxification, *Res. Chem. Intermed.,* 20, 953–992, 1994.

18. (a) Wilkinson, F. and Brummer, J. G., Rate constants for the decay and reactions of the lowest excited singlet state of molecular oxygen, *J. Phys. Chem. Ref. Data.,* 10, 804–999, 1981. (b) Bielski, H. J., Cabelli, D. E., and Arudi, R. L., Reactivity of HO_2/O_2^- radicals in aqueous solution, *J. Phys. Chem. Ref. Data,* 14, 1041–1100, 1985. (c) Buxton, G. V., Greenstock, C. L., Helman, W. P., and Ross, A. B., Critical review of rate constants for reactions of hydrated electrons, hydrogen atoms and hydroxyl radicals ($\cdot OH/\cdot O^-$) in aqueous solution, *J. Phys. Chem. Ref. Data,* 17, 513–886, 1988. (d) Neta, P., Huie, R. E., and Ross, A. B., Rate constants for reactions of inorganic radicals in aqueous solutions, *J. Phys. Chem. Ref. Data,* 17, 1027–1284, 1988.

19. Janzen, E. G., Spin trapping, *Acc. Chem. Res.,* 4, 31–40, 1971.

20. Jaeger, C. D. and Bard, A., Spin trapping and electron spin resonance detection of radical intermediates in the photodecomposition of water at titanium dioxide particulate systems, *J. Phys. Chem.,* 83, 3146–3152, 1979.

21. Wolfrum, E. J., Ollis, D. F., Lim, P. K., and Fox, M. A., The UV-H_2O_2 process. Quantitative epr determination of radical concentrations, *J. Photochem. Photobiol. A: Chem.,* 78, 259–265, 1994.

22. Noda, H., Oikawa, K., Ohya-Nishiguchi, H., and Kamada, H., Efficient hydroxyl radical production and their reactivity with ethanol in the presence of photoexcited semiconductors, *Bull. Chem. Soc. Jpn.,* 67, 2031–2037, 1994.

23. Kochany, J. and Lipczynska-Kochany, E., Application of the EPR spin-trapping technique for the investigation of the reactions of carbonate, bicarbonate, and phosphate anions with hydroxyl radicals generated by the photolysis of H_2O_2, *Chemosphere,* 25, 1769–1782, 1992.

24. Zafiriou, O. C., Blough, N. V., Micinski, E., Dister, B., Kieber, D., and Moffett, J., Molecular probe systems for reactive transients in natural waters, *Mar. Chem.,* 30, 45–70, 1990.

25. Gollnick, K., Chemical aspects of photodynamic action in the presence of molecular oxygen, in: *Radiation Research: Biochemical, Chemical and Physical Perspectives,* Nygaard, O. F., Adler, H. I., and Sinclair, W. K. (Eds.), Academic Press, New York, 1975, p. 590.

26. Gerischer, H., Conditions for an efficient photocatalytic activity of TiO_2 particles, in: *Photocatalytic Purification and Treatment of Water and Air,* Ollis, D. F. and Al-Ekabi, H. (Eds.), Elsevier Science Publishers, Amsterdam, 1993, pp. 1–17.

27. Howe, R. F. and Grätzel, M., EPR observation of trapped electrons in colloidal TiO_2, *J. Phys. Chem.,* 89, 4495–4499, 1985.

28. Rothenberger, R., Moser, J., Grätzel, M., Serpone, N., and Sharma, D. K., Charge carrier trapping and recombination dynamics in small semiconductor particles, *J. Am. Chem. Soc.,* 107, 8054–8059, 1985.

29. Lawless, D., Serpone, N., and Meisel, D., Role of $\cdot OH$ radicals and trapped holes in photocatalysis. A pulse radiolysis study, *J. Phys. Chem.,* 95, 5166–5170, 1991.

30. Serpone, N., Pelizzetti, E., and Hidaka, H., Identifying primary events and the nature of intermediates during the photocatalyzed oxidation of organics mediated by irradiated semiconductors, in: *Photocatalytic Purification and Treatment of Water and Air,* Ollis, D. F. and Al-Ekabi, H. (Eds.), Elsevier Science Publishers, Amsterdam, 1993, pp. 225–250.

31. Fox, M. A., The role of hydroxyl radicals in the photocatalyzed detoxification of organic pollutants: pulse radiolysis and time-resolved diffuse reflectance measurements, in: *Photocatalytic Purification and Treatment of Water and Air,* Ollis, D. F. and Al-Ekabi, H. (Eds.), Elsevier Science Publishers, Amsterdam, 1993, pp. 163–167.

32. Goldstein, S., Czapski, G., and Rabani, J., Oxidation of phenol by radiolytically generated $\cdot OH$ and chemically generated $SO_4{}^{\cdot-}$. A distinction between $\cdot OH$ transfer and hole transfer oxidation in the photolysis of TiO_2 colloid solution, *J. Phys. Chem.,* 98, 6586–6591, 1994.

33. Tanaka, K., Hisanaga, T., and Rivera, A. P., Effect of crystal form of TiO_2 on the photocatalytic degradation of pollutants, in: *Photocatalytic Purification and Treatment of Water and Air,* Ollis, D. F. and Al-Ekabi, H., Elsevier Science Publishers, Amsterdam, 1993, pp. 169–178.

34. Rivera, A. P., Tanaka, K., and Hisanaga, T., Photocatalytic degradation of pollutant over TiO_2 in different crystal structures, *Appl. Catal. B: Environ.,* 3, 37–44, 1993.

35. Schenk, M., Color reactions of trivalent and quadrivalent titanium, *Helv. Chim. Acta,* 19, 625–639, 1936.

36. Schwarzenbach, G., Mühleback, J., and Müller, K., Peroxo complexes of titanium, *Inorg. Chem.,* 9, 2381–2390, 1970.

37. Pelizzetti, E., Minero, C., and Maurino, V., The role of colloidal particles in the photodegradation of organic compounds of environmental concern in aquatic systems, *Adv. Colloid Interface Sci.,* 32, 271–316, 1990.

38. Stumm, W. (Ed.), *Aquatic Surface Chemistry, Chemical Processes at the Particle-Water Interface,* Wiley-Interscience, New York, 1987.

39. Sulzberger, B., Suter, D., Siffert, C., Banwart, S., and Stumm, W., Dissolution of Fe(III) (hydr) oxides in natural waters; laboratory assessment on the kinetics controlled by surface coordination, *Marine Chem.,* 28, 127–144, 1989.

40. Matthews, R. W., Kinetics of photocatalytic oxidation of organic solutes over titanium dioxide, *J. Catal.,* 111, 264–272, 1988.

41. Turchi, C. S. and Ollis, D. F., Mixed reactant photocatalysis: intermediates and mutual rate inhibition, *J. Catal.,* 119, 483–496, 1989.

42. Turchi, C. S. and Ollis, D. F., Photocatalytic degradation of organic water contaminants: mechanisms involving hydroxyl radical attack, *J. Catal.,* 122, 178–192, 1990.

43. Pruden, A. L. and Ollis, D. F., Photoassisted heterogeneous catalysis: the degradation of trichloroethylene in water, *J. Catal.,* 82, 404–417, 1983.

44. Ollis, D. F., Contaminant degradation in water, *Environ. Sci. Technol.,* 19, 480–484, 1985.

45. Abdullah, M., Low, G. K.-C., and Matthews, R. W., Effects of common inorganic anions on rates of photocatalytic oxidation of organic carbon over illuminated titanium dioxide, *J. Phys. Chem.,* 94, 6820–6825, 1990.

46. Mills, A. and Morris, S., Photomineralization of 4-chlorophenol sensitized by titanium dioxide: a study of the initial kinetics of carbon dioxide photogeneration, *J. Photochem. Photobiol. A: Chem.,* 71, 75–83, 1993.

47. Cunningham, J. and Al-Sayyed, G., Factors influencing efficiencies of TiO_2-sensitized photodegradation. Part 1. Substituted benzoic acids — Discrepancies with dark-adsorption parameters, *J. Chem. Soc. Faraday Trans. 1,* 86, 3935–3941, 1990.

48. Cunningham, J., Al-Sayyed, G., and Srijaranai, S., Adsorption of model pollutants onto TiO_2 particles in relation to photoremediation of contaminated water, in: *Aquatic and Surface Photochemistry,* Helz, G. R., Zepp, R. G., and Crosby, D. G. (Eds.), Lewis Publishers, Boca Raton, FL, 1994, pp. 317–347.

49. Cunningham, J. and Sedlák, P., Initial rates of TiO_2-photocatalysed degradations of water pollutants: influences of adsorption, pH and photon flux, in: *Photocatalytic Purification and Treatment of Water and Air,* Ollis, D. F. and Al-Ekabi, H. (Eds.), Elsevier Science Publishers, Amsterdam, 1993, pp. 67–81.

50. D'Oliveira, J.-C., Minero, C., Pelizzetti, E., and Pichat, P., Photodegradation of dichlorophenols and trichlorophenols in TiO_2 aqueous suspensions: kinetic effects of the positions of the Cl atoms and identification of the intermediates, *J. Photochem. Photobiol. A: Chem.,* 72, 261–267, 1993.

51. Davis, A. P. and Huang, C. P., A kinetic model describing photocatalytic oxidation using illuminated semiconductors, *Chemosphere,* 26, 1119–1135, 1993.

52. Okamoto, K., Yamamoto, Y., Tanaka, H., Tanaka, M., and Itaya, A., Heterogeneous photocatalytic decomposition of phenol over anatase powder, *Bull. Chem. Soc. Jpn.,* 58, 2015–2022, 1985.

53. Okamoto, K., Yamamoto, Y., Tanaka, H., and Itaya, A., Kinetics of heterogeneous photocatalytic decomposition of phenol over anatase TiO_2 powder, *Bull. Chem. Soc. Jpn.,* 58, 2023–2028, 1985.

54. Zeltner, W. A., Hill, C. G., and Anderson, M. A., Supported titania for photodegradation, *ChemTech,* 32, 21, 1993.

55. Hill, C. G., Jr., *An Introduction to Chemical Engineering. Kinetics and Reactor Design,* John Wiley & Sons, New York, 1977, Chapter 6.

56. Cunningham, J. and Srijaranai, S., Isotope-effect evidence for hydroxyl radical involvement in alcohol photo-oxidation sensitized by TiO_2 in aqueous suspension, *J. Photochem. Photobiol. A: Chem.,* 43, 329–335, 1988.

57. Gerischer, H. and Heller, A., The role of oxygen in photooxidation of organic molecules on semiconductor particles, *J. Phys. Chem.,* 95, 5261–5267, 1991.

58. Gerischer, H. and Heller, A., Photocatalytic oxidation of organic molecules at TiO_2 particles by sunlight in aerated water, *J. Electrochem. Soc.,* 139, 113–118, 1992.

59. Gerischer, H., Photoelectrochemical catalysis of the oxidation of organic molecules by oxygen on small semiconductor particles with TiO_2 as an example, *Electrochim. Acta,* 38, 3–9, 1993.

60. Henrich, V. E., Dresselhaus, D., and Zeiger, H. J., Observation of two-dimensional phases associated with defect states on the surface of titanium dioxide, *Phys. Rev. Lett.,* 36, 1335–1339, 1976.

61. Grätzel, M., Photocatalysis with colloidal semiconductors, *NATO ASI Sr., Ser. C,* 237, 367–398, 1988.

62. Grätzel, M., *Heterogeneous Photoelectrochemical Electron Transfer,* CRC Press, Boca Raton, FL, 1988.

63. Zhao, J., Hidaka, H., Takamura, A., Pelizzetti, E., and Serpone, N., Photodegradation of surfactants. 11. ζ-Potential measurements in the photocatalytic oxidation of surfactants in aqueous TiO_2 dispersions, *Langmuir,* 9, 1646–1650, 1993.

64. Mpandou, A. and Siffert, B., Sodium carboxylate adsorption onto TiO_2: shortest chain length allowing hemimicelle formation and shearplane position in the electric double layer, *J. Colloid Interface Sci.,* 102, 138–145, 1984.

65. Parks, G. A., The isoelectric point of solid oxides, solid hydroxides and aqueous hydroxo complex systems, *Chem. Rev.,* 65, 177–198, 1965.

66. Pichat, P., Guillard, C., Maillard, C., Amalric, L., and D'Oliveira, J.-C., TiO_2 photocatalytic destruction of water aromatic pollutants: intermediates; properties–degradability correlation; effects of inorganic ions and TiO_2 surface area; comparison with H_2O_2 processes, in: *Photocatalytic Purification and Treatment of Water and Air,* Ollis, D. F. and Al-Ekabi, H. (Eds.), Elsevier Science Publishers, Amsterdam, 1993, pp. 207–223.

67. Kazarinov, V. E., Andreev, V. N., and Mayorov, A. P., Investigation of the adsorption properties of the TiO_2 electrode by the radioactive tracer method, *J. Electroanal. Chem.,* 130, 277–285, 1981.

68. Tanaka, K., Capule, M. F. V., and Hisanaga, T., Effect of crystallinity of TiO_2 on its photocatalytic action, *Chem. Phys. Lett.,* 187, 73–76, 1991.

69. Mills, A. and Morris, S., Photomineralization of 4-chlorophenol by titanium dioxide. A study of annealing the photocatalyst at different temperatures, *J. Photochem. Photobiol. A: Chem.,* 71, 285–289, 1993.

70. Mills, A. and Sawunyama, P., Photocatalytic degradation of 4-chlorophenol mediated by TiO_2: a comparative study of the activity of laboratory made and commercial TiO_2 samples, *J. Photochem. Photobiol. A: Chem.,* 84, 305–309, 1994.

71. Palmisano, L., Schiavello, M., Sclafani, A., Matra, G., Borello, E., and Coluccia, S., Photocatalytic oxidation of phenol on TiO_2 powders. A Fourier transform study, *Appl. Catal. B: Environ.,* 3, 117–132, 1994.

72. Sclafani, A., Palmisano, L., and Schiavello, M., Phenol and nitrophenol photodegradation using aqueous TiO_2 dispersions, in: *Aquatic and Surface Photochemistry,* Helz, G. R., Zepp, R. G., and Crosby, D. G. (Eds.), Lewis Publishers, Boca Raton, FL, 1994, pp. 419–425.

73. Palmisano, L., Schiavello, M., Sclafani, A., Martin, I., and Rives, V., Surface properties of iron-titania photocatalysts employed for 4-nitrophenol photodegradation in aqueous TiO_2 dispersion, *Catal. Lett.,* 24, 303–315, 1994.

74. Serpone, N., Lawless, D., Disdier, J., and Herrmann, J.-M., Spectroscopic, photoconductivity, and photocatalytic studies of TiO_2 colloids: naked and with the lattice doped with Cr^{3+}, Fe^{3+}, and V^{5+} cations, *Langmuir,* 10, 643–652, 1994.

75. Larson, S. A. and Falconer, J. L., Characterization of TiO_2 photocatalysts used in trichloroethene oxidation, *Appl. Catal. B: Env.,* 4, 325–342, 1994.

76. Zang, L., Liu, C.-Y., and Ren, X.-M., Adsorption of cations on TiO_2 particles. A method to determine the surface density of OH groups, *J. Chem. Soc. Chem. Commun.,* 1865–1866, 1994.

77. Serpone, N., Terzian, R., Lawless, D., Kennepohl, P., and Sauvé, G., On the usage of turnover numbers and quantum yields in heterogeneous photocatalysis, *J. Photochem. Photobiol. A: Chem.,* 73, 11–16, 1993.

78. Formenti, M., Juillet, F., Mériaudeau, P., and Teichner, S. J., Heterogeneous photocatalysis for partial oxidation of paraffins, *Chem. Technol.,* 1, 680–686, 1971.

79. Schiavello, M., Augugliaro, V., and Palmisano, L., An experimental method for the determination of the photon flow reflected and absorbed by aqueous dispersions containing polycrystalline solids in heterogeneous photocatalysis, *J. Catal.,* 127, 332–341, 1991.

80. Augugliaro, V., Schiavello, M., and Palmisano, L., Rate of photon absorption and turnover number. 2. Parameters for the comparison of heterogeneous photocatalytic systems in a quantitative way, *Coord. Chem. Rev.,* 125, 173–182, 1993.

81. Childs, L. P. and Ollis, D. F., Is photocatalysis catalytic? *J. Catal.,* 66, 383–390, 1980.

82. Palmisano, L., Augugliaro, V., Campostrini, R., and Schiavello, M., A proposal for the quantitative assessment of heterogeneous photocatalytic processes, *J. Catal.,* 143, 149–154, 1993.

83. Lepore, P., Langford, C. H., Víchová, J., and Vlcek, A., Photochemistry and picosecond absorption spectra of aqueous suspensions of a polycrystalline titanium dioxide optically transparent in the visible spectrum, *J. Photochem. Photobiol. A: Chem.,* 75, 67–75, 1993.

84. Lepore, G. P., Pant, B. C., and Langford, C. H., Limiting quantum yield measurements for the disappearance of 1-propanol and propanal: an oxidation reaction study employing a TiO_2 based photoreactor, *Can. J. Chem.,* 71, 2051–2059, 1993.

85. Valladares, J. E. and Bolton, J. R., A method for determination of quantum yields in heterogeneous systems: the TiO_2 photocatalyzed bleaching of methylene blue, in: *Photocatalytic Purification and Treatment of Water and Air,* Ollis, D. F. and Al-Ekabi, H. (Eds.), Elsevier Science Publishers, Amsterdam, 1993, pp. 111–120.

86. Mills, A., Morris, S., and Davies, R., Photomineralisation of 4-chlorophenol sensitized by titanium dioxide: a study of the intermediates, *J. Photochem. Photobiol. A: Chem.,* 70, 183–191, 1993.

87. Sabate, J., Anderson, M. A., Kikhawa, H., Edwards, M., and Hill, C. G., A kinetic study of the photocatalytic degradation of 3-chlorosalicylic acid over TiO_2 membranes supported on glass, *J. Catal.,* 127, 167–177, 1991.

88. Sabate, J., Anderson, M. A., Aguado, M. A., Gimenez, J., Cevera-March, S., and Hill, C. G., Comparison of TiO_2 powder suspensions and TiO_2 ceramic membranes supported on glass as photocatalytic systems in the reduction of chromium(VI), *J. Mol. Catal.,* 71, 57–68, 1992.

89. Chester, G., Anderson, M., Read, H., and Esplugas, S., A jacketed annular membrane photocatalytic reactor for wastewater treatment. Degradation of formic acid and atrazine, *J. Photochem. Photobiol. A: Chem.,* 71, 291–297, 1993.

90. Aguado, M. A. and Anderson, M. A., Degradation of formic acid over semiconducting membranes supported on glass: effects of structure and electronic doping, *Solar Energy Mater. Solar Cells,* 28, 345–361, 1993.

91. Yue, P. L., Modelling, scale-up and design of multiphasic photoreactors, in: *Photocatalytic Purification and Treatment of Water and Air,* Ollis, D. F. and Al-Ekabi, H. (Eds.), Elsevier Science Publishers, Amsterdam, 1993, pp. 495–510.

92. Jakob, L., Oliveros, E., Legrini, O., and Braun, A. M., Titanium dioxide photocatalytic treatment of water. Reactor design and optimization experiments, in: *Photocatalytic Purification and Treatment of Water and Air,* Ollis, D. F. and Al-Ekabi, H. (Eds.), Elsevier Science Publishers, Amsterdam, 1993, pp. 511–532.

93. Findeling, C., Viriot, M. L., Carre, M. C., and Andre, J. C., Industrial photochemistry. 22. The best use of photons in water purification, *J. Photochem. Photobiol, A: Chem.,* 71, 191–193, 1993.

94. Nargiello, M. and Herz, T., Physical-chemical characteristics of P-25 making it extremely suited as the catalyst in photodegradation of organic compounds, in: *Photocatalytic Purification and Treatment of Water and Air,* Ollis, D. F. and Al-Ekabi, H. (Eds.), Elsevier Science Publishers, Amsterdam, 1993, pp. 801–807.

95. Kato, K., Photosensitized oxidation of ethanol on alkoxy-derived TiO_2 powders, *Bull. Chem. Soc. Jpn.,* 65, 34–38, 1992.

96. Kato, K., Synthesis of TiO_2 photocatalysts with high activity by the alkoxide method, in: *Photocatalytic Purification and Treatment of Water and Air,* Ollis, D. F. and Al-Ekabi, H. (Eds.), Elsevier Science Publishers, Amsterdam, 1993, pp. 809–813.

97. Aurian-Blajeni, B., Halmann, M., and Manassen, J., Photoreduction of carbon dioxide and water into formaldehyde and methanol on semiconductor materials, *Solar Energy,* 25, 165–170, 1980.

98. Serpone, N., Borgarello, E., Harris, R., Cahill, P., Borgarello, M., and Pelizzetti, E., Photocatalysis over TiO_2 supported on a glass substrate, *Solar Energy Mater.,* 14, 121–127, 1986.

99. Al-Ekabi, H. and Serpone, N., Kinetic studies in heterogeneous photocatalysis. 1. Photocatalytic degradation of chlorinated phenols in aerated aqueous solutions over TiO_2 supported on a glass matrix, *J. Phys. Chem.,* 92, 5726–5731, 1988.

100. Matthews, R. W., Photooxidation of organic impurities in water using thin films of titanium oxide, *J. Phys. Chem.,* 91, 3328–3333, 1987.

101. Matthews, R. W., Solar-electric water purification using photocatalytic oxidation with TiO_2 as a stationary phase, *Solar Energy,* 38, 405–413, 1987.

102. Matthews, R. W., Kinetics of photocatalytic oxidation of organic solutes over titanium dioxide, *J. Catal.,* 111, 264–272, 1988.

103. Matthews, R. W., Abdullah, M., and Low, G. K.-C., Photocatalytic oxidation for total organic carbon analysis, *Anal. Chim. Acta,* 233, 171–179, 1990.

104. Murabayashi, M., Itoh, K., Kuroda, S., Huda, R., Masuda, R., Takahashi, W., and Kawashima, K., Photocatalytic degradation of chloroform with TiO_2 coated glass fiber cloth, *Denki Kagaku,* 60, 741–742, 1992.

105. Kato, K., Tsuzuki, A., Torii, Y., Taoda, H., Kato, T., and Butsugan, Y., TiO_2 coatings tailored for photocatalytic degradation of organic compounds, *Abstr. 2nd Int. Symp. on New Trends in Photoelectrochemistry,* University of Tokyo, March 1994, p. 62.

106. Taoda, H., Watanabe, E., Horiuch, T., Ohta, K., and Iseda, K., Photocatalytic degradation of tetrachloroethylene in water using thin films of titanium dioxide, *Abstr. 2nd Int. Symp. New Trends Photoelectrochemistry,* University of Tokyo, March 1994, p. 76.

107. Matthews, R. W., Adsorption photocatalytic oxidation: a new method of water purification, *Chem. Ind. (London),* 28–30, 1988.

108. Zhang, Y., Crittenden, J. C., Hand, D. W., and Perram, D. L., Fixed-bed photocatalysts for solar decontamination, *Environ. Sci. Technol.,* 28, 435–442, 1994.

109. Bellobono, I. R., Bonardi, M., Castellano, L., Selli, E., and Righetto, L., Photosynthetic membranes. 23. Degradation of some chloroaliphatic water contaminants by photocatalytic membranes immobilizing titanium dioxide, *J. Photochem. Photobiol., A: Chem.,* 67, 109–115, 1992.

110. Bellobono, I. R., Carrara, A., Barni, B., and Gazzotti, A., Laboratory- and pilot-plant-scale photodegradation of chloroaliphatics in aqueous solutions by photocatalytic membranes immobilizing titanium dioxide, *J. Photochem. Photobiol. A: Chem.,* 84, 83–90, 1994.

111. Kato, K., Photocatalytic property of TiO₂ anchored on porous alumina ceramic support by the alkoxide method, *J. Ceramic Soc. Jpn.,* 1101, 245–249, 1993.

112. Tomonari, M., Takaoka, Y., Aketa, I., and Murasawa, S., Nano-structured TiO₂ photocatalysts for advanced applications, *Abstr. 2nd Internat. Symp. New Trends Photoelectrochemistry,* University of Tokyo, March 1994, p. 67.

113. O'Regan, B. and Grätzel, M., A low-cost, high efficiency solar cell based on dye-sensitized colloidal TiO₂ films, *Nature,* 353, 737–740, 1991.

114. Sopyan, I., Murasawa, S., Hashimoto, K., and Fujishima, A., Highly efficient TiO₂ film photocatalyst. Degradation of gaseous acetaldehyde, *Chem. Lett.,* 723–726, 1994.

115. Hofstadler, K., Bauer, R., Novalic, S., and Heisler, G., New reactor design for photocatalytic wastewater treatment with TiO₂ immobilized on fused-silica glass fibers: photomineralization of 4-chlorophenol, *Environ. Sci. Technol.,* 28, 670–674, 1994.

116. Dagan, G. and Tomkiewicz, M., TiO₂ aerogels for photocatalytic decontamination of aquatic environments, *J. Phys. Chem.,* 97, 12651–12655, 1993.

117. Tomkiewicz, M., Dagan, G., and Zhu, Z., Morphology and photocatalytic activity of TiO₂ aerogels, *Res. Chem. Intermed.,* 20, 701–710, 1994.

118. Turchi, C. S. and Ollis, D. F., Photocatalytic reactor design: An example of mass-transfer limitations with an immobilized catalyst, *J. Phys. Chem.,* 92, 6852–6853, 1988.

119. Sclafani, A., Brucato, A., and Rizzuti, L., Mass transfer limitations in a packed bed photoreactor used for phenol removal, in: *Photocatalytic Purification and Treatment of Water and Air,* Ollis, D. F. and Al-Ekabi, H. (Eds.), Elsevier Science Publishers, Amsterdam, 1993, pp. 533–545.

120. Yatmaz, H. C., Howarth, C. R., and Wallis, C., Photocatalysis in a falling film reactor, in: *Photocatalytic Purification and Treatment of Water and Air,* Ollis, D. F. and Al-Ekabi, H. (Eds.), Elsevier Science Publishers, Amsterdam, 1993, pp. 795–800.

121. Christensen, P. A., Hamnett, A., He, R., Howarth, C. R., and Shaw, K. E., Fundamental photocatalytic studies on immobilized films of TiO₂, in: *Photocatalytic Purification and Treatment of Water and Air,* Ollis, D. F. and Al-Ekabi, H. (Eds.), Elsevier Science Publishers, Amsterdam, 1993, pp. 765–770.

122. Notthakun, S., Crittenden, J. C., Hand, D. W., Perram, D. L., and Mullins, M. E., Regeneration of adsorbents using heterogeneous advanced oxidation, *J. Environ. Eng.,* 119, 695–714, 1993.

123. Fantoni, R., Giardiniguidoni, A., Mele, A., Pizzella, G., and Teghil, R., Laser degradation of pollutants. Polychlorobiphenyls, triazines and polycyclic aromatic hydrocarbons, *Proc. Indian Acad. Sci. — Chem. Sci.,* 105, 735–746, 1993.

124. Eliasson, B., Kogelschatz, U., and Stein, H. J., New trends in high intensity UV generation, *EPA Newslett.,* 32, 29–40, 1988.

125. Kogelschatz, U., Silent-discharge driven excimer UV sources and their applications, *Appl. Surf. Sci.,* 54, 410–423, 1992.

126. Jakob, L., Hashem, T. M., Bürki, S., Guindy, N. M., and Braun, A. M., Vacuum-ultraviolet (VUV) photolysis of water: oxidative degradation of 4-chlorophenol, *J. Photochem. Photobiol. A: Chem.,* 75, 97–103, 1993.

127. Oppenländer, T., Novel incoherent excimer UV irradiation units for the application in photochemistry, photobiology/-medicine and for waste water treatment, *EPA Newslett.,* 50, 2–8, 1994.

128. Oppenländer, T., Baum, G., and Egle, W., Comparison of advanced oxidation processes (AOP's) for waste water treatment, *EPA Newslett.,* 52, 53–57, 1994.

129. Loraine, G. A., Short wavelength ultraviolet photolysis of aqueous carbon tetrachloride, *Haz. Waste Haz. Mater.,* 10, 185–194, 1993.

130. Sczechovski, J. G., Koval, C. A., and Noble, R. D., Evidence of critical illumination and dark recovery times for increasing the photoefficiency of aqueous heterogeneous photocatalysis, *J. Photochem. Photobiol. A: Chem.,* 74, 273–278, 1993.

131. Kawaguchi, H., Dependence of photocatalytic reaction rate on titanium dioxide concentration in aqueous suspensions, in: *Photocatalytic Purification and Treatment of Water and Air,* Ollis, D. F. and Al-Ekabi, H. (Eds.), Elsevier Science Publishers, Amsterdam, 1993, pp. 665–682.

132. Kawaguchi, H., Dependence of photocatalytic reaction rate on titanium dioxide concentration in aqueous suspensions, *Environ. Technol.,* 15, 183–188, 1994.

133. Kim, D. H. and Anderson, M. A., Photoelectrolytic degradation of formic acid using a porous TiO_2 thin-film electrode, *Environ. Sci. Technol.,* 28, 479–483, 1994.

Chapter II

Inorganic Ions and Molecules

Inorganic ions and molecules include the water pollutants nitrite, nitrate, and sulfide, the highly toxic cyanide, heavy metal ions such as Hg, Pb, and Cr, and some extremely toxic metal-organic compounds.

Direct photoexcitation of inorganic anions may be either by "charge transfer to solvent" (CTTS) transitions, involving the excitation of an electron from the hydrated or solvated ion to an orbital in the solvation shell, or alternatively by an internal transition, with excitation of an electron to a higher orbital of the ion, such as in n–π* or π–π* transitions.[1]

I. WATER AND HYDROGEN PEROXIDE

In the photodegradation of water pollutants, the interactions of water molecules with the substrates, sensitizers, catalysts, and radical intermediates are important steps in the overall reactions. The direct photolysis of water occurs only by far-ultraviolet (vacuum UV) irradiation. Photolysis by near-UV or visible ($\lambda > 300$ nm) light requires sensitizers. In homogeneous solutions, water was photooxidized using Prussian blue as a sensitizer and $Na_2S_2O_8$ as sacrificial electron acceptor.[2]

Di- and trihydroxybenzene derivatives, such as gallic acid (3,4,5-trihydroxybenzoic acid), tannic acid (tannin, which is a mixture of gallate esters of glucose, is a cancer suspect agent), and various catechols in weakly alkaline aqueous media (pH 7.0 to 8.5) in the presence of oxygen or air are known to undergo a thermal reaction causing the production of hydrogen peroxide. A similar photochemical oxidation of water to H_2O_2 was shown to occur by illumination with UV light, even in mildly acidic solutions (pH 6.0 or below), in which the thermal reaction was negligible. Highest quantum yields for hydrogen peroxide production were obtained under illumination at 254 nm with air-saturated aqueous solutions (4 mM) of 3,5-dihydroxybenzoic acid, gallic acid, and pyrogallol, reaching 3.3, 2.4, and 2.2%, respectively.[3] Such catechol derivatives are ubiquitous decomposition products of plant materials and are commonly found in all natural waters. With their near-UV absorption bands (in the 300 to 400 nm region) they may thus be one of the possible sensitizers involved in the production of hydrogen peroxide in natural waters.

Heterogeneous photocatalytic oxidation of water to hydrogen peroxide or oxygen has attracted much attention.[4,5] Hydrogen peroxide was also efficiently formed in aerated aqueous solutions illuminated with light shorter than 380 nm,

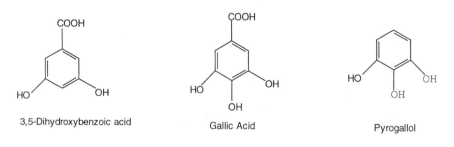

3,5-Dihydroxybenzoic acid   Gallic Acid   Pyrogallol

SCHEME II.1

in the presence of cobalt(II) tetrasulfophthalocyanine covalently linked to TiO_2.[6]

Remarkably high quantum yields of hydrogen peroxide were produced by UV illumination of quantum-sized (Q-sized) colloidal particles of ZnO suspended in oxygenated solutions in the presence of hole scavengers. With Q-sized ZnO (4 to 5 nm), the initial rate of production of hydrogen peroxide was up to 1000 times more rapid than with regular bulk ZnO (particle sizes 10 to 1000 μM). The colloidal particles of ZnO were prepared by the hydrolysis of zinc acetate in isopropanol. The quantum yields were highest at low illumination intensities, reaching up to $\Phi = 0.3$, and depended on the inverse square root of the light intensity. The formation of hydrogen peroxide was due to the two-electron reduction of oxygen with conduction band electrons,

$$O_2 + 2e_{cb}^- + 2H^+ \rightarrow H_2O_2 \tag{1}$$

Efficient hydrogen peroxide production required the presence of hole scavengers. Among various electron donors tested, the order of reactivity was formate > oxalate > acetate > water > citrate, with which the quantum yields of peroxide production were 5.4, 4.2, 3.1, 2.8, and 2.3%, respectively. The only oxidation product from citrate, oxalate, and formate was CO_2. From acetate as hole scavenger, the oxidation products were glyoxylate, glycolate, and formate, which suggested that the oxidation of acetate involved surface-adsorbed $\cdot$OH radicals.[7-9]

The rate of the true photocatalytic decomposition of H_2O_2 to oxygen over aqueous suspensions of TiO_2 was determined in a kinetic study in which the photochemical (noncatalytic) decomposition of H_2O_2 was determined in a similar experiment, but with Cr-doped TiO_2, which has no photocatalytic activity. This true photocatalytic rate depended on the initial concentration of H_2O_2 in accordance with the Langmuir–Hinshelwood mechanism. The observed oxygen evolution was smaller than expected from the stoichiometry,

$$H_2O_2 \rightarrow H_2O + {}^1\!/_2 O_2 \tag{2}$$

This was accounted for by the decomposition of hydrogen peroxide to hydroxyl radicals, which hydroxylate the TiO_2 particle surface layer according to

$$\cdot OH + Ti^{4+}O^{2-} \rightarrow Ti^{3+}(OOH)^{-} \tag{3}$$

The same mechanism of hydroxylation of TiO_2 was proposed to explain the photoinduced uptake of oxygen by aqueous suspensions of this semiconductor, that is, reduction of oxygen, followed by protonation and dissociation of the hydrogen peroxide to hydroxyl radicals:[10]

$$O_2 + 2e^- + 2H^+ \rightarrow H_2O_2 \rightarrow \cdot OH \tag{4}$$

II. HALIDE IONS

The halide ions are transparent in the visible region and absorb light only in the ultraviolet or far ultraviolet regions. The absorption maxima (and molar absorptivities) of the first two absorption bands of Br^- in aqueous solutions are 196 and 188 nm (12,000 and 13,000 mol^{-1} cm^{-1}), respectively, and those of iodide are 226 and 195 nm (13,700 and 14,000 L mol^{-1} cm^{-1}), respectively.[11] These far-ultraviolet absorption bands have been assigned to "charge transfer to solvent" transitions.[12]

The photooxidation of bromide ions in dilute aqueous solutions has been the subject of many studies. The generally accepted mechanism involves hydroxyl radicals as the primary oxidant,

$$\cdot OH + Br^- \rightarrow BrOH^- \rightarrow Br + OH^- \tag{5}$$

followed by several reactions of the $BrOH^-$ radical anion,

$$BrOH^- + Br^- \rightarrow Br_2^- + OH^- \tag{6}$$

$$Br + Br^- \rightarrow Br_2^- \tag{7}$$

The transient Br_2^- radical anion, with a strong absorption band at $\lambda_{max} = 350$ nm, and also the Cl_2^- and I_2^- radical anions, had been produced and detected by flash photolysis[13-15] and pulse radiolysis[16] techniques.

A useful application of photocatalyzed oxidation is in the recovery of oxidizable species from solution. Thus, in steady-state photolysis, the photooxidation of chloride or bromide with near-UV light in dilute aqueous solutions has been achieved by sensitization either with suspended or colloidal semiconductor particles such as TiO_2 or CdS,[17-21] or in homogeneous solutions, by sensitization with ferric chloride,[22] with anthraquinone sulfonates,[23–27] with an iridium bipyridine complex[28] and with aluminum-, copper-, or cobalt-phthalo-cyanine tetrasulfonic acids.[29]

III. NITROGEN COMPOUNDS

A. Nitrate

The mechanism of direct photolysis of the nitrate ion has been studied by flash photolysis, with conductometric detection.[30] The primary reactions in the photoexcitation of nitrate are the formation of the nitrite ion and the radical anion $\cdot O^-$, which is rapidly protonated to the highly reactive hydroxyl radical,

$$NO_3^- + h\nu \rightarrow NO_2^- + O(^3P) \tag{8}$$

$$NO_3^- + h\nu \rightarrow NO_2 + \cdot O^- \tag{9}$$

$$\cdot O^- + H_2O \rightarrow \cdot OH + OH^- \tag{10}$$

In dilute aqueous solutions of nitrate in the pH range 6 to 8 illuminated at 313 nm, the quantum yield for the production of $\cdot OH$ radicals was measured to be 0.013 and 0.017 at 20°C and 30°C, respectively.[31]

B. Nitrite

The direct UV photolysis of the nitrite ion in anaerobic aqueous solutions at its absorption maximum (360 nm) resulted in the formation of hydroxyl and $\cdot NO_2$ radicals, formed by the following sequence of reactions:[32–35]

$$\left(NO_2^-\right)^* + H_2O \rightarrow \cdot OH + \cdot NO + OH^- \tag{11}$$

$$\cdot OH + \cdot NO \rightarrow HNO_2 \tag{12}$$

$$\cdot OH + NO_2^- \rightarrow \cdot NO_2 + OH^- \tag{13}$$

$$\cdot NO + \cdot NO_2 \rightarrow N_2O_3 \tag{14}$$

$$N_2O_3 + H_2O \rightarrow 2HNO_2 \tag{15}$$

Thus, in oxygen-free solutions of nitrite, photolysis does not result in any apparent chemical change. In air- or oxygen-containing solutions, an additional reaction is oxidation of the nitrogen oxide radicals,

$$\cdot NO + \cdot NO_2 + O_2 + H_2O \rightarrow 2HNO_3 \tag{16}$$

The formation of the $\cdot OH$ radical during the anaerobic photolysis of the nitrite ion in aqueous solution was proven by electron spin resonance (ESR) identification of the spin trap radical formed when irradiated at 360 nm in the presence of the spin trap 5,5-dimethyl-1-pyrroline *N*-oxide (DMPO). The $\cdot NO_2$ radical did not react with DMPO, but a spin-trap radical was formed by irradiating nitrite in aqueous solutions containing nitromethane. In order to confirm that the resulting spin-trap adduct was indeed derived from the nitrite ion, ^{15}N-labeled nitrite was used. The resulting ESR spectrum indicated the formation of the spin adduct $^-O_2$ $^{15}N=\cdot CH=^{14}NO_2^-$. In air-saturated solutions of pure $NaNO_2$ (0.05 *M*), UV irradiation at 360 nm caused only very minor oxygen consumption, with a quantum yield of 0.0007. Addition of organic solutes resulted in very marked increases in the rates of oxygen consumption, with quantum yields ranging from 0.031 in the presence of phenol to 0.066 in the presence of propionaldehyde. Presumably, the hydroxyl radicals formed in the photolysis of nitrite reacted with the organic compounds, producing carbon-centered radicals, which scavenged oxygen molecules, leading to peroxy radicals. The formation of carbon-centered radicals was proven by photolysis of anaerobic solutions of sodium nitrite containing DMPO as well as several organic substrates. In the presence of methanol, ethanol, dimethyl sulfoxide (DMSO), formate, and formamide, the spin adducts identified were $\cdot CH_2OH$, $\cdot CH(CH_3)OH$, $\cdot CH_3$, $\cdot CO_2^-$, and $\cdot CONH_2$, respectively. The very high rates of consumption of oxygen and oxidation of organic substrates in near-UV-illuminated aqueous solutions containing nitrite suggests that such a process may be important in the natural decontamination of organic pollutants in environmental waters.[35]

C. Nitrite to Nitrate

Nitrite contamination in drinking water is an environmental problem because of the health hazard of nitrosation of body fluids, leading to carcinogens such as nitrosamines.[36] Thus, there have been some efforts to study the conversion of nitrite to the less dangerous nitrate. The photocatalytic oxidation of nitrite to nitrate was achieved in the presence and absence of oxygen, in heterogeneous mixtures containing aqueous slurries of powdered semiconductors, such as Ag_2O, PbO, TiO_2, Si, ZnO, SnO_2, CdS, and Bi_2O_3. Most rapid oxidation was with TiO_2 (anatase) in the presence of oxygen.[37] The rate of the TiO_2- promoted photooxidation of nitrite to nitrate was found to increase with rising pH.[38] The mechanism proposed involves interaction of hydroxyl radicals with the nitrite ion,

$$2 \cdot OH + NO_2^- \rightarrow NO_3^- + H_2O \qquad (17)$$

The photooxidation of nitrite ions on anatase TiO_2 was shown to proceed in a zero-order reaction with respect to O_2.[39]

The photoassisted oxidation of NO_2^- to NO_3^- in aqueous solutions open to air was performed by illumination with a 250-W xenon lamp in the presence of dispersions of several powdered semiconductors. Optimal yields of photo-oxidation were achieved with 2 g/L TiO_2 and with 4 g/L ZnO. WO_3 and Fe_2O_3 were less active. From illuminated ZnO suspensions, hydrogen peroxide was also produced, presumably formed via reduction of oxygen by photogenerated electrons. No H_2O_2 was released when using TiO_2 as photocatalyst, possibly because any peroxide formed remained bound to the TiO_2 surface. The yields of nitrite photooxidation decreased with rising pH of the solutions. This was explained to be due to negative charges on these semiconductor particles at pH values higher than the point of zero charge (PZC), which are at pH 6 and 8.5 for TiO_2 and ZnO, respectively. These negative charges hinder the approach and adsorption of NO_2^- to the semiconductor surface. At lower pH than the PZC, the semiconductors are positively charged, thus attracting and facilitating the adsorption of the nitrite ions. The photooxidation of nitrite with both TiO_2 and ZnO followed first-order kinetics and was in agreement with the Langmuir–Hinshelwood model. The proposed mechanism involved oxidation of surface-adsorbed nitrite ions by hydroxyl radicals.[40] Enhanced photoactivity in the photooxidation of nitrite to nitrate was obtained by iron doping of the TiO_2. The most active photocatalyst was TiO_2 doped with 0.5 wt% Fe. With the Fe-doped catalysts, the nitrite photooxidation observed zero-order kinetics.[41]

D. Nitrate to Nitrite, Hydroxylamine, and Ammonia

The reduction of nitrate ions to nitrite and ammonia has also been achieved, using several photocatalysts in aqueous suspensions, and in the absence of added oxygen. Illumination of solutions of 1 M $NaNO_3$ containing 0.3 wt% Pt-TiO_2 with a 500-W xenon arc through a Pyrex® window ($\lambda > 300$ nm) resulted in the formation of NH_3, NO_2^-, and O_2 at rates of 1.1, 0.1, and 1.1 μmol/h, respectively. In acidic solutions (1 M HNO_3) with the same catalyst, the production of nitrite was suppressed, and the rate of ammonia formation was 2.4 μmol/h. On the other hand, in alkaline solutions (1 M $NaNO_3$ + 0.1 M NaOH), the ammonia production was suppressed, while the nitrite formation reached 2.8 μmol/h. Presumably, hydrogen produced on the Pt particles of the catalyst served to reduce nitrate to ammonia. Bare TiO_2, without platinum, was ineffective in the reduction of nitrate to ammonia.[42]

Illumination of 2.2 M $NaNO_3$ in deoxygenated aqueous suspensions of TiO_2 (Degussa P25) and $SrTiO_3$ (Ventron) in the presence of 2 M H_3PO_4 for 24 h with a 150-W xenon lamp resulted in the formation of 1.6 and 2.0 μM hydroxylamine and of 160 and 400 μM ammonia, respectively. The overall chemical yield was minute (less than 0.1%). Substantial yields of hydroxylamine and extremely high yields of ammonia were obtained in the photoelectrochemical reduction of KNO_3 (2.2 M, in 2 M H_3PO_4) at an illuminated p-GaP photoelectrode (−0.09 V vs. SCE) and a platinum counterelectrode in a two-compartment cell, with the current maintained constant at 0.05 mA. The faradaic yield (current conversion efficiency) for hydroxylamine production according to the equation,

$$HNO_3 + 6H^+ + 6e^- \rightarrow NH_2OH + 2H_2O \qquad (18)$$

after 0.5 h of illumination with a 45-W high-pressure Hg lamp was 77%. The faradaic yields for ammonia were much larger (more than 100%) than those calculated for the eight-electron reduction of nitrate,

$$NO_3^- + 10H^+ + 8e^- \rightarrow NH_4^+ + 3H_2O \qquad (19)$$

which may indicate the occurrence of a photocatalytic reaction on the p-GaP surface.[43]

E.   Nitrite to Ammonia

The photoassisted reduction of the nitrite anion to ammonia was achieved in deoxygenated alkaline (1 M KOH) aqueous suspensions of semiconductors in the presence of sodium sulfide (0.1 M) as electron donor under illumination with a 150-W xenon lamp. No reaction occurred in the absence of the sulfide, and sodium sulfite was ineffective as electron donor. The order of photoactivity of different catalysts was $SrTiO_3$ > TiO_2 > CdS ~ CdS-ZnS (coprecipitated on SiO_2). Hydroxylamine production was not observed. Nitrate ions were not reduced to ammonia under these conditions.[44] The reduction of nitrite to ammonia was also carried out in a photoelectrochemical cell, using a CdSe/CdS photoanode and a stainless steel counterelectrode, in a single-compartment cell with 1 M KOH, 1 M potassium polysulfide, 1 M sulfur, and 0.2 M KNO_2, in an argon atmosphere. The cell was operated as a regenerative liquid junction solar cell, with simultaneous production of electrical energy (initially about 5% efficiency, which, however, decreased markedly within several days). Under illumination of the photoanode with a W-halogen lamp (100 mW cm^{-2}), the NH_3 production rate per illuminated area reached 18 μmol h^{-1} cm^{-2}.[45]

In the photocatalyzed reduction of nitrite to ammonia in aqueous dispersions of ZnS, the catalyst was effective only when pretreated by heating in air at about 600°C for 2 h, followed by etching with concentrated HNO_3 to leach out any oxide or sulfate formed. The yields of nitrite reduction were enhanced by metallization of the ZnS, optimally with about 1% metal loading. The order of catalytic activity of the metals was Rh > Ru > Pt > Pd. The yield of nitrite reduction to ammonia was increased in the presence of the sacrificial electron donor sulfite. Nitrate ion reduction to ammonia was also achieved in aqueous suspensions of ZnS, although at a lower rate than that of nitrite.[46]

See also Chapter III, Section IV, on the photochemical nitrosation and nitration of secondary amines, aromatic amino acids, and polycyclic aromatic compounds by aqueous nitrite and nitrate ions, with formation of mutagenic nitroso and nitro derivatives. See Chapter VIII, Section I on the nitrate-induced photooxidation of organic compounds in natural waters.

F. Ammonia to Nitrate

Near-UV illumination of NH_4^+ ions in aqueous dispersions of TiO_2 resulted in oxidation to NO_3^-. The dependence of initial reaction rates on initial ammonium ion concentrations was in accordance with the Langmuir–Hinshelwood model.[47]

G. Cyanides, Hexacyanoferrate(II) and -(III) Ions

Cyanide ions are widely used in precious metals processing, for example, to recover gold-mill tailings. In impoundments of gold-mill tailings, cyanide often is complexed as hexacyanoferrate(II) and hexacyanoferrate(III). These cyanoferrate complexes may undergo photolysis, releasing "free" cyanide ions. While cyanide ions themselves can be oxidized by alkaline chlorination or with hydrogen peroxide, the thermodynamically stable cyanide complexes of iron, cobalt, and nickel are resistant to these treatments. In the United States, the land disposal of solid wastes containing cyanide has been banned (RCRA Act of 1987, Section 3004).[48]

The photocatalytic oxidation of cyanide ions in alkaline aqueous solutions was accomplished by Frank and Bard[49,50] in heterogeneous media, using semiconductor powders, such as TiO_2, ZnO, and CdS, yielding the nontoxic cyanate ion, CNO^-, as the product. Fe_2O_3 and WO_3 were inactive as photocatalysts.[50] In a further study it was found that CNO^- was the product of the photooxidation of cyanide only during the first minutes of irradiation, but that after long illumination times the cyanide carbon was completely converted to carbonate. The reaction was applied to a continuous flow process by impregnating the TiO_2 onto 3-Å molecular sieve, which was packed into a glass column. Using this flow reactor, photodegradation resulted in up to 97% removal of cyanide.[51] While both TiO_2 and ZnO were effective as photocatalyst for the oxidation of cyanide, Fe_2O_3, WO_3, ZnS, and CdS were inactive. Initial rates of cyanide photooxidation were about four times higher with TiO_2 than with ZnO. CdS was found to dissolve within 15 min in cyanide solutions, forming a cyano complex, $Cd(CN)_4^{2-}$. The chemical yield of photooxidation of cyanide increased with the concentration of ZnO as photocatalyst. At initially 0.1 mM

cyanide solutions, the limiting value was reached at about 8 g/L ZnO in suspension, with which 68% oxidation was reached after 9 min of irradiation with a 125-W high-pressure Hg lamp. In strongly alkaline solution (pH > 12), ZnO is unstable, dissolving to form zincates. On the other hand, TiO_2 is stable even in strongly alkaline solutions (pH 13) and can be reused repeatedly. This was applied to the elimination of cyanide from aqueous copper cyanide solutions. The presence of copper ions was found to enhance the rate of cyanide photooxidation, with Cu^{2+} trapping conduction band electrons, and being reduced to $Cu°$.[52]

Removal of cyanide from water was also achieved by solar irradiation, using ZnO as the photocatalyst. Optimal rates of oxidation of cyanide were achieved at pH 11. The intermediate product was cyanate, but after 2 h of exposure of initially 0.1 mM CN^-, 87% of the carbon atoms was converted to CO_3^{2-}.[53] In the presence of rhodium-loaded cadmium sulfide in alkaline sulfide media, illumination with visible light resulted in the transformation of cyanide to the much less toxic thiocyanate, with a quantum efficiency of more than 25%.[54]

Alkaline aqueous solutions (pH 10.5) of potassium hexacyanoferrate(II) (potassium ferrocyanide, $K_4[Fe(CN)_6]$) and of potassium hexacyanoferrate(III) (potassium ferricyanide, $K_3[Fe(CN)_6]$), under sunlight irradiation in an open vessel underwent slow photolysis, releasing cyanide, which was oxidized to cyanate. However, after 17 d of such exposure, a large fraction of the original hexacyanoferrate was still either unconverted or remained as free cyanide.[55] This direct photolysis yielded as primary products aquapentacyanoferrate(II) and -(III):[56]

$$\left[Fe(CN)_6\right]^{4+} + H_2O = \left[Fe(CN)_5H_2O\right]^{3-} + CN^- \tag{20}$$

$$\left[Fe(CN)_6\right]^{3+} + H_2O = \left[Fe(CN)_5H_2O\right]^{2-} + CN^- \tag{21}$$

The primary step in the excitation of the hexacyanoferrate(II) ion by illumination at $\lambda < 313$ nm was found to be a CTTS transition, with formation of the hydrated electron. This was followed by a photoaquation reaction leading to the pentacyanoaqua complex. Laser pulse excitation at 337 nm of aqueous hexacyanoferrate(II) caused formation of the pentacyanoaqua complex within ≤ 1 ns. The pentacyanoaqua complex was also formed by steady state illumination in the wavelength range $\lambda = 254$ to 365 nm, and was identified by its yellow color (λ_{max} about 440 nm). When the light was interrupted this color disappeared, due to the back reaction of the pentacyanoaqua complex with cyanide, reforming the hexacyanoferrate(II) ion.[56]

In the presence of TiO_2 (10 g/L), the oxidation was much more rapid. Starting with an initial concentration of 1 mM of either $K_3[Fe(CN)_6]$ or

$K_4[Fe(CN)_6]$, within 11 or 12 days of sunlight exposure the total amounts of cyanide and hexacyanoferrate had been decreased to practically zero concentration. The main products of this oxidation were cyanate and nitrate. This photocatalytic method could be a practical approach to the treatment of cyanide-containing wastewaters.[55]

In another study, potassium ferricyanide (~1 mM), in the presence of a dilute TiO_2 sol (~10 mM, stabilized with 1 wt% polyvinyl alcohol) at pH 10 in a fused silica cuvette was illuminated by UV light or sunlight. Using a 4-W Hg lamp, 93% of the $[Fe(CN)_6]^{3-}$ had disappeared after 9 h. Complete destruction of ferricyanide was achieved by sunlight exposure of only 1.5 h.[57]

H. Thiocyanate

Laser flash photolysis of aqueous solutions of thiocyanate in the presence of colloidal TiO_2 led to the formation of a transient, detected by its absorption spectrum, with $\lambda_{max} = 480$ nm.[58] This same transient had previously been observed by pulse radiolysis of thiocyanate solutions, and was assigned as the dimer radical anion. Its formation was shown to be due to attack by hydroxyl radicals, followed by reaction of the thiocyanate radical with the excess thiocyanate ion. The dimer radical anion then decomposed to form the stable product, $(SCN)_2$:[59]

$$\cdot OH + SCN^- \rightarrow \cdot SCN + OH^- \tag{22}$$

$$\cdot SCN + SCN^- \rightarrow \cdot (SCN)_2^- \tag{23}$$

$$2 \cdot (SCN)_2^- \rightarrow (SCN)_2 + 2SCN^- \tag{24}$$

The same transient, with $\lambda_{max} = 480$ nm, was also observed by laser flash photolysis of thiocyanate in aqueous solution containing suspended TiO_2 (Degussa P25), but using diffuse reflectance detection.[60]

I. Azide

Illumination of aqueous solutions of N_3^- at 254 nm resulted in the formation of N_2, NH_2OH, H_2, N_2H_4, and NH_3. In dilute solution (≤1 mM) the quantum yield for the photodecomposition was 0.27.[61]

IV. PHOSPHORUS OXYANIONS

Orthophosphate is nontoxic and is a natural component of environmental waters. However, excessive loading of orthophosphate and of polyphosphates (e.g., from sewage and from detergents) in fresh water and coastal waters causes eutrophication and the appearance of algal blooms. Photochemical effects with inorganic phosphorus oxyanions occur mainly in the far-UV region, below 200 nm.

Aqueous solutions of salts of $H_2PO_4^-$, HPO_4^{2-}, pyrophosphate, tripolyphosphate, trimetaphosphate, hypophosphite, phosphite, and hypophosphate have broad shallow UV absorption bands above 200 nm, probably due to internal electronic transitions, and steeply rising absorption edges below 200 nm. Solvent and temperature effects on the short-wavelength absorption edges indicated that these are due to charge transfer to solvent (CTTS) transitions, thus resulting in the formation of hydrated electrons, e_{aq}^-, as well as of phosphorus oxyanion radicals.[62,63] This conclusion was confirmed by flash photolysis[64] and pulse radiolysis[65] experiments on aqueous solutions of mono-, di-, and tribasic orthophosphate and of pyrophosphate anions. The transient absorption bands observed included that of the hydrated electron at $\lambda_{max} = 700$ nm, and of the $\cdot H_2PO_4$ and $\cdot HPO_4^-$ radicals, for both of which λ_{max} was at 500 nm.[64]

V. SULFUR OXYANIONS AND SULFIDE

The SO_3^{2-} ion in aqueous solution has a strong absorption band below 265 nm, which was assigned to a charge transfer to solvent transition,[66] leading to the hydrated electron and to the sulfite radical anion $\cdot SO_3^-$. In the presence of oxygen, the predominant subsequent reaction was the formation of the peroxymonosulfate radical anion,[67]

$$\cdot SO_3^- + O_2 \rightarrow \cdot SO_5^- \qquad (25)$$

Under illumination at 254 nm in the absence of oxygen, the main products were sulfate and dithionate, produced in a very rapid chain reaction,[68] with a quantum yield of about 500.[69]

$$SO_3^{2-} \rightarrow SO_4^{2-} + S_2O_6^{2-} \qquad (26)$$

In heterogeneous photocatalysis, sulfite ions in aqueous solutions were photooxidized by illumination in the presence of the suspended powdered semiconductors TiO_2, ZnO, CdS, and Fe_2O_3, with the order of activity $Fe_2O_3 \cong ZnO \cong CdS > TiO_2$.[50] Enhanced photooxidation of aqueous aerated sulfur dioxide was obtained using cobalt(II) tetrasulfophthalocyanine covalently linked to the surface of TiO_2 particles — resulting in the conversion of SO_2 to sulfate. The mechanism proposed involved valence-band holes reacting with adsorbed S(IV) to produce S(VI), while electrons were channeled via the cobalt–titanium complex to molecular oxygen, producing hydrogen peroxide.[70] Aqueous colloidal suspensions of α-Fe_2O_3 were also found to be effective as photocatalysts in the oxidation of sulfur dioxide, in the pH range of 2 to 10.5. A maximal quantum yield of 0.3 was observed at pH 5.7.[71]

Photolysis of aqueous solutions of Na_2S (pH ~ 11) under illumination at λ = 248 nm resulted in hydrogen and sulfane, H_2S_2, as the primary products. The initial rate of hydrogen generation was proportional to the light intensity, and the quantum yield of hydrogen production was about 0.04. During illumination the absorption maximum of NaS_2 at 230 nm decreased and new absorption bands appeared at 305 and 370 nm. The solution changed from colorless to yellow, due to production of sodium sulfane, Na_2S_2, which has λ_{max} = 370 nm and ε = 450 M^{-1} cm^{-1}. The mechanism proposed involved homolytic cleavage of the H-S bond, yielding $\cdot S^-$ radical anions, which dimerize to sulfane ions:[72]

$$HS^- + h\nu \rightarrow \cdot H + \cdot S^- \tag{27}$$

$$HS^- + \cdot H \rightarrow H_2 + \cdot S^- \tag{28}$$

$$\cdot S^- + \cdot S^- \rightarrow S_2^{2-} \tag{29}$$

The photooxidation of aqueous sulfide and sulfite solutions in the presence of platinized microcrystals of CdS and in the absence of oxygen was accompanied by the formation of hydrogen. In solutions containing both S^{2-} and SO_3^{2-}, thiosulfate was produced in a quantum yield of 0.25.[73] In the similar reaction using illuminated ZnS as photocatalyst, sulfite was oxidized to sulfate and dithionate, with concomitant formation of hydrogen.[74]

Enhanced photooxidation of H_2S was achieved by interparticle electron transfer (IPET), using two coupled semiconductors, such as CdS/TiO_2. By combining the two semiconductors, the extent of e^-/h^+ recombination was decreased, and charge separation was obtained by the transfer of electrons from the CdS to the TiO_2 particles.[75] A severe drawback of the use of CdS may be the release of toxic cadmium to the treated wastewater.

VI. HEAVY METAL IONS

A. Photooxidation of Fe(II)

The photooxidation of ferrous to ferric ions at λ < 300 nm in aqueous solutions, in the absence of oxygen, was shown to result in the production of the hydrated electron as the primary transient species:[76,77]

$$Fe^{2+} + h\nu \rightarrow Fe^{3+} + e_{aq}^- \tag{30}$$

At 253.7 nm, the quantum yield of this reaction was determined as 0.08.[78] In acid solutions, the complete process resulted in hydrogen production,[77]

$$2Fe^{2+}(aq) + h\nu \rightarrow 2Fe^{3+}(aq) + H_2 \tag{31}$$

On the other hand, at pH > 6.5, the main active species was shown to be $[Fe(OH)]^+_{aq}$, which was photooxidized to Fe^{3+} even with light of $\lambda > 400$ nm, without formation of solvated electrons. In this region, the proposed primary step is[76]

$$[Fe(OH)]^+_{aq} + h\nu \rightarrow [Fe(OH)_2]^+_{aq} + H\cdot \tag{32}$$

The photochemical oxidation of the tris (2,2′-bipyridine)–iron(II) complex by the periodate anion was studied in acidic aqueous solution. The mechanism proposed included the formation of an outer sphere complex in a thermal equilibrium reaction,

$$Fe(bpy)_3^{2+} + IO_4^- \leftrightarrow \left[Fe(bpy)_3^{2+} \cdots IO_4^-\right] \tag{33}$$

followed by photochemical excitation with irreversible intramolecular electron transfer,

$$2Fe(bpy)_3^{2+} + IO_4^- + h\nu \rightarrow Fe(bpy)_3^{3+} + IO_3^- \tag{34}$$

as well as the decomposition of the iron complex. In contrast to the very reactive photooxidation of the $Fe(bpy)_3^{2+}$ complex by periodate ions, the persulfate anions, $S_2O_8^{2-}$, were quite inactive. This was attributed to the bulky geometry of the persulfate ions, which prevented access to the internal d-orbitals of $Fe(bpy)_3^{2+}$. The ready access of the much smaller periodate ion to the Fe d shell facilitated electron transfer from the iron atom to the periodate ion.[79]

See also Section III.G on the photodegradation of hexacyanoferrate(II) and -(III) ions.

B. Photoreduction of Metal Salts

An attractive approach to the decontamination of toxic metal salts or the recovery of noble metals involves reductive photocatalytic reactions. The relative reactivity for the recovery of heavy metals observed the following pattern:[80]

$$Ag > Pd > Au > Pt \gg Rh \gg Ir \gg Cu = Ni = Fe$$

1.   Chromium

Chromium(VI), in chromate or bichromate salts, is water soluble, very stable, only weakly adsorbed to inorganic solids, carcinogenic, and extremely toxic. Chromium(III) is much less toxic, and in neutral and alkaline solutions in the absence of complexing agents is precipitated as $Cr(OH)_3$. Chromium(VI) pollution originates from a large variety of industries, such as of chrome plating, leather tanning, and metallurgical processes.

In a comparison of the photocatalytic activity of aqueous suspensions of several semiconductor materials for the reduction of Cr(VI) to Cr(III), the order of decreasing activity was CdS > ZnS > WO_3. The presence of oxygen caused decreased yields.[81] Another very active material for the reduction of Cr(VI) under sunlight exposure was ZnO.[82] A different series of photocatalytic activities was observed by Yoneyama et al.[83] in strongly acid solutions: WO_3 > TiO_2 (rutile) > TiO_2 (anatase) > $\alpha - Fe_2O_3$ > $SrTiO_3$. Using WO_3, under illumination at $\lambda = 400$ nm, the quantum yield of the reduction of Cr(VI) was 0.02.[83] The TiO_2-photocatalyzed reduction was applied to the experimental remediation of wastes from a metal-surface treatment plant, containing initially 2 mM Cr(VI). After dilution to 0.5 mM Cr(VI) and addition of TiO_2, the sample was illuminated with a 125-W high-pressure Hg lamp. Within 20 min, about 90% of the Cr(VI) was reduced.[84]

The photocatalytic reduction of Cr(VI) in acidic aqueous solutions (initially 200 µg/mL; pH 5) over dispersions of CdS, under illumination with a 250-W tungsten-halogen lamp, resulted in more than 90 and 99.9% conversion within 5 and 20 min, respectively. The optimal amount of CdS was reported to be 5 g/L. Under sunlight, the photoreduction of Cr(VI) was somewhat slower — 99.3% was converted within 1 h. In these experiments, untreated CdS (C.P. grade) was used. If the CdS had been heat treated (6 h at 400°C under argon), the photoactivity decreased.[85] In the photocatalytic reduction of Cr(VI) on aqueous dispersions of titanium dioxide, a basic medium was shown to be more favorable than an acidic medium, because in the basic solution the TiO_2 was less subject to photocorrosion, and also the $Cr(OH)_3$ product was strongly adsorbed on the photocatalyst and thus removed from the solution. The overall process, which is thermodynamically downhill, was described by the two half-cell reactions,[86]

$$CrO_4^{2-}(aq) + 4H_2O + 3e^- + h\nu \rightarrow Cr(OH)_3(ads) + 5OH^-(aq) \qquad (35)$$

$$4OH^- + 4h^+ \rightarrow O_2 + 2H_2O \qquad (36)$$

in which photogenerated conduction-band electrons from the TiO_2 particles reduce the soluble chromate ions to the insoluble and adsorbed Cr(III), while valence-band holes oxidize hydroxide ions to oxygen. For efficient reduction of Cr(VI), the reaction was performed in N_2-purged solutions. In the presence of oxygen, the reduction of Cr(VI) was considerably inhibited, presumably because of competition with O_2 molecules for the conduction-band electrons. Under illumination with a 400-W medium-pressure Hg lamp, an initially 0.5 mM solution of Na_2CrO_4 (at pH 10, containing 4.44 g/L of TiO_2) within 3 h completely lost its original yellow color (λ_{max} = 370 nm), while the originally white TiO_2 turned pale green. The used TiO_2 could be regenerated by washing off the $Cr(OH)_3$ with 3 M NaOH:

$$Cr(OH)_3 + OH^- \rightarrow Cr(OH)_4^- \qquad (37)$$

The initial rates of the photoreduction of Cr(VI) on dispersed TiO_2 were linearly related to the square root of the initial chromate concentration. The concentration dependence could also be fitted to Langmuir–Hinshelwood kinetics.[86]

The efficiency of suspended particles of TiO_2 for the reduction of Cr(VI) was about four times higher than that of TiO_2 ceramic membranes supported on glass. The general form of the kinetics was similar in both systems, with rising rates at lower pH. At pH < 2, the kinetic order with respect to Cr(VI) was 0.5.[87] The reduction of Cr(VI) was also performed in a flow system, using either CdS or TiO_2 as photocatalysts. Highest reactivity was obtained using CdS fixed onto borosilicate pearls. However, the CdS gradually dissolved, releasing Cd^{2+} ions. Best overall results were with TiO_2 (P 25).[52]

A pilot-plant study was made at the Plataforma Solar de Almería, Spain, on the photocatalytic reduction of Cr(VI) in aqueous suspensions of TiO_2 (Degussa P25). Using concentrated sunlight illumination in a loop of continuous flow-through tubular photoreactors in the focus of parabolic trough mirrors, the process was operated both in single-pass and in recirculation mode. The residence time distribution in the loop corresponded to a plug flow model. The flow rate, temperature, and partial pressure had little influence on the rate of Cr(VI) reduction. On the other hand, the rate was strongly affected by the pH, with best results obtained at pH 1. The reaction kinetic order was found to be $^1/_2$ with respect to the Cr(VI) concentration, and increased with the TiO_2 concentration, reaching a plateau at about 2 g/L.[88]

See also Chapter VIII, Section I, on the photoreduction of Cr(VI) to Cr(III) in natural waters.

2. Mercury, Lead, Manganese, Thallium, Cobalt, and Uranium

Illumination of aerated aqueous solutions of the metal ions Pb^{2+}, Mn^{2+}, Tl^+, Co^{2+}, and Hg^{2+} in the presence of Pt-loaded TiO_2 led to deposition of oxides on the photocatalyst. Thus, Pb^{2+} led to PbO_2 deposition, and Tl^+ led to Tl_2O_3 deposition. Bare TiO_2 (not Pt-loaded) was much less effective than Pt/TiO_2.[89] Aqueous suspensions of ZnO, TiO_2, and WO_3 were good photocatalysts for the elimination of Hg(II), which was reduced to metallic Hg deposited on the catalyst particles and could thus be recovered. The presence of large concentrations of Cl⁻ caused a decrease in the efficiency of reduction of Hg(II), while the presence of SO_4^{2-} or NO_3^- had no effect.[90,91] Uranium from aqueous solutions of uranyl ions, in the presence of isopropanol, acetate, or formate as hole scavengers, could be photodeposited on TiO_2 as U_3O_8.[92]

While mercury ions (Hg^{2+} and Hg^+) undergo photoreduction to metallic Hg in deoxygenated aqueous dispersions of TiO_2 containing a reducing agent, no photoreduction occurred in the presence of oxygen, because oxygen molecules very rapidly trap the conduction band electrons.[93] In the absence of oxygen, the conduction-band electrons on the semiconductor surface are available for the reduction of heavy metal ions.[94]

Improved photocatalytic reduction of Hg(II) in aerated aqueous solutions was achieved with dispersions of two-component semiconductor powders. Almost complete elimination of initially 100 µg/L of Hg(II) was obtained with ZnO/WO_3 as catalyst by irradiation for 1 h with a 250-W tungsten-halogen lamp. The catalyst was prepared by grinding together the two components. Optimal rate of photoreduction was obtained with a Zn/W atomic ratio of 0.5. The rate of photoreduction was independent of pH in the range of pH 3 to 4.5, and decreased below pH 3. An alternative photocatalyst system involved mixtures obtained by grinding together WO_3 with metal-phthalocyanines (MeP_cs). Among Ni, Co, and Cu-P_cs, best results were obtained with WO_3/Ni-P_c. In a comparison, for the photoreduction of Hg(II) by 2 h of exposure to sunlight, the ZnO/WO_3 and $WO_3/5wt\%$ Ni-P_c catalysts resulted in 99.1 and 84.4% reduction, respectively.[95]

3.   Copper

Reiche et al.[96] observed the photocatalytic reduction of Cu(II) to copper metal in aqueous suspensions of TiO_2, both in the presence and absence of acetate. Bideau et al.[97] found that in the presence of formic acid and TiO_2, Cu(II) was photodeposited as a red material on the TiO_2, presumably a mixture of Cu(0) and Cu(I), while formic acid was oxidized to carbon dioxide.[97]

Foster et al.[98] developed the photocatalytic reduction of Cu(II) to a process of concentration and volume reduction of Cu(II) waste streams. In a typical experiment, solutions at pH 3.6, containing 2 g/L TiO_2 and 0.1 M sodium formate (which is the usual waste organic compound from the oxidation of formaldehyde in electroless copper plating baths), were purged with nitrogen (to prevent the reduction of oxygen) and were illuminated with a 15-W black-light (mainly 365 nm). Within 5 min of illumination, the concentration of Cu(II) in solution, initially 51 mg/L, decreased to ≤0.018 mg/L. The originally white TiO_2 turned purple. X-ray photoelectron (XPS) and electron spin resonance (EPR) spectroscopy measurements were consistent with the tentative assignment of the purple species as Cu(I). In addition to formic acid, oxalic acid, citric acid, and ethylenediamine tetraacetic acid (EDTA) were also effective "hole traps" in the photoreduction of Cu(II). The useful pH range was 1.8

to 6.6. In highly acidic solution (pH 0.6 to 1.4) the photoreduction did not occur, possibly because of competition with the photoreduction of water to hydrogen. The purple copper/TiO_2 could be separated by decanting off the supernatant and resuspending in a smaller volume of water. The copper could then be redissolved simply by bubbling air through the dispersion, while the TiO_2 regained its white color. This "dark" reoxidation was very rapid, and was 98% complete within ~2 min. Such a concentration step (with volume reduction by an order of magnitude) could be a practical preliminary step for the electrolytic recovery of copper.[98,99]

The photodeposition of metallic copper from Cu(II) ions onto ZnO particles in aqueous suspensions was effected under sunlight exposure. The yield of the reduction of Cu(II) was found to decrease with increasing initial Cu(II) concentration.[100]

See also Chapter I, Section IV.E, on the adsorption of Cu^{2+}, Cd^{2+}, and Zn^{2+} on TiO_2 particles as a method to determine the surface density of OH groups.[101]

4. Mercury, Gold, Platinum, Silver, and Chromium

Recovery of gold from solutions of Au^{3+} salts was achieved by UV illumination in the presence of suspensions of either TiO_2 or WO_3, but TiO_2 provided more rapid reduction. Au recovery from Au^{3+} was most efficient at pH 3 to 6. The protonated form of the gold complex $HAuCl_4$ at pH 0 was more difficult to reduce, but the addition of the hole scavenger methanol enabled gold reduction even at this very acidic pH.[102]

The initial rate of the photoreduction of Ag^+ ions to metallic silver on aqueous dispersions of TiO_2 depended on the starting concentration of the silver ions in accordance with the Langmuir–Hinshelwood relationship. Under illumination with a 125-W Hg lamp ($\lambda = 300$ to 400 nm), with initially 1 mM Ag^+ in an oxygen-free solution, silver was quantitatively removed from the solution within about 10 min. The initial rate of deposition of Ag was proportional to the radiant flux (at least up to 7.6×10^{15} photons s^{-1} cm^{-2}), and the initial apparent quantum yield was 0.16. The presence of thiosulfate salts in the silver ion solutions did not interfere in the photodeposition of silver on TiO_2, and thus this method could be applicable for the recovery of silver from photographic waste solutions.[94]

The mechanism of the photocatalytic reduction of Au^{3+} cations on colloidal TiO_2 was studied by *in situ* extended X-ray absorption fine structure (EXAFS) spectroscopy, using for analysis the intense photon beam from high-energy storage rings (synchrotron radiation). The Au^{3+}–TiO_2 colloid containing about 4 wt% Au^{3+} was prepared from $HAuCl_4$ and colloidal TiO_2 (obtained by hydrolyzing titanium tetraisopropoxide in a mixture of isopropanol and hydrochloric acid). The argon-purged reaction mixture was circulated through a reaction cell and through an analysis cell. In the reaction cell, the Au^{3+}–TiO_2 was illuminated with a 125-W high-pressure Hg lamp, resulting in reduction of Au^{3+} to atomic gold. In the analysis cell, the mixture underwent EXAFS spectroscopy. The results indicated a two-step mechanism of the photoreduction of Au^{3+}. In the initial nucleation step, very small (10 Å) colloidal Au particles were formed. This was followed by the growth of metallic crystallites.[103]

The photoreduction of hexachloroplatinate in aqueous dispersions of TiO_2 resulted in the deposition of platinum on the photocatalyst. The initial rate of platinum deposition was proportional to the square root of the light intensity.[104]

Hg(II), Au(III), and Ag(I) salts in illuminated aqueous suspensions of TiO_2 were reduced to metallic crystallites on the photocatalyst. As noted earlier, Cr(VI) salts were reduced to Cr(III) salts, which remained soluble at acidic pH but could be precipitated at alkaline pH. These reactions proceeded very slowly when only water was available as the reductant, but were very markedly accelerated when certain organic reductants were added. Good results were reported with salicylic acid (0.22 mM) as the reductant, and also with EDTA and with citric acid. Methanol, ethanol, and acetic acid were quite ineffective as reductants. The presence of oxygen slightly inhibited the reduction of the metal ions, presumably by competition of the oxygen molecules for conduction-band electrons. These photocatalytic reduction reactions occurred only with those metal salts that had standard reduction potentials more positive than 0.3 V (vs. the normal hydrogen electrode). Cu(II) and Pt(IV) salts were only partly reduced, while Cd(II) and Ni(II) salts were not affected by this photocatalytic treatment.

The photocatalytic method could be useful for the detoxification of waters polluted with the highly toxic mercury and the carcinogenic chromate salts, which would be converted to Hg metal adsorbed on the TiO_2 particles and to the less toxic Cr(III) salts.[105,106]

5.   Iron(III)

The photoreduction of iron(III) to iron(II) in natural waters is discussed in detail in Chapter VIII, Section I.

VII. ORGANO-METALLIC COMPOUNDS

The serious environmental hazard of spills of mercury and of other heavy metals (particularly lead and cadmium) into natural waters is due to the formation of organo-metallic compounds, some of which are extremely toxic.

A. Mercurochrome

As an example of the photocatalytic degradation of organo-mercury compounds, the photodecomposition of the highly toxic dye mercurochrome (merbromin) was studied. Under illumination with a 400-W medium-pressure Hg lamp, in aerated aqueous dispersions of TiO_2, the intense color of mercurochrome ($\lambda_{max} = 498$ nm) gradually faded and disappeared within a few hours. An almost

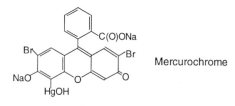

Mercurochrome

SCHEME II.2

stoichiometric amount of CO_2 was released, and the TiO_2 turned ash colored and sank to the bottom of the reactor, due to the deposition of metallic mercury. This result suggested that it may perhaps be possible to decontaminate also other organo-metallic water pollutants, such as methyl-Hg or methyl-Pb compounds, by photocatalytic degradation.[107]

B. Tributyltin Chloride

The direct photolysis of the highly toxic and corrosive tributyltin chloride (TBT) was studied in aerated aqueous solution under irradiation with a 100-W medium-pressure Hg lamp through its quartz envelope ($\lambda \geq 185$ nm). The disappearance of TBT followed pseudo first-order kinetics, and was accompanied by the gradual formation and decay of dibutyltin dichloride, and eventual mineralization to tin(IV). Surprisingly, the photodegradation of TBT was slower in oxygen-flushed solutions than in aerated or nitrogen- or argon-flushed solutions. On the other hand, the presence of oxygen enhanced the photodegradation of dibutyltin dichloride.[108]

C. Bis(tributyltin) Oxide

Bis(tributyltin) oxide, TBTO, $\{[CH_3(CH_2)_3]_3Sn\}_2O$, is an important component of marine and freshwater antifouling paints. It is extremely toxic to aquatic organisms, and at concentrations above several milligrams per liter it is also quite resistant to microbial degradation. Its direct photolysis in aerated aqueous solutions occurred by illumination under sunlight and more rapidly with a 300-W UV lamp, leading to the highly toxic dibutyltin oxide, $[CH_3(CH_2)_3]_2SnO$, DBTO, as the main photoproduct. Further photodegradation to the nontoxic tin oxide was not observed during the period of the experiment (300 h). The photolysis of TBTO was more rapid in seawater or in distilled water containing NaCl than in pure water, and also was strongly photosensitized by the presence of anthraquinone. It was therefore concluded that natural photodegradation of TBTO in the marine environment may result in the detoxification of this pollutant.[109]

D. Phenylmercury Compounds

Organomercury compounds are highly toxic but are still used in various applications, particularly in marine antifouling paints. The direct photolysis of diphenylmercury (DPM) and of monophenylmercury chloride (MPM) was performed by illumination of their aerated solutions in acetonitrile–water (1:1) with a 100-W medium-pressure Hg lamp through its quartz envelope. Photoexcitation occurred mainly at the absorption maximum of DPM at 225 nm, and at the absorption shoulder of MPM at 212 nm. The photodegradation of DPM resulted in the intermediate formation and subsequent decay of MPM. Other identified intermediates and products were phenol, *p*-hydroxyphenylmercury, bis(trihydroxyphenyl)mercury, and *p*-hydroxyacetophenone mercury, as well as Hg^{2+} and metallic mercury. Since UV light below 270 nm was required for the photodegradation of MPM and DPM, the natural sunlight photodecomposition of such compounds in the environment may be expected to be very slow.[110]

REFERENCES

1. Strickler, S. J. and Kasha, M., Solvent effects on the electronic absorption spectrum of nitrite ion, *J. Am. Chem. Soc.,* 85, 2899–2901, 1963.
2. Christensen, P. A., Harriman, A., Neta, P., and Richoux, M. C., Photooxidation of water using Prussian Blue as catalyst, *J. Chem. Soc., Faraday Trans. 1,* 81, 2461–2466, 1985.
3. Clapp, P. A., Du, N., and Evans, D. F., Thermal and photochemical production of hydrogen peroxide from dioxygen and tannic acid, gallic acid and other related compounds in aqueous solution, *J. Chem. Soc., Faraday Trans.,* 86, 2587–2592, 1990.
4. Rives-Arnau, V., Remarks on "Photocatalytic oxidation of water to hydrogen peroxide by irradiation of aqueous suspensions of TiO_2", *J. Electroanal. Chem.,* 190, 279–281, 1985.
5. Tanaka, K., Murata, S., and Harada, K., Oxygen evolution by the photooxidation of water, *Solar Energy,* 34, 303–308, 1985.
6. Hong, A. P., Bahnemann, D. W., and Hoffmann, M. R., Cobalt(II) tetrasulfo phthalocyanine on titanium dioxide: a new efficient electron relay for the photocatalytic formation and depletion of hydrogen peroxide in aqueous suspensions, *J. Phys. Chem.,* 91, 2109–2117, 1987.
7. Hoffman, A. J. and Hoffmann, M. R., Photocatalytic oxidation of organic compounds and reduction of dioxygen with quantum-sized semiconductors, in: *Photocatalytic Purification and Treatment of Water and Air,* Ollis, D. F. and Al-Ekabi, H. (Eds.), Elsevier Science Publishers, Amsterdam, 1993, pp. 155–162.
8. Hoffman, A. J., Carraway, E. R., and Hoffmann, M. R., Photocatalytic oxidation of organic acids on quantum-sized semiconductor colloids, *Environ. Sci. Technol.,* 28, 786, 1994.

9. Carraway, E. R., Hoffman, A. J., and Hoffmann, M. R., Photocatalytic production of H_2O_2 and organic peroxides on quantum-sized semiconductor colloids, *Environ. Sci. Technol.,* 28, 776, 1994.

10. Jenny, B. and Pichat, P., Determination of the actual photocatalytic rate of H_2O_2 decomposition over suspended TiO_2. Fitting to the Langmuir–Hinshelwood form, *Langmuir,* 7, 947–954, 1991.

11. Halmann, M. and Platzner, I., Deuterium oxide solvent effects on ultraviolet absorption of anions, *Proc. Chem. Soc.,* 261–262, 1964.

12. Burak, I. and Treinin, A., Solvent scale for charge-transfer-to-solvent spectra of anions, *Trans. Faraday Soc.,* 59, 1490–1496, 1963.

13. Grossweiner, L. I. and Matheson, M. S., Short-lived species from the photolysis of aqueous alkali halide and halogen solutions, *J. Chem. Phys.,* 23, 2443–2444, 1955.

14. Wong, D. and DiBartolo, B., Evolution of the dihalide ion Br_2^- in aqueous solutions, *J. Photochem.,* 4, 249–268, 1975.

15. Wagner, I. and Strehlow, H., On the flash photolysis of bromide ions in aqueous solutions, *Ber. Bunsen Ges. Phys.* 91, 1317–1321, 1987.

16. Zehavi, D. and Rabani, J., Oxidation of aqueous bromide ions by hydroxyl radicals. Pulse radiolytic investigation, *J. Phys. Chem.,* 76, 312–319, 1972.

17. Herrmann, J.-M. and Pichat, P., Heterogeneous photocatalysis. Oxidation of halide ions by oxygen in ultraviolet irradiated aqueous suspensions of titanium oxide, *J. Chem. Soc. Faraday Trans. I,* 76, 1138–1146, 1980.

18. Reichman, B. and Byvik, C. E., Photoproduction of I_2, Br_2 and Cl_2 on n-semiconducting powder, *J. Phys. Chem.,* 85, 2255–2258, 1981.

19. Dabestani, R., Wang, X., Bard, A. J., Campion, A., Fox, M. A., Webber, S. E., and White, J. M., Photoinduced oxidation of bromide to bromine on irradiated platinized TiO_2 powders and platinized particles supported in Nafion films, *J. Phys. Chem.,* 90, 2729–2732, 1986.

20. Moser, J. and Grätzel, M., Photoelectrochemistry with colloidal semiconductors; laser studies of halide oxidation in colloidal dispersions of TiO_2 and α-Fe_2O_3, *Helv. Chim. Acta,* 65, 1436–1444, 1982.

21. Henglein, A., Colloidal TiO_2 catalyzed photo and radiation chemical processes in aqueous solution, *Ber. Bunsenges. Phys. Chem.* 86, 241–246, 1982.

22. Chen, S.-N., Lichtin, N. N., and Stein, G., Fuel formation from aqueous ferric bromide by photolysis in the visible, *Science,* 190, 879–880, 1975.

23. Eckert, A., About oxidation in light, *Ber. Dtsch. Chem. Ges.,* 58B, 313–320, 1925.

24. Eckert, A., Oxidations in light, *Ber. Dtsch. Chem. Ges.,* 60B, 1691–1693, 1927.

25. Scharf, H.-D. and Weitz, R., Photooxidation of chloride to chlorine with anthraquinone derivatives in aqueous solution. A contribution to the storage of solar energy as chemical energy, *Tetrahedron,* 35, 2255–2262, 1979.

26. Metcalfe, J., Photochemistry of anthraquinone-2-sodium methyl suphonate in aqueous solutions, *J. Chem. Soc., Faraday Trans. 1,* 79, 1721–1731, 1983.

27. Fuchs, B., Mayer, W. J. W., and Abramson, S., Photoassisted oxidation of hydrogen bromide to bromine, *J. Chem. Soc., Chem. Commun.,* 1711–1713, 1985.

28. Slama-Schwok, A., Gershuni, S., Rabani, J., Cohen, H., and Meyerstein, D., An iridium–bipyridine complex as a sensitizer for the bromide oxidation to bromine by oxygen, *J. Phys. Chem.,* 89, 2460–2464, 1985.

29. Grätzel, M. and Halmann, M., Photosensitized oxidation of bromide to bromine with phthalic acid derivatives in aqueous solutions, *Solar Energy Mater.,* 20, 117–129, 1990.

30. Wagner, I., Strehlow, H., and Busse, G., Flash photolysis of nitrate ions in aqueous solution, *Z. Physik. Chem., Neue Folge,* 123, 1–33, 1980.

31. Zepp, R. G., Hoigné, J., and Bader, H., Nitrate-induced photooxidation of trace organic chemicals in water, *Environ. Sci. Technol.,* 21, 443–450, 1987.

32. Treinin, A. and Hayon, E., Absorption spectra and reaction kinetics of NO_2, N_2O_2, and N_2O_4 in aqueous solution, *J. Am. Chem. Soc.,* 92, 5821–5828, 1970.

33. Grätzel, M., Taniguchi, S., and Henglein, A., Pulse radiolytic study of NO oxidation and of the equilibrium $N_2O \leftrightarrow NO + NO_2$ in aqueous solution, *Ber. Bunsenges. Phys. Chem.,* 74, 488–492, 1970.

34. Strehlow, H. and Wagner, I., Flash photolysis in aqueous nitrite solutions, *Z. Phys. Chem., Neue Folge,* 132, 151–160, 1982.

35. Bilski, P., Chignell, C. F., Szychlinski, J., Borkowski, A., Oleksy, E., and Reszka, K., Photooxidation of organic and inorganic substrates during UV photolysis of nitrite anion in aqueous solution, *J. Am. Chem. Soc.,* 114, 549–556, 1992.

36. Glidewell, C., The nitrate/nitrite controversy, *Chem. Br.,* 26, 137–140, 1990.

37. Hori, Y., Nakatsu, A., and Suzuki, S., Heterogeneous catalytic oxidation of NO_2^- in aqueous suspensions of various semiconductor powders, *Chem. Lett.,* 1429–1432, 1985.

38. Tennakone, K., Punchihewa, S., and Tantrigoda, R. U., Photocatalytic oxidation of nitrite in water to nitrate, *Environ. Pollut.,* 57, 299–305, 1989.

39. Hori, Y., Bandoh, A., and Nakatsu, A., Electrochemical investigation of photocatalytic oxidation of NO_2^- at TiO_2 (anatase) in the presence of O_2, *J. Electrochem. Soc.,* 137, 1155–1161, 1990.

40. Milis, A., and Domènech, X., Photoassisted oxidation of nitrite to nitrate over different semiconducting oxides, *J. Photochem. Photobiol. A: Chem.,* 72, 55–59, 1993.

41. Milis, A., Peral, J., Domènech, X., and Navio, J. A., Heterogeneous photocatalytic oxidation of nitrite over iron-doped TiO_2 samples, *J. Catal.,* 87, 67–74, 1994.

42. Kudo, A., Domen, K., Maruya, K.-I., and Onishi, T., Reduction of nitrate ions into nitrite and ammonia over some photocatalysts, *J. Catal.,* 135, 300–303, 1992.

43. Halmann, M., Tobin, J., and Zuckerman, K., Photoassisted and electrochemical reduction of nitric acid to hydroxylamine and ammonia, *J. Electroanal. Chem.,* 209, 405–411, 1986.

44. Halmann, M. and Zuckerman, K., Nitrite reduction to ammonia in illuminated aqueous suspensions of powdered semiconductors in alkaline sulphide solutions, *J. Chem. Soc. Chem. Commun.,* 455–457, 1986.

45. Halmann, M., Taouk, M., and Haneman, D., Nitrite reduction to ammonia in cadmium-chalcogenide-polysulfide solar cells, *Solar Cells,* 18, 55–60, 1986.

46. Ranjit, K. T., Krishnamoorthy, R., and Viswanathan, B., Photocatalytic reduction of nitrite and nitrate on ZnS, *J. Photochem. Photobiol. A: Chem.,* 81, 55–58, 1994.

47. Bravo, A., Garcia, J., Domènech, X., and Peral, J., Some aspects of the photocatalytic oxidation of ammonium ion by titanium dioxide, *J. Chem. Res. S. (Synopses),* 376–377, 1993.

48. Code of Federal Regulations 40, Part 261, revised as of July 1, 1987. Wild, S. R., Rudd, T., and Neller, A., Fate and effects of cyanide during waste-water treatment processes, *Sci. Total Environ.*, 156, 93–107, 1994.

49. Frank, S. N. and Bard, A. J., Heterogeneous photocatalytic oxidation of cyanide ion in aqueous solution at TiO_2 powder, *J. Am. Chem. Soc.*, 99, 303–304, 1977.

50. Frank, S. N. and Bard, A. J., Heterogeneous photocatalytic oxidation of cyanide and sulfite in aqueous solutions at semiconductor powders, *J. Phys. Chem.*, 81, 1484–1488, 1977.

51. Peral, J., Muñoz, J. and Domènech, X., Photosensitized CN^- oxidation over TiO_2, *J. Photochem. Photobiol. A: Chem.*, 55, 251–257, 1990.

52. Domènech, X., Photocatalysis for aqueous phase decontamination: Is TiO_2 the better choice? in: *Photocatalytic Purification and Treatment of Water and Air*, Ollis, D. F. and Al-Ekabi, H. (Eds.), Elsevier Science Publishers, Amsterdam, 1993, pp. 337–351.

53. Doménech, J. and Peral, J., Removal of toxic cyanide from water by heterogeneous photocatalytic oxidation over zinc oxide, *Solar Energy*, 41, 55–59, 1988.

54. Borgarello, E., Terzian, R., Serpone, N., Pelizzetti, E., and Barbeni, M., Photocatalyzed transformation of cyanide to thiocyanate by rhodium-loaded cadmium sulfide in alkaline aqueous sulfide media, *Inorg. Chem.*, 25, 2135–2137, 1986.

55. Rader, W. S., Solujic, L., Milosavijevic, E. B., and Hendrix, J. L., Sunlight-induced photochemistry of aqueous solutions of hexacyanoferrate(II) and -(III) ions, *Environ. Sci. Technol.*, 27, 1875–1879, 1993.

56. Shirom, M. and Stein, G., Excited state chemistry of the ferrocyanide ion in aqueous solution. II. Photoaquation, *J. Chem. Phys.*, 55, 3379–3382, 1971.

57. Bhakta, D., Shukla, S. S., Chandras, M. S., and Margrave, J. L., A novel photocatalytic method for detoxification of cyanide wastes, *Environ. Sci. Technol.*, 26, 625–626, 1992.

58. Duonghong, D., Ramsden, J., and Grätzel, M., Dynamics of interfacial electron-transfer processes in colloidal semiconductor systems, *J. Am. Chem. Soc.*, 104, 2979–2985, 1982.

59. Baxendale, J. H., Bevan, P. L. T., and Stott, D. A., Pulse radiolysis of aqueous thiocyanate and iodide solutions, *Trans. Faraday Soc.*, 64, 2389–2397, 1968.

60. Draper, R. B. and Fox, M. A., Titanium dioxide photooxidation of thiocyanate $(SCN)_2 \cdot^-$ studied by diffuse reflectance flash photolysis, *J. Phys. Chem.*, 94, 4628–4634, 1990.

61. Burak, I. and Treinin, A., Photochemistry of N_3^- in aqueous solution at 254 nm, *J. Am. Chem. Soc.*, 87, 4031–4036, 1965.

62. Halmann, M. and Platzner, I., The photochemistry of phosphorus compounds. Part II. Far-ultraviolet absorption spectra of some phosphorus oxyanions in aqueous solutions, *J. Chem. Soc.*, 1440–1449, 1965.

63. Benderly, H. and Halmann, M., The photochemistry of phosphorus compounds. V. Far-ultraviolet absorption spectra of hypophosphite, phosphite, and hypophosphate ions in aqueous solution, *J. Phys. Chem.*, 71, 1053–1060, 1967.

64. Huber, J. R. and Hayon, E., Flash photolysis in the vacuum ultraviolet region of the phosphate anions $H_2PO_4^-$, HPO_4^{2-}; and $P_2O_7^{4-}$ in aqueous solutions, *J. Phys. Chem.*, 72, 3820–3827, 1968.

65. Black, E. D. and Hayon, E., Pulse radiolysis of phosphate anions $H_2PO_4^-$, HPO_4^{2-}, PO_4^{3-} and $P_2O_7^{4-}$ in aqueous solutions, *J. Phys. Chem.,* 74, 3199–3203, 1970.

66. Hayon, E., Treinin, A., and Wilf, J., Electronic spectra, photochemistry, and autoxidation mechanisms of the sulfite–bisulfite–pyrosulfite system, *J. Am. Chem. Soc.,* 94, 47–57, 1972.

67. Anbar, M., Bambenek, M., and Ross, A. B., Selected Specific Rates of Reactions of transients from water in aqueous solution: 1. hydrated electron; 2. hydrogen atom; *Natl. Bur. Stand.* 43 and 51; U.S. Government Printing Office, Washington, DC, 1973 and 1975.

68. Bäckstrom, H. L. J., The chain mechanism in the autoxidation of sodium sulfite solutions, *Z. Physik, Chem.,* B25, 122–138, 1934.

69. Deister, U. and Warneck, P., Photooxidation of SO_3^{2-} in aqueous solution, *J. Phys. Chem.,* 94, 2191–2198, 1990.

70. Hong, A. P., Bahnemann, D. W., and Hoffmann, M. R., Cobalt(II) tetrasulfo phthalocyanine on titanium dioxide. 2. Kinetics and mechanisms of the photocatalytic oxidation of aqueous sulfur dioxide, *J. Phys. Chem.,* 91, 6245–6251, 1987.

71. Faust, B. C., Hoffmann, M. R., and Bahnemann, D. W., Photocatalytic oxidation of sulfur dioxide in aqueous suspensions of α-Fe_2O_3, *J. Phys. Chem.,* 93, 6371–6381, 1989.

72. Dzhabiev, T. S. and Tarasov, B. B., Photochemical decomposition of an aqueous solution of sodium sulphide, *J. Photochem. Photobiol. A: Chem.,* 72, 23–27, 1993.

73. Buhler, N., Meier, K., and Reber, J.-F., Photochemical hydrogen production with cadmium sulfide suspensions, *J. Phys. Chem.,* 88, 3261–3268, 1984.

74. Reber, J.-F. and Meier, K., Photochemical production of hydrogen with zinc sulfide suspensions, *J. Phys. Chem.,* 88, 5903–5913, 1984.

75. Serpone, N., A decade of heterogeneous photocatalysis in our laboratory: pure and applied studies in energy production and environmental detoxification, *Res. Chem. Intermed.,* 20, 953–992, 1994.

76. Airey, P. L. and Dainton, F. S., The photochemistry of aqueous solutions of Fe(II). I. Photoelectron detachment from ferrous and ferrocyanide ions, *Proc. Roy. Soc., Ser. A,* 291, 340–352, 1966.

77. Braterman, P. S., Cairns-Smith, A. G., Sloper, R. W., Truscott, T. G., and Craw, M., Photooxidation of iron(II) in water between pH 7.5 and 4.0, *J. Chem. Soc., Dalton Trans.,* 1441–1445, 1984.

78. Logan, S. R., Effects of temperature and wavelength on the photooxidation of the ferrous ion, *XIII IUPAC Symp. on Photochemistry,* Univ. Warwick, England, Abstract P3, 1990.

79. Gerasimov, O. V., Parmon, V. N., and Lymar, S. V., Photooxidation of tris(2,2′-bipyridine)-iron(II) complex by periodate in aqueous solution, *J. Photochem. Photobiol. A: Chem.,* 73, 21–29, 1993.

80. Herrmann, J.-M., Guillard, G., and Pichat, P., Heterogeneous photocatalysis: an emerging technology for water treatment, *Catal. Today,* 17, 7–20, 1993.

81. Domènech, J. and Muñoz, J., Photochemical reduction of chromium(VI) over cadmium sulfide, zinc sulfide, and tungsten(VI) oxide, *J. Chem. Res. Synop.,* 106–107, 1987; *Chem. Abstr.,* 106, 224248g.

82. Domènech, J. and Muñoz, J., Photocatalytic reduction of chromium(VI) over zinc oxide powder, *Electrochim. Acta,* 32, 1383–1386, 1987.

83. Yoneyama, H., Yamashita, Y., and Tamura, H., Heterogeneous photocatalytic reduction of dichromate on n-type semiconductor catalysts, *Nature,* 282, 817–818, 1979.

84. Muñoz, J. and Domènech, X.,TiO$_2$ catalyzed reduction of Cr(VI) in aqueous solutions under ultraviolet illumination, *J. Appl. Electrochem.,* 20, 518–521, 1990.

85. Wang, S., Wang, Z., and Zhuang, Q., Photocatalytic reduction of the environmental pollutant Cr^{VI} over cadmium sulphide powder under visible light illumination, *Appl. Catal. B: Environ.,* 1, 257–270, 1992.

86. Lin, W. Y., Wei, C., and Rajeshewar, K., Photocatalytic reduction and immobilization of hexacovalent chromium at titanium dioxide in aqueous basic media, *J. Electrochem. Soc.,* 140, 2477–2482, 1993.

87. Sabate, J., Anderson, M. A., Aguado, M. A., Gimenez, J., Cevera-March, S., and Hill, C. G., Comparison of TiO$_2$ powder suspensions and TiO$_2$ ceramic membranes supported on glass as photocatalytic systems in the reduction of chromium(VI), *J. Mol. Catal.,* 71, 57–68, 1992.

88. Cervera-March, S., Giménez-Farreras, J., Aguado, M. A., Borrell, L., Curcó, D., and Queral, M. A., Kinetic and radiation studies for the photoreactor design in photocatalytic detoxification processes using TiO$_2$, in: *Photocatalytic Purification and Treatment of Water and Air,* Ollis, D. F. and Al-Ekabi, H. (Eds.), Elsevier Science Publishers, Amsterdam, 1993, pp. 633–638.

89. Tanaka, K., Harada, K., and Murata, S., Photocatalytic deposition of metal ions onto titanium dioxide powder, *Solar Energy,* 36, 159–161, 1986.

90. Domènech, J. and Andres, M., Elimination of mercury(II) from aqueous solutions by photocatalytic reduction over zinc oxide powder, *N. J. Chem.,* 11, 443–447, 1987.

91. Domènech, J. and Andres, M., Photocatalytic reduction of mercury(II) ions in aqueous suspensions of titanium dioxide and tungsten trioxide, *Gazz. Chim. Ital.,* 117, 495–498, 1987.

92. Amadelli, R., Maldotti, A., Sostero, S., and Carassitti, V., Photodeposition of uranium oxides onto TiO$_2$ from aqueous uranyl solutions, *J. Chem. Soc. Faraday Trans.,* 87, 3267–3273, 1991.

93. Tennakone, K., Photoreduction of carbonic acid by mercury coated n-titanium dioxide, *Sol. Energy Mater.,* 10, 235–238, 1984.

94. Herrmann, J. M., Disdier, J., and Pichat, P., Photocatalytic deposition of silver on powder titania: consequences for the recovery of silver, *J. Catal.,* 113, 72–78, 1988.

95. Wang, Z.-H. and Zhuang, Q.-X., Photocatalytic reduction of pollutant Hg(II) on doped WO$_3$ dispersion, *J. Photochem. Photobiol. A: Chem.,* 75, 105–111, 1993.

96. Reiche, H., Dunn, W. W., and Bard, A. J., Heterogeneous photocatalytic and photosynthetic deposition of copper on titanium dioxide and tungsten(VI) oxide powders, *J. Phys. Chem.,* 83, 2248–2251, 1979.

97. Bideau, M., Claudel, B., Faure, L., and Rachimoellah, M., Photooxidation of formic acid by oxygen in the presence of titanium dioxide and dissolved copper ions: oxygen transfer and reaction kinetics, *Chem. Eng. Commun.,* 93, 167–197, 1990.

98. Foster, N. S., Noble, R. D., and Koval, C. A., Reversible reductive deposition and oxidative dissolution of copper ions in titanium dioxide aqueous suspensions, *Environ. Sci. Technol.,* 27, 350–356, 1993.

99. Foster, N. S., Brown, G. N., Noble, R. D., and Koval, C. A., Use of TiO_2 photocatalysis in the treatment of spent electroless copper plating baths, in: *Photocatalytic Purification and Treatment of Water and Air,* Ollis, D. F. and Al-Ekabi, H. (Eds.), Elsevier Science Publishers, Amsterdam, 1993, pp. 365–373.

100. Domènech, J., Prieto, A., and Franco, C., Photodeposition of metallic copper on powdered zinc oxide under solar irradiation, *Affinidad,* 44, 330–332, 1987; *Chem. Abstr.,* 107, 187050j.

101. Zang, L., Liu, C.-Y., and Ren, X.-M., Adsorption of cations on TiO_2 particles. A method to determine the surface density of OH groups, *J. Chem. Soc. Chem. Commun.,* 1865–1866, 1994.

102. Borgarello, E., Harris, R., and Serpone, N., Photochemical deposition and photorecovery of gold using semiconductor dispersions. A practical application of photocatalysis, *Nouv. J. Chim.,* 9, 743–747, 1985.

103. Caballero, A., González-Elipe, A. R., Fernández, A., Herrmann, J.-M., Dexpert, H., and Villain, F., Experimental set-up for *in-situ* absorption spectroscopy analysis of photochemical reactions: the photocatalytic reduction of gold on titania, *J. Photochem. Photobiol. A: Chem.,* 78, 169–172, 1994.

104. Curran, J. S., Doménech, J., Jaffrezie-Renault, N., and Phillipe, R., Kinetics and mechanism of platinum deposition by photoelectrolysis in illuminated suspensions of semiconducting titanium dioxide, *J. Phys. Chem.,* 89, 957–963, 1985.

105. Prairie, M. R., Evans, L. R., Stange, B. M., and Martinez, S. L., An investigation of TiO_2 photocatalysis for the treatment of water contaminated with metals and organic chemicals, *Environ. Sci. Technol.,* 27, 1776–1782, 1993.

106. Prairie, M. R., Stange, B. M., and Evans, L. R., TiO_2 photocatalysis for the destruction of organics and the reduction of heavy metals, in: *Photocatalytic Purification and Treatment of Water and Air,* Ollis, D. F. and Al-Ekabi, H. (Eds.), Elsevier Science Publishers, Amsterdam, 1993, pp. 353–363.

107. Tennakone, K., Thaminimulle, C. T. K., Senadeera, S., and Kumarasinghe, A. R., TiO_2-catalysed oxidative photodegradation of mercurochrome: an example of an organo-mercury compound, *J. Photochem. Photobiol. A: Chem.,* 70, 193–195, 1993.

108. Navio, J. A., Marchena, F. J., Cerrillos, C., and Pablos, F., UV photocatalytic degradation of butyl tin chlorides in water, *J. Photochem. Photobiol. A: Chem.,* 71, 97–102, 1993.

109. El Anba, F. and Mille, G., Kinetic study of the photodegradation of bis (tributyltin) oxide (TBTO) in aqueous solution, *C. R. Acad. Sci., Ser. II,* 318, 317–322, 1994.

110. Cerrillos, C., Pradera Adrian, M. A., and Navio, J. A., UV photocatalytic degradation of phenylmercury compounds in water–acetonitrile (1:1) media, *J. Photochem. Photobiol. A: Chem.,* 84, 299–303, 1994.

Chapter III

Hydrocarbon Derivatives

Hydrocarbon derivatives discussed in this chapter include various harmful aliphatic and aromatic compounds, such as the extremely toxic nonylphenol, and the treatment of oil spills, surfactants, and pulp and paper industry effluents.

Partial photooxidation of hydrocarbon compounds is useful for the synthesis of oxidized derivatives, such as aldehydes and carboxylic acids.[1–3] For the degradation of organic contaminants in freshwater or wastewater, the preferred target is the complete mineralization of the organic compounds.

I. ALIPHATIC COMPOUNDS

A. Homogeneous Photolysis

1. Phosphotungstates and Phosphomolybdates as Sensitizers

One approach to the photoassisted oxidation of organic compounds in homogeneous solutions has been by utilization of the heteropoly compounds of molybden and tungsten. These compounds were described by the general formula $A_a X_x O_m^{n-}$, where A = P, Si, Fe, or H_2, X = Mo or W, and x/a = 12 or 6. In the presence of easily oxidizable organic compounds, these heteropoly compounds underwent photoreduction with visible and ultraviolet (UV) light to the heteropoly blues. The organic compounds that were thus oxidized included alcohols, glycols, hydroxyacids, and carboxylic acids. Most effective were phosphotungstates, PW_{12}^{3-}, and phophomolybdates, PMo_{12}^{3-}. The interaction of the heteropoly compounds with the organic molecules involved either electron or hydrogen transfer, resulting in oxidation of the organic species and formation of the heteropoly blue. In the presence of molecular oxygen, the heteropoly blue compounds underwent rapid reoxidation, thus enabling the photocatalytic oxidation of organic compounds.[4]

2. Hg Salts as Sensitizers

Alkanes in aqueous solutions undergo photooxidation in the presence of Hg(II) salts and O_2. These reactions were carried out with mixtures of $HgSO_4$ and Hg_2SO_4 (1 mM each) in 0.1 M H_2SO_4 on linear and cyclic hydrocarbons from C_1 to C_7, including cyclopentane, cyclohexane, and cycloheptane. The mechanism proposed involved excitation of Hg(II) as the primary step:[5]

$$Hg(II) + h\nu \rightarrow Hg^*(II) \tag{1}$$

$$Hg^*(II) + RH = RHg(II) + H^+ \qquad (2)$$

$$RHg(II) + h\nu = R + Hg(I) \qquad (3)$$

$$2Hg(I) \rightarrow Hg_2^{2+} \qquad (4)$$

The alkyl radicals formed, then presumably reacted with oxygen to form peroxyl radicals, which decomposed to release carbon dioxide, producing lower alkanes:

$$R_1CH_2 + O_2 \rightarrow R_1C(O)OH \qquad (5)$$

$$R_1C(O)OH + h\nu \rightarrow CO_2 + R_1H \qquad (6)$$

3. Benzophenones as Sensitizers

Ultraviolet irradiation of ethanol in dilute aqueous solutions in the presence of water-soluble benzophenones, with either positive (tetraalkyl ammonium) or negative (sulfonate) side chains, resulted in dehydrogenation of the alcohol to acetaldehyde, with condensation of the benzophenones to pinacols. In the presence of colloidal Pt as catalyst, the intermediate ketyl and 1-hydroxyethyl radicals caused the reduction of water to hydrogen.[6] This method could have been an attractive approach to photolytic water splitting, but for the fact that the photosensitizing benzophenones absorb strongly only at or below 300 nm.

4. Formic Acid with H_2O_2

Irradiation of aqueous solutions of formic acid or of sodium formate with a 10-W low-pressure Hg lamp ($\lambda = 254$ nm) did not result in decomposition, since both HCOOH and HCOO⁻ do not absorb at this wavelength. However, in the presence of added hydrogen peroxide, quite rapid photolysis occurred, which was due to light absorption by H_2O_2 at 254 nm and its photodissociation to hydroxyl radicals. Within 3 h, with initially 0.01 M H_2O_2, formate ion solutions (pH 9.5) had mineralized completely, while formic acid solutions (pH 1.5) had decomposed to about 60%. At low initial concentrations of both hydrogen peroxide and formic acid or formate, in the range of 1 mM, the decreases in concentration of formic acid or formate were not accompanied by decreases in the concentration of hydrogen peroxide, which remained stable. This surprising result was explained by the following mechanism, in which the primary step was photolysis of hydrogen peroxide,[7]

$$H_2O_2 + h\nu \rightarrow 2 \cdot OH \qquad (7)$$

followed by a hydrogen atom abstraction reaction of the hydroxyl radical,

$$\cdot OH + HCOOH \rightarrow \cdot COOH + H_2O \tag{8}$$

$$\cdot COOH + O_2 \rightarrow CO_2 + \cdot HO_2 \tag{9}$$

$$\cdot HO_2 + \cdot HO_2 \rightarrow H_2O_2 + O_2 \tag{10}$$

Thus, for each mole of H_2O_2 that was photolyzed, 1 mol of H_2O_2 was regenerated, and the concentration stayed constant. The "new" H_2O_2 according to this mechanism was therefore derived from dissolved oxygen. The higher photooxidation rate of formate ion relative to that of formic acid is understood considering the rate constants for reaction of $\cdot OH$ radicals with HCOOH and HCOO$^-$, which are 0.8 to 1.4×10^8 and 2.4×10^9 L mol^{-1} s^{-1}, respectively.[7]

5. Acetic Acid with H_2O_2

Support for the preceding mechanism of photodecomposition of formic acid in the presence of hydrogen peroxide came from an analogous study on acetic acid. Illumination with a low-pressure Hg lamp of aqueous solutions of acetic acid (initially 1 mM) not containing added hydrogen peroxide caused the formation of hydrogen peroxide, together with methane and carbon dioxide. The production of H_2O_2 could be accounted for by the following reaction steps:[8]

$$CH_3COOH + h\nu \rightarrow \cdot CH_3 + \cdot COOH \tag{11}$$

$$\cdot COOH + O_2 \rightarrow \cdot HO_2 + CO_2 \tag{12}$$

$$2 \cdot HO_2 \rightarrow H_2O_2 + O_2 \tag{13}$$

6. Formic Acid with Fe(III)

The photooxidation of formic acid in oxygenated aqueous solution is catalyzed by iron ions. The overall reaction is represented by

$$HCOOH + {}^1/_2 O_2 \rightarrow CO_2 + H_2O \tag{14}$$

This reaction was shown to depend on two photochemical steps,

$$2Fe^{3+} + HCOOH + h\nu_1 \rightarrow 2Fe^{2+} + CO_2 + 2H^+ \tag{15}$$

$$2Fe^{2+} + {}^1\!/_2 O_2 + 2H^+ + h\nu_{II} \rightarrow 2Fe^{3+} + H_2O \tag{16}$$

At wavelengths above 300 nm (at which formic acid is transparent) this reaction depends on light absorption by the iron ions. Thus, at 366 nm, the absorptivities for Fe(II) are $\varepsilon = 1.6$ cm^{-1} M^{-1} and for Fe(III), $\varepsilon = 125$ cm^{-1} M^{-1}. Under O_2, in aqueous solutions of formic acid containing either ferrous or ferric ions, the formic acid was completely degraded to carbon dioxide, while the Fe(II)/Fe(III) ratio approached a stationary state plateau. The rate of photooxidation of formic acid was found to be proportional to the absorbed light intensity.[9]

7. 2-Propanol with H_2O_2

2-Propanol (isopropanol) is used in the electronics industry for the cleaning of semiconductors. It thus contributes to the pollution of wastewater. Under illumination with a 10-W low-pressure Hg lamp ($\lambda = 254$ nm), with an initial concentration of 1 mM 2-propanol and 3 mM H_2O_2, the 2-propanol had nearly disappeared within 12 min. This photolysis resulted in the formation of acetone as primary intermediate, followed by formation of acetic acid and formic acid in a later stage of degradation, and complete mineralization within 9 h.[8]

B. Heterogeneous Photodegradation

The heterogeneous photocatalytic oxidation of organic compounds on powdered semiconductors suspended in aqueous media was pioneered by Bard et al.[10,11] Using platinized TiO_2 powders, aliphatic hydrocarbons such as hexane, heptane, and nonane, cyclohexane, and kerosene, as well as aromatic compounds such as benzene, were degraded to oxidized intermediates and eventually to carbon dioxide.[10]

1. Carboxylic Acids

The photocatalytic decarboxylation of carboxylic acids has been shown by Kraeutler and Bard[11] to be a photochemical variant of the Kolbe reaction. Acetic acid or acetate ions in oxygenated aqueous dispersions of platinized TiO_2 (1 to 5 wt% Pt on anatase) under illumination with an Xe–Hg lamp or with sunlight produced CO_2 as well as the hydrocarbons methane and ethane (in the ratio 12:1). In the absence of oxygen, the photodecomposition of acetic acid or acetate was even more quantitative toward the production of methane, in addition to small yields (5 to 10%) of ethane and hydrogen:

$$CH_3COOH \rightarrow CH_4 + CO_2 \tag{17}$$

$$2CH_3COOH \rightarrow C_2H_6 + 2CO_2 + H_2 \tag{18}$$

The photo Kolbe reaction thus differs from the thermal Kolbe reaction, in which the main product from acetic acid/acetate is ethane. This photocatalytic decarboxylation was extended to other carboxylic acids: propionic, *n*-butyric, *n*-valeric, pivalic, and adamantane-1-carboxylic acids, with production of the corresponding hydrocarbons as the major products. The proposed mechanism involved photoproduced holes and electrons trapped in shallow surface traps. These trapped holes react with adsorbed carboxylate anions, while electrons react with carboxylic acids or with oxygen:[11]

$$e^- + O_2 \rightarrow O_2^- \tag{19}$$

$$e^- + RCOOH \rightarrow H_{ads} + RCOO^- \tag{20}$$

$$h^+ + RCOO^- \rightarrow R\cdot + CO_2 \tag{21}$$

$$e^- + R\cdot + RCOOH \rightarrow RH + RCOO^- \tag{22}$$

$$R\cdot + H_{ads} \rightarrow RH \tag{23}$$

$$2R\cdot \rightarrow R_2 \tag{24}$$

$$2H_{ads} \rightarrow H_2 \tag{25}$$

In a further detailed kinetic study of the photodegradation with TiO_2 of acetic acid and of the chloroacetic acids, Chemseddine and Boehm[12] concluded that the preceding photo Kolbe mechanism was of only secondary importance. The major contribution to the degradation was proposed to be attack at the α-C–H bonds, resulting in hydrogen abstraction. The reactive species carrying out this abstraction is the ·OH radical, and in the presence of oxygen also the superoxide ion, O_2^- (or in acidic solution its conjugate acid, $HO_2\cdot$). This explained the very low reactivity of trichloroacetic acid. In a comparison of the photocatalytic degradation of acetic acid with that of the chloroacetic acids in oxygenated aqueous dispersions of TiO_2, the order of the initial rates of CO_2 release was dichloroacetic acid = chloroacetic acid > acetic acid $\gg$ trichloroacetic acid.[12]

A reconsideration of the mechanism of photocatalytic degradation of acetate over TiO_2 was based on detailed radiation chemical studies. These had revealed that in basic solutions the acetate anion was attacked by hydroxyl radicals mainly at the methyl group, resulting in hydrogen abstraction and formation of a ·CH_2COO^- radical. This reacted rapidly with oxygen, forming a peroxy radical, ·$OOCH_2COO^-$, which then broke down, producing glyoxylate, glycolate, and formaldehyde. If methyl radicals were released from acetate ions by direct electron transfer via the Kolbe reaction (Equation 21), such radicals should, as known from radiation chemistry, react with oxygen to form the peroxy radical $\cdot OOCH_3$, which would break down to produce formaldehyde, methanol, and formate. The TiO_2 photocatalyzed degradation of acetate at pH 10.6 yielded mainly glycolate and formate, with smaller amounts of glyoxylate. This is thus in agreement with a mechanism involving hydroxyl radicals. On the other hand, in acidic suspensions of TiO_2, the photooxidation of acetic acid yielded only formaldehyde and formate. In such media, the predominant mechanism may therefore be the direct oxidation of acetic acid by valence-band holes.[13]

The decarboxylation of carboxylic acids was studied in the presence of light acceptors such as UO_2^{2+}, Fe^{2+}, and Cu^{2+}. Dissolved Fe^{2+} and Cu^{2+} ions considerably enhanced the activity of TiO_2 (anatase form) as photocatalyst for the decarboxylation of formic acid.[13,14]

In the presence of suspended n-ZnS, sodium formate in aqueous solutions underwent photooxidation with production of carbonate, carbon monoxide, and hydrogen. A quantum yield of 0.24 was obtained under illumination at 290 nm.[16]

Saturated carboxylic acids in aqueous solutions were photodecomposed by size-quantized Fe_2O_3 microcrystallites incorporated in sodium montmorillonite clay. The clay contained about 36 wt% Fe_2O_3, and photocatalysis was performed in deaerated dispersion of this clay (10 mg/10 ml solution) under illumination with a 500-W high-pressure Hg lamp. The photodecomposition of acetic, propionic, and n-butyric acids resulted in the formation of methane, ethane, and propane, respectively, indicating a photo Kolbe mechanism. In the pH range 2 to 7, the rates of photodegradation had a maximum at pH 4.8, which was also the point of zero charge (pzc) of the Fe_2O_3/clay. In contrast to the photoactivity of this iron oxide clay, α-Fe_2O_3 was practically inactive as photocatalyst for the decarboxylation of carboxylic acids.[17]

The photodegradation of formic acid in aqueous solutions was performed on semiconducting membranes supported on glass. A sol-gel method was applied for coating TiO_2 on glass. Doping of the TiO_2 with MgO, Li_2O, or Pt-metal enhanced the rate of mineralization of formic acid, relative to that with undoped TiO_2. The reaction rate as a function of the formic acid concentration could be fitted to a Langmuir–Hinshelwood model.[18] Using ceramic membranes of TiO_2, quantum efficiencies as high as 30% were achieved for the photocatalytic oxidation of aqueous solutions of formic acid.[19] Illumination in the presence of suspended TiO_2 or Pt/TiO_2 of aqueous solutions (pH 1) of acetic acid, acetamide, and acetonitrile, as well as of benzoic acid, benzamide, and benzonitrile, caused mineralization to CO_2.[20]

As described in detail in Chapter I, Section V.E, photoelectrochemical degradation of aqueous formic acid has been achieved using TiO_2-coated tin-oxide glass slides biased at +0.3V (vs. SCE), with a current of 1 to 2 mA.

Applying the bias voltage increased the rate of oxidation of formic acid by 20 to 30%.[21,22]

In a study of the photocatalytic oxidation of oxalic acid in aqueous solution, the effect of transition metal dopants in colloidal TiO_2 was investigated. Addition of Cr^{3+}, Fe^{3+}, and V^{5+} cations did not enhance the rate of photooxidation.[23]

2. Methanol and Ethanol

Illumination of a mixture of methanol and water (9:1 v/v) in a quartz cell with a high-pressure Hg lamp caused the formation of hydrogen, ethylene glycol, and formaldehyde. When carried out in an alkaline medium (pH 12) at 50°C, the selectivity to the photoproduction of ethylene glycol reached 95%. Spin-trapping experiments with the spin trap PBN (phenyl-*N-tert*-butyl nitrone) and electron spin resonance measurements indicated the intermediate formation of the radical $\cdot CH_2OH$, which presumably dimerized to form ethylene glycol.[24] However, in deaerated aqueous mixtures of methanol and ethanol (e.g., alcohols to water = 4:1, v/v, and MeOH to EtOH = 5:1, v/v) containing colloidal ZnS, illumination resulted not in degradation but in the production of hydrogen as well as 1,2-propanediol, 2,3-butanediol, and formaldehyde. Presumably, methanol and ethanol reacted with the photoexcited ZnS to produce the intermediate radicals $\cdot CH_2OH$ and $\cdot CHOH\text{-}CH_3$, which then combined to produce 1,2-propanediol and 2,3-butanediol:[25]

$$CH_3OH + CH_3CH_2OH + h\nu \rightarrow \cdot CH_2OH + \cdot CHOH{-}CH_3 \qquad (26)$$

$$\cdot CH_2OH + \cdot CHOH - CH_3 \rightarrow$$
$$HOCH_2{-}CHOH{-}CH_3 + CH_3{-}CHOH{-}CHOH{-}CH_3 \qquad (27)$$

The photocatalytic oxidation of ethanol to acetaldehyde was also achieved using TiO_2-coated hollow glass microbeads.[26]

See also Section II.C on the use of such TiO_2-coated microbeads for the treatment of oil spills.

3. Direct ESR Identification of Radical Intermediates

Direct identification of the organic radicals formed by the photooxidation of acetic acid, methanol, ethanol, and 2-propanol in aqueous dispersions of TiO_2 loaded with platinum group metals was achieved by electron spin resonance (ESR) spectroscopy. Experiments were made at room temperature, under illumination with a 500-W super-high-pressure Hg lamp ($\lambda = 300$ to 400 nm). Among the metals, Pt, Pd, and Rh were most effective in generation of the intermediate radicals, while Ir, Os, and Ru were less effective. The optimal loading of the metal for effective radical formation was metal/TiO_2 = 0.5 to 1.5 × 10^{-3} molar ratio. No radicals could be detected when "bare" TiO_2 was used. The radical intermediates identified from acetic acid were the methyl ($\cdot CH_3$) and the carboxymethyl radicals ($\cdot CH_2COOH$). From methanol, ethanol, and 2-propanol, the intermediate species detected were α-hydroxymethyl ($\cdot CH_2OH$), α-hydroxyethyl ($\cdot CH[CH_3]OH$), and α-hydroxypropyl ($\cdot CH[CH_3]_2OH$), respectively.[27] The formation of the methyl radicals from acetic acid could be explained by a photo Kolbe reaction of the holes (h[+]) on the semiconductor particle surface with the acetate anion:[11]

$$CH_3COO^- + h^+ \rightarrow \cdot CH_3 + CO_2 \tag{28}$$

$$2 \cdot CH_3 \rightarrow C_2H_6 \tag{29}$$

The other radical species observed could be formed by hydrogen abstraction reactions of hydroxyl radicals ($\cdot OH$), such as from acetic acid and from methanol:[27]

$$\cdot OH + CH_3COOH \rightarrow H_2O + \cdot CH_2COOH \tag{30}$$

$$\cdot OH + CH_3OH \rightarrow H_2O + \cdot CH_2OH \tag{31}$$

4. 1-Propanol and Propanal

Limiting quantum yields Φ' for the degradation of 1-propanol and of propanal (propionaldehyde) were measured in three systems: in regular aqueous dispersions of TiO_2, with TiO_2 small-size microcrystallite particles, and in a flow reactor with immobilized TiO_2 on a fiberglass mesh. The limiting quantum yield was defined as a parameter that approaches the true quantum yield Φ when all the light entering the reactor was absorbed by the photocatalyst. Thus, Φ' is smaller or equal to Φ. For 1-propanol, the value $\Phi' \sim 1.0$ was measured in all three types of reactors. The high quantum yield indicated that surface-adsorbed hydroxyl radicals reacted more rapidly with the alcohol molecules than with surface-trapped electrons (or Ti^{3+} centers). Thus, interfacial electron transport to adsorbed molecules may be rapid compared with the recombination of trapped charge carriers.[28,29] Picosecond laser spectroscopy of the microcrystallite TiO_2 revealed differences from the ps spectral transients observed previously in TiO_2 colloids,[30–32] indicating different photophysical processes.

See Lepore et al.,[28] Chapter I, Section IV.H, on the method of extraction of microcrystallites of TiO_2.

5. Methyl Vinyl Ketone

Methyl vinyl ketone is an industrial pollutant that inhibits biological wastewater treatment. Its photodegradation in oxygen purged solutions was compared in the presence of TiO_2 either suspended or immobilized on a glass surface. Both the disappearance of the methyl vinyl ketone and the appearance of CO_2 were substantially faster in the case of the suspended TiO_2.[33]

6. Polyvinyl Alcohol

Polyvinyl alcohol (PVA) is released to wastewater during the washing of sized warp yarn in the weaving industry. The photocatalytic degradation of an aqueous slurry of PVA was performed in oxygenated dispersions of TiO_2 (10 g/L) under illumination with a 100-W high-pressure Hg lamp ($\lambda > 300$ nm). Acetic acid and acetone were detected as intermediate products, and within 2 to 3 h the COD (chemical oxygen demand) had essentially disappeared, indicating complete mineralization.[37]

II. AROMATIC AND OTHER CYCLIC COMPOUNDS

A. Homogeneous Photolysis

1. Toluene and Phenol — Direct Photolysis

The direct photolysis of aromatic hydrocarbons in aqueous solution is very inefficient. The pseudo first-order rate constants for the photodegradation of toluene and phenol at 254 nm were 0.0058 and 0.0019 min^{-1}, respectively.[35] This direct photolysis was due to absorption of light below 280 nm. Thus, in neutral and acidic solutions of phenol, $\lambda_{max} = 271$ nm; $\varepsilon = 2.19 \times 10^3$ M^{-1} cm^{-1}. Under illumination with a 400-W high-pressure Hg lamp, in the presence of hydrogen peroxide (initially 14 mM), the COD (chemical oxygen demand) of a solution of phenol (initially 1 mM) decreased to nearly zero within about 45 min. No reaction took place in the dark. The photodegradation resulted in the stoichiometric conversion of phenol and hydrogen peroxide, resulting in practically complete mineralization according to the equation[36]

$$C_6H_5OH + 14H_2O_2 \rightarrow 6CO_2 + 17H_2O \tag{32}$$

2. Flash Photolysis of Phenol

Illumination of aqueous solutions of phenol with 10-μs flashes from a xenon lamp resulted in the formation of *p*-benzoquinone as the primary identified product, in addition to smaller amounts of hydroquinone. After repeated flashing, 2-hydroxyhydroquinone was also detected.[37]

3. Toluene and Phenol with H_2O_2

The rates of photolysis of toluene and phenol were considerably enhanced in the presence of dilute hydrogen peroxide (8 mM), to 0.154 and 0.199 min^{-1}, respectively. The primary step was the photolysis of hydrogen peroxide to hydroxyl radicals by light below 310 nm:

$$H_2O_2 + h\nu \rightarrow 2 \cdot OH \qquad (33)$$

The hydroxylation of aromatic compounds was proposed to result initially in the formation of cyclohexadienyl radicals, followed by their oxidation to phenols, or by dissociation to radical cations. The rates of reaction of hydroxyl radicals with aromatic compounds are close to those of diffusion-controlled reactions. Values for the second-order rate constants of hydroxyl radicals reacting with benzene, chlorobenzene, nitrobenzene, and phenol (at pH 7) are 7.8×10^9, 5.5×10^9, 3.9×10^9, and 6.6×10^9 M^{-1} s^{-1}, respectively. Hydroxyl radical reactions are much less efficient with halogenated alkanes. With chloroform and dichloromethane, the rate constants for reactions with hydroxyl radicals are only 1.5×10^7 and 5.8×10^7 M^{-1} s^{-1}, respectively.[35]

While the rate of phenol photooxidation was considerably enhanced by the presence of hydrogen peroxide, the rate of the photodecomposition of hydrogen peroxide was not affected by the presence of phenol. An even greater acceleration of phenol photooxidation was observed in the presence of both hydrogen peroxide and oxygen. This acceleration was explained to be due to the formation of $\cdot HO_2$ radicals:[38,39]

$$H_2O_2 + h\nu \rightarrow H_2O_2^* \qquad (34)$$

$$H_2O_2^* + O_2 \rightarrow 2 \cdot HO_2 \qquad (35)$$

$$\cdot HO_2 + C_6H_5OH \rightarrow H_2O_2 + \cdot C_6H_4OH \qquad (36)$$

$$\cdot C_6H_4OH + O_2 \rightarrow \text{organic peroxide} \qquad (37)$$

See also Chapter VIII, Section I, on the humic acid photosensitized degradation of 2,4,6-trimethylphenol.[40]

4. Photo Fenton Reaction

By illuminating aqueous solutions of Fe^{2+} and H_2O_2 (initial pH 3.3) at $\lambda > 320$ nm, a "photo Fenton" type reaction was applied to the degradation of several aromatic and alicyclic compounds. The remaining TOC (total organic carbon) after 5 h of illumination of the aromatic compounds phenol, hydroquinone,

4-chlorophenol, 4-chloroaniline, 4-nitroaniline, and 3-nitroaniline was 6, 2, 2, 8, 24, and 26%, respectively. In the dark, the degradation was much slower. For the alicyclic compounds cyclohexanol and cycohexanone, the remaining TOC after 5 h of illumination was 98 and 86%. The lability of the aromatic compounds was explained by attack of hydroxyl radicals (formed in the photo Fenton reaction) on the aromatic ring, resulting in ring opening and further degradation. The alicyclic compounds were merely hydroxylated, without ring cleavage.[41,42]

5. Dye Sensitization and QSAR Analysis of $^1O_2$ Reaction with Phenols

Dye sensitizers cause enhanced photodegradation of phenols, which are common pollutants in natural waters. The rates of disappearance under sunlight illumination of both undissociated phenols and of phenolate anions in the presence of rose bengal as sensitizer were correlated with the half-wave potentials, $E_{1/2}$, and with the substituent constant σ. A QSAR (quantitative structure–activity relationship) analysis of the reaction of substituted phenols with singlet oxygen was performed in order to determine the importance of this reaction for the natural abiotic degradation of such pollutants. Both with *ortho*- and multisubstituted phenols a significant quantitative structure–activity relationship was observed. A good linear correlation was found between the logarithm of the second-order rate constants and the half-wave potentials or Hammett σ constants for those phenols that react at high rates. Such QSAR data may be useful for predicting the unknown reaction rates of other substituted phenols with $^1O_2$. Much lower rates than expected from the half-wave potentials were observed for phenols with substituents that are through-resonance acceptors, such as *ortho*- and *para*-substituted cyano, nitro, and acetyl groups. The involvement of singlet oxygen in these reactions was proven by the inhibitory effects of added azide ion (5 mM) or of DABCO (1,4-diazabicyclo[2.2.2]octane), or the stimulatory effect of replacing H_2O as solvent by D_2O (in which $^1O_2$ is quenched more slowly). The proposed mechanism involved a rate-limiting formation of a precursor complex with a small extent of charge-transfer character.[43–45] The rose bengal-photosensitized oxidation by singlet oxygen of 2,4–6-trimethylphenol was very rapid and could be described as the sum of reactions of the protonated phenol and of the phenoxy anion. The second-order rate constants for these species with $^1O_2$ in water at 27°C were 6.2 × 10^7 M^{-1} s^{-1} and 1.1 × 10^9 M^{-1} s^{-1}, respectively.[46]

The photolytic decomposition of phenols was considerably accelerated in the presence of riboflavin as sensitizer.[47] Under illumination with a 200-W medium-pressure Hg lamp ($\lambda > 300$ nm), in aqueous solutions containing riboflavin (5 μM), phenol underwent rapid decomposition, at rates that decreased with rising pH of the solution. The decomposition did not follow simple first-order kinetics. Thus, at pH 5, during the first 2 min, riboflavin served as the main sensitizer. Subsequently, as riboflavin was decomposed (with $t_{1/2} = 1.5$ min) to lumichrome, the degradation of phenol slowed down, while lumichrome served as the main sensitizer. Phenol was resistant to direct unsensitized photolysis under these conditions. However, 2,4-dichlorophenol, 2,4,6-trichlorophenol, pentachlorophenol, *p*-cresol, and 1-naphthol underwent quite rapid direct photolysis, with half-lives of 12, 21, 9.0, 45, and 29 min, respectively. The rates of photodegradation of all these phenols were substantially enhanced in the presence of riboflavin. Since the photolysis rates were larger in the absence than in the presence of oxygen, it was proposed that the mechanism did not involve singlet oxygen as an intermediate. Rather, a "Type I" sensitized reaction,[48] by a mechanism of direct energy or electron transfer between the flavin sensitizer and the phenol, had to be assumed.[47]

The rose bengal dye-sensitized photodegradation of phenol under sunlight was performed in an annular bubble-column photoreactor. The rate of reaction of phenol with singlet oxygen increased with rising pH, and was quenched by the triplet state of the sensitizer, $^3D^*$. The primary intermediate product was *p*-benzoquinone, which was further degraded to carboxylic acids.[49]

See also Chapter VIII on the photochemistry of riboflavin and lumichrome in natural waters.[50]

6. Nonylphenol

The very toxic 4-nonyl phenol (C_9H_{19}–C_6H_4OH) and nonylphenol polyethoxylates are the products of the biodegradation of the nonionic alkylphenol polyethoxylated surfactants during sewage treatment (see below, Chapter III, Section III.A). These compounds, which are lethal to aquatic organisms, are released in secondary effluents and were found in relatively high concentrations in Swiss lakes.[51] The direct photolysis ($\lambda > 280$ nm) of nonylphenol and of nonylphenol polyethoxylates in distilled water was very slow. Much higher rates of photolysis were observed in filtered lake water. The estimated half-life of nonylphenol under noon sunlight illumination in surface natural waters was estimated at 10 to 15 h. The nonylphenol polyethoxylates underwent much slower photodegradation. Experiments on the photosensitized oxidation of nonylphenol in the presence of rose bengal ($\lambda > 530$ nm) showed that at pH 10.7 the rate in D_2O was 4.6 times faster than in H_2O as solvent. At this alkaline pH, the photolysis presumably occurred mainly via a singlet oxygen mechanism. On the other hand, at pH 9, which is more representative of natural waters, the ratio of photolysis rates D_2O/H_2O was only 1.8. This result seemed to indicate that singlet oxygen was not significantly involved in the photooxidation of nonylphenol in natural waters.[51]

7. Phthalic and Maleic Anhydrides

Phthalic anhydride is released from the organic chemicals industry in Slovenia into an adjacent river, and subsequently into the Gulf of Trieste (Northern

Adriatic). The direct photolysis of phthalic anhydride was studied in N_2-purged distilled water, as well as in river water, and in artificial and natural seawater, using illumination by a 125-W Hg lamp or by sunlight. The photochemical transformation could be described by steps of hydrolysis to phthalic acid, followed by decarboxylation, producing small amounts of benzoic acid, and mainly polymerization to polyphenyl. Nuclear magnetic resonance (NMR) data indicated that in this polymer, 1,4-bonding was predominant, with a minor degree also of 1,2-bonding:[52]

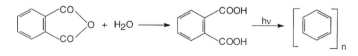

SCHEME III.1

The average molecular weight of this polymer was distributed in the quite narrow range of 2800 to 3600. In the comparison of the different types of waters, the degradation of phthalic anhydride decreased in the order distilled water > river water ~ artificial seawater > natural seawater. Half-lives in these waters under Hg-lamp exposure were about 3.9, 6.3, 6.8, and 9.6 h, respectively. Under sunlight the reaction was much slower, and in distilled water only 4% of the phthalic anhydride had disappeared after 15 days of exposure.[52] The slower rates of reaction in the natural waters may be due to the deactivation or capture of radical intermediates.

The direct photolysis of both phthalic anhydride and maleic anhydride in distilled water was considerably enhanced in the presence of oxygen, resulting in first-order rate constants of 1.0×10^{-4} and 6.8×10^{-5} s^{-1}, respectively.[53]

See also in Section II.B.11 on the semiconductor-photocatalyzed degradation of phthalic and maleic anhydrides.[53]

8. α-Pinene

α-Pinene is a typical representative of the terpene compounds that are released in large amounts by terrestrial plants into the atmosphere and by blue-green algae into natural waters. The photolysis of aerated solutions of α-pinene (in CH_3CN–H_2O, 3:2 v/v) was found to be sensitized by humic acid and by methylene blue. The quantum yield of the photoconversion of α-pinene by humic acid was about 60 times smaller than that sensitized by methylene blue. Since the main photoproduct, *trans*-3-hydroperoxidepin-2-(10)-ene,[54] was obtained using either humic acid or methylene blue as sensitizers, it was proposed that the mechanism of oxygenation was the same, that is, by the involvement of singlet oxygen. Such abiotic degradation of a terpene may account for the low observed concentration of terpene compounds in natural waters.[54]

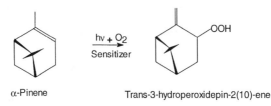

α-Pinene   Trans-3-hydroperoxidepin-2(10)-ene

SCHEME III.2

9. Hydroquinone

Hydroquinone is released in the effluents of many industries, such as from photographic processing and in paper production. Due to its toxicity, it is a considerable environmental problem. The direct photolysis in alkaline aqueous solutions (pH 8) in the absence of oxygen resulted in the formation of 2,5,2′,5′-tetrahydroxybiphenyl (THBP) as the major photoproduct, formed with a quantum yield of 0.002 at 296 nm. The rate of conversion of hydroquinone into THBP was much enhanced if benzoquinone was added. The proposed mechanism of formation of THBP was excitation of hydroquinone to its triplet state $^3QH_2$, followed by reaction with benzoquinone to form an exciplex. In the absence of quenchers, the half-life of the triplet state of hydroquinone was found to be 0.9 μs. In aerated and oxygenated solutions, the quantum yields at 296 nm of the phototransformation of hydroquinone were $\Phi = 0.048$ and 0.053, respectively, and the main photoproducts were hydrogen peroxide, benzoquinone, and hydroxybenzoquinone.[55]

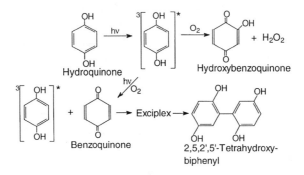

SCHEME III.3

Hydroquinone in deoxygenated aqueous solution was also photooxidized by nitrate ions, resulting mainly in the production of benzoquinone (85 to 90%) and also of some hydroxybenzoquinone and nitrohydroquinone. These transformations, when carried out under illumination at 296 nm, must be due to excitation of the hydroquinone, since the NO_3^- ion has very little absorptivity at this wavelength. At 296 nm, in the absence of oxygen, this phototransformation of hydroquinone had the very high quantum yield of 0.25 ± 0.05. The photo-oxidation by nitrate ions was inhibited by the presence of oxygen. Nanosecond laser flash photolysis at 266 nm of neutral aqueous solutions of hydroquinone indicated the appearance of transient absorption bands at $\lambda_{max} = 312$, 405, and 428 nm, identified as due to the semiquinone anion $Q^{\cdot-}$. Addition of nitrate ions resulted in enhanced formation of this transient. The dependence of the transient absorbance at 312 nm on the laser pulse energy indicated that the production of the semiquinone anion was due to one-photon processes. The quantum yield of the formation of the semiquinone anion at $\lambda = 266$ nm was evaluated to be $\Phi = 0.13$, 0.31, and 0.52 in deoxygenated, oxygen-saturated, and nitrate-containing solutions, respectively.[55]

10. Ascorbic Acid Synthesis

An interesting example of homogeneous photooxidation is that applied to the synthesis of ascorbic acid (vitamin C). Ascorbic acid is produced in nature by plants in a biosynthetic pathway starting with D-glucose, which is oxidized to D-gluconic acid, then undergoes ring closure to D-glucono-γ-lactone, and is further dehydrogenated to 1-ascorbic acid. In an abiological photochemical analogue, D-sorbitol (the reduction product of D-glucose) or L-sorbose was oxidized to ascorbic acid in the presence of oxygen, UV light from a high-pressure Hg lamp, and as sensitizer the natural flavanoid hesperidin (which is found in many plants, and in particularly high concentrations in grapefruit and other citrus fruits).[56]

B. Heterogeneous Photodegradation

Heterogeneous photooxidation in slurries of semiconductor catalysts has been studied extensively, in reactions in which hydroxyl radicals were implicated as the main oxidant.[10,38,57–67,72–74] In the photooxidation of aromatic compounds with TiO_2, the primary reaction intermediates were due to hydroxylation of the aromatic ring, often accompanied by cleavage of side chains, if present. Substituents on the ring with electron-donating property, such as HO and CH_3O groups, caused hydroxylation at *ortho* and *para* positions. On the other hand, electron-withdrawing substituents, as in nitrobenzene and in benzamide, had little effect on the orientation of hydroxylation. The intermediate products formed, hydroxyaromatics, quinones, were quite unstable and readily underwent further photodegradation.[68]

Proposed intermediates in the TiO_2-photocatalyzed degradation of aromatic derivatives were peroxycyclohexadienyl derivatives and mucondialdehyde-type compounds.[57] Rapid destruction of salicylic acid, phenol, 2-chlorophenol, 4-chlorophenol, benzoic acid, 2-naphthol, naphthalene, and fluorescein was achieved by circulating their aqueous solutions through a tubular glass reactor coated with a thin film of TiO_2 and illuminating with a 20-W blacklight UV fluorescent lamp. The times for 50% disappearance of 500 mL of initially 10 μM solutions were 7.1, 7.2, 8.2, 8.7, 6.9, 8.5, 4.3, and 6.4 min, respectively. The degradations observed first order kinetics. From the temperature dependence for the photodegradation of salicylic acid, an activation energy of 11.0 kJ mol^{-1} was derived.[66] For the destruction of phenol in UV-illuminated aqueous dispersions of TiO_2, the activation energy was 10 kJ mol^{-1}.[58,59]

In the photocatalyzed oxidation of salicylic acid in oxygen-saturated suspensions of ZnO (2 g/L), the intermediate products identified were 2,5-dihydroxybenzoic acid, 2,3-dihydroxybenzoic acid, and pyrocatechol. In the presence of the hydroxyl radical scavenger 2-propanol, the only observed product was 2,5-dihydroxybenzoic acid. This product was formed even when azide ions, which are effective scavengers of $^1O_2$, were present. Also, the ZnO-promoted phototransformation of salicylic acid was slower in D_2O than in H_2O. If a singlet oxygen mechanism were involved, the rate should have been faster in D_2O, in which the lifetime of $^1O_2$ is longer than in normal water. Thus, the production of 2,5-dihydroxybenzoic acid from salicylic acid was proposed to be due neither to the action of $\cdot OH$ radicals, nor to that of $^1O_2$, but instead was attributed to oxidation by positive holes on the semiconductor surface.[60]

1. Ethylbenzene

The photocatalytic degradation of ethylbenzene (initially 3 to 50 μM) in aqueous dispersions of TiO_2 (Degussa P25, 1 g/L) was performed in a recirculating reactor under illumination with a 1600-W xenon lamp ($\lambda > 300$ nm). The disappearance of ethylbenzene followed pseudo first-order kinetics, and the dependence on the initial concentration of the compound could be fitted to the Langmuir–Hinshelwood model. The degradation intermediates identified were 4-ethylphenol, acetophenone, 2-methylbenzyl alcohol, 2-ethylphenol, and 3-ethylphenol. The formation of the ethylphenols could be understood to be due to electrophilic substitution reactions by $\cdot OH$ radicals on the aromatic ring,with preferential orientation at the *para* and *ortho* positions. Substitution reaction at the ethyl side chain was proposed to be by free-radical hydroxylation, with preferential attack at the benzylic hydrogen atoms, thus forming the intermediates acetophenone and 2-methylbenzyl alcohol. Since the phenols are much more toxic than ethylbenzene (the European Community-recommended maximal concentrations in drinking water for ethylbenzene, acetophenone, and phenol are 0.1, 0.1, and 0.005 mg/L, respectively), it is imperative in the photocatalytic reaction to oxidize all these intermediates. At an initial pH of 4.5, about 65 min was required for the complete mineralization.[69]

2. Polynuclear Aromatic Hydrocarbons

Polynuclear aromatic hydrocarbons are products of the incomplete combustion of fossil fuels. By entering water bodies they become priority pollutants. Polynuclear aromatics accumulate in aquatic organisms, undergoing partial transformations. The chemical and photochemical oxidation of these hydrocarbons has been carefully reviewed.[70] The polynuclear aromatics absorb strongly in the near ultraviolet region (>290 nm). With increasing size of these molecules, their absorption maxima are shifted to longer wavelengths and into the visible part of the solar spectrum. Thus, direct photolysis is a major pathway for their abiotic degradation in the environment. Estimated half-lives for the photolysis by sunlight of polynuclear aromatic hydrocarbons range from 0.1 to 5 h. Intermediates observed were peroxides, while quinones were the main photoproducts.[70]

The photocatalytic degradation of acenaphthene, anthracene, fluorene, and naphthalene in oxygenated aqueous suspensions of TiO_2 (1 g/L; Degussa P25) under illumination either with a 500-W super-high-pressure Hg lamp or with sunlight was followed both by the disappearance of the fluorescence of these hydrocarbons and by the release of CO_2. With naphthalene, after 3 h of artificial illumination ($\lambda > 345$ nm), most of the fluorescence had disappeared, while the CO_2 release indicated 27% mineralization. The relatively stable intermediates identified were 2-formyl-3-hydroxycinnamaldehyde and 5,8-dihydroxy-1,4-naphthoquinone. From anthracene solutions, the photoproduct identified was anthraquinone. The proposed mechanism involved the action of hydroxyl and superoxide radicals.[71]

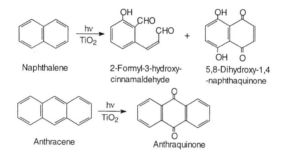

| Naphthalene | 2-Formyl-3-hydroxy-cinnamaldehyde | 5,8-Dihydroxy-1,4-naphthaquinone |

Anthracene Anthraquinone

SCHEME III.4

3. Benzoic Acid

In the photooxidation of benzoic acid or of sodium benzoate in UV-illuminated oxygenated aqueous dispersions of TiO_2, the products were *ortho-*, *meta-*, and *para*-hydroxybenzoic acids.[72–74] These products were formed in the same isomeric distribution previously found for γ radiolysis of benzoic acid, which is known to involve ·OH radical attack.[75]

4. Phenol

In the photooxidation of phenol, the anatase form of TiO_2 was reported to be much more effective than the rutile form. The presence of oxygen was a necessary requirement for this oxidation, which also proceeded, albeit more slowly, in a homogeneous reaction in the presence of oxygen and hydrogen peroxide (see Section II.A.1).[38] In the TiO_2-promoted reaction, the degradation rate was found to decrease at higher initial phenol concentration. In a further study, the effects of added hydrogen peroxide and of either ferric or cupric ions were determined. In the absence of added H_2O_2, ferric ions caused a slight enhancement in the rate of phenol photodegradation, while cupric ions caused retardation. However, in the presence of both added hydrogen peroxide and either Fe^{3+} or Cu^{2+} ions, a dramatic rate increase was observed. This synergistic effect of H_2O_2 and the metal ions was explained by a radical chain reaction:[62,63]

$$H_2O_2 + O_2^- \rightarrow \cdot OH + O_2 + OH^- \tag{38}$$

$$H_2O_2 + h\nu \rightarrow 2 \cdot OH \tag{39}$$

$$H_2O_2 + e^- \rightarrow \cdot OH + OH^- \tag{40}$$

$$H_2O_2 + Fe^{3+} \rightarrow H^+ + \cdot HO_2 + Fe^{2+} \tag{41}$$

$$H_2O + h^+ \rightarrow \cdot OH + H^+ \tag{42}$$

with the ferrous ion causing the production of additional hydroxyl radicals by the normal Fenton reaction,

$$Fe^{2+} + H_2O_2 \rightarrow Fe^{3+} + OH^- + \cdot OH \tag{43}$$

Wei and Wan[63] developed a theoretical model, based on rate-determining generation of $\cdot OH$ radicals, to account for the kinetics of the TiO_2-photocatalyzed degradation of phenol, without and with added H_2O_2 and Fe^{3+} ions.

While the direct photolysis of phenol in oxygenated aqueous solutions, under illumination with a 500-W high-pressure Hg lamp, became appreciable only above pH 6.5, the TiO_2-photocatalyzed degradation was rapid both in acidic and alkaline media, with a maximal rate at pH 3, and resulted in complete mineralization. Oxygen and hydrogen peroxide molecules were proposed to adsorb in competition on the same adsorption site on the TiO_2 surface, but phenol molecules adsorbed on a different site. In the presence of both oxygen and hydrogen peroxide, the adsorbed oxygen acted mainly as an electron trap, while the adsorbed hydrogen peroxide acted mainly as a hole trap. Under these conditions, the adsorbed phenol molecules underwent oxidation by two parallel mechanisms. A Langmuir–Hinshelwood type model was proposed to fit the experimental results, using separate second-order rate constants for the reactions of the adsorbed oxygen and hydrogen peroxide species with the adsorbed phenol.[38,39,64]

The preferred orientation for attack by ·OH radicals in the TiO_2-photocatalyzed oxidation of phenol and substituted phenols is *o*-hydroxylation, in addition to some *p*-hydroxylation, as well as oxidation of the *p*-substituent. On the other hand, the preferred orientation for h[+] oxidation is at the *p*-position in phenols. The oxidation by h[+] could be revealed by scavenging the ·OH radicals with isopropanol.[76]

Efforts to sensitize the TiO_2-induced photodegradation of phenol to visible light by doping the semiconductor with chromium were unsuccessful. Although doping TiO_2 with Cr did shift the absorption into the visible range, low levels of doping the TiO_2 with Cr had no effect on the rate of destruction of phenol, while high levels of Cr doping were detrimental.[77,78]

In the photooxidation of phenol in the presence of suspended ZnO, the main products were hydroquinone, catechol, *p*-benzoquinone, maleic acid, and carbon dioxide.[79]

A significant step toward the practical application of photocatalyzed oxidation of contaminants in open ponds was made by experiments on the natural sunlight-promoted degradation of phenol in open shallow dishes with TiO_2 immobilized on beach sand. The catalyst was prepared by sonicating a suspension of TiO_2 (Degussa P25) in water, adding to the sand, and drying in a rotary evaporator under vacuum at 95°C. With 0.5 g TiO_2 supported on 100 g of sand, in 500 mL of 0.1 mM phenol, circulated at a flow rate of about 250 mL/min in a shallow glass dish, 90% of the phenol had disappeared within 70 min. Mineralization to CO_2 was slower, and about 150 min was required for 90% removal of TOC (total organic carbon). Roughly 3 times higher rates of photooxidation had been observed using a 1% suspension of pure TiO_2. However, for large-scale operation in shallow lagoons, the TiO_2 in low concentration supported on sand will probably be more economic.[80,81]

The reaction mechanism in the semiconductor promoted photooxidation of phenol was proposed by Okamoto et al.[58] to involve stepwise hydroxylation, via di-, tri-, and tetrahydroxybenzenes (mixtures of the various isomers), leading to formic acid, which was finally oxidized to carbon dioxide.[58] The overall stoichiometry is

$$C_6H_5OH + 7O_2 \rightarrow 6CO_2 + 3H_2O \qquad (44)$$

Phenol in oxygenated aqueous solutions in the presence of colloidal Fe_2O_3, under irradiation with either sunlight or a 450-W xenon lamp ($\lambda > 380$ nm), was degraded to hydroquinone and catechol as initial products. Highest rates of photodegradation were obtained at the natural pH of the phenol solution, about pH 5.5. Good results were also obtained with a photocatalyst of colloidal Fe_2O_3 immobilized on beach sand.[82]

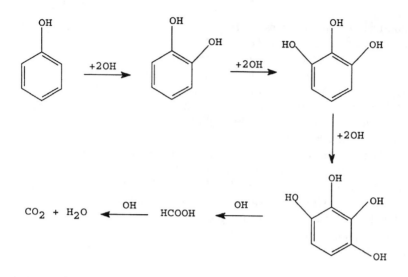

SCHEME III.5

5. *o-, m-,* and *p*-Cresol

In aerated aqueous dispersions of TiO_2 (2 g/L) at pH 3 illuminated with a 1000-W Hg/Xe lamp ($\lambda \geq 300$ nm), *ortho-*, *meta-*, and *para*-cresol underwent total mineralization within about 4 h. At shorter irradiation times, the major intermediates identified from *p*-cresol were 4-methylcatechol and 4-methylresorcinol, from *o*-cresol were 3-methylcatechol, 3-methylresorcinol, methylhydroquinone, and 2-methylresorcinol, and from *m*-cresol were 4-methylcatechol, 3-methylcatechol, orcinol, and methylhydroquinone. Methylhydroquinone disappeared and was completely mineralized within 1 h, while 4-methylcatechol disappeared within 20 min, and was totally mineralized after 1.5 h.[83,84] The photodegradation of *o-* and *p*-cresol occurred by first-order kinetics, but that of *m*-cresol was by zero-order kinetics. The degradation and mineralization of the cresols in oxygen-saturated solutions was faster than in aerated solution. For *m*-cresol, the change from aeration to oxygenation decreased the half-life of degradation from 174 to 59 min. The initial rate of photodegradation of *m*-cresol was linearly related to the radiant power level of the light source, up to 36 mW cm^{-2}. The photochemical efficiencies (defined as the number of molecules degraded per incident photon, or the lower limits of the quantum yields) for the disappearance at 365 nm of *o-*, *m-*, and *p*-cresol were 0.0096, 0076, and 0.010, respectively. The kinetic data could be fitted to a Langmuir–Hinshelwood type model. It was suggested that adsorption or photoadsorption of molecular oxygen occurred on Ti(III) sites, while organic substrates such as the cresols adsorbed on surface hydroxyl Ti(IV)-OH$^-$ sites.[83–85] In another study, irradiation of *p*-cresol with a medium-pressure Hg lamp in aerated aqueous suspensions of TiO_2 resulted initially in the formation of *p*-hydroxybenzaldehyde and *p*-hydroxybenzoic acid, and only traces of 4-methylcatechol. After 1 h of irradiation, the amount of CO_2 evolved corresponded to 95% mineralization of the *p*-cresol. The rate of oxidation of *p*-cresol was considerably enhanced if also hydrogen peroxide was present, and in this case significant amounts of 4-methylcatechol were produced.[86]

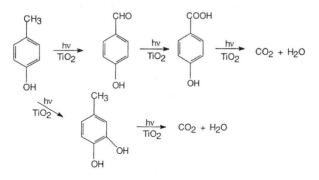

SCHEME III.6

The intermediate formation of hydroxyl radicals during the photocatalytic reactions of *p*-cresol in aqueous dispersions of TiO_2 was shown by spin trapping of the hydroxyl radicals with DMPO (5,5-dimethyl-1-pyrrolline *N*-oxide). The adsorption of DMPO was about three times higher than that of *p*-cresol. The reciprocal values of the concentration of the DMPO–·OH adduct was linearly related to the initial *p*-cresol concentration, in agreement with a Langmuir–Hinshelwood model of adsorption-controlled reaction. The kinetics in the presence of the spin trap could be accounted for by assuming competitive adsorption by both DMPO and *p*-cresol on one type of active site on the TiO_2 surface.[87,88]

6. Phenoxyacetic Acid

Phenoxyacetic acid, $C_6H_5OCH_2COOH$, may be considered the parent compound of the chlorophenoxy herbicides 2,4-D and 2,4,5-T (see Chapter IV, Section II.B). The rate of its photocatalytic degradation in aqueous dispersions of TiO_2 under illumination with a 125-W high-pressure Hg lamp observed Langmuir–Hinshelwood kinetics. After 30 min of illumination, about 72% of the initial phenoxyacetic acid was removed from the solution. The reaction rate increased with the TiO_2 concentration, reaching a limiting value at a concentration of 2 g/L. The rate of removal was proportional to the square root of the light intensity, indicating that the electron–hole recombination on the catalyst surface effectively competed with the photodegradation of the substrate,

$$e^- + h^+ \rightarrow \text{Heat} \qquad (45)$$

The temperature dependence of the reaction rate could be fitted to an Arrhenius law relationship, resulting in the rather low activation energy of 9.3 kJ mol[-1]. The main identified product of the photodegradation of phenoxyacetic acid was phenol, presumably formed by reaction of ·OH radicals on the alkyl chain of the molecule. Hydroquinone was obtained as a minor product, possibly by attack of ·OH radicals at the *para* position of phenol.[89]

7. Dimethoxybenzenes

The three isomeric dimethoxybenzenes in aqueous solutions did not undergo direct photolysis at $\lambda > 340$ nm, since the absorption maxima of 1,2-dimethoxybenzene (the antiseptic veratrole) and of 1,3-dimethoxybenzene (dimethylresorcinol) were $\lambda = 270$ nm, and that of 1,4-dimethoxybenzene (hydroquinone dimethylether) was $\lambda = 285$ nm. In the presence of dispersed TiO_2, in oxygenated solutions illuminated with a 125-W high-pressure Hg lamp, the disappearance of these dimethoxybenzenes was very rapid, and followed pseudo first-order kinetics. The order of disappearance was *meta > para > ortho,* and the first-order rate constants of these isomers were 0.087, 0.069, and 0.023 min[-1], respectively. The aromatic intermediates identified were due to *para* and *ortho* orientations and included mainly hydroxylation of the ring, with or without displacement of methoxy groups. These intermediates readily underwent further photocatalytic degradation. Complete mineralization was slower, requiring for 1,2-dimethoxybenzene about 9 h of illumination at $\lambda > 340$ nm.[90]

In an ingenious mechanistic study of the photocatalytic destruction of 1,2-dimethoxybenzene in aqueous dispersions of either TiO_2 or ZnO, Amalric et al.[91] determined the relative importance of H_2O_2 and O_2^- as reactive intermediates. The effects of H_2O_2 could be suppressed in the presence of catalase, which specifically decomposed hydrogen peroxide ($2H_2O_2 \rightarrow 2H_2O + O_2$), while the superoxide anion was rapidly decomposed in the presence of superoxide dismutase (SOD) ($2O_2^- + 2H^+ \rightarrow O_2 + H_2O_2$). Catalase had no effect on the ZnO-catalyzed photodegradation of the dimethoxybenzene, and only minor effects on the TiO_2-catalyzed process. On the other hand, SOD substantially inhibited the photocatalytic destruction of 1,2-dimethoxybenzene by both TiO_2 and ZnO. These results indicated that hydrogen peroxide formed *in situ* by photoexcitation of these semiconductors played a rather small role in the overall reaction, while the superoxide radical-anion played a major role in the destruction of the pollutant. In a tentative mechanism, dimethoxybenzene may capture a surface-generated hole, forming a radical cation (DMB[+]·) The superoxide anion was proposed to act as a Brönsted base, acting on the dimethoxybenzene radical cation, which thus acted as a Brönsted acid.[91]

8. Toluenes

A "photo Fenton" type reaction was observed in the photocatalytic oxidation of toluenes, in the presence of Cu^{2+} ions and suspended TiO_2, by the hydrogen peroxide produced from dissolved O_2. Side-chain oxidation prevailed over cresol formation, except at high Cu^{2+} concentrations in acidic solutions, in which cresol formation was the preferred reaction.[92] This effect of enhanced photooxidation rate of toluene, as well as of chlorobenzene, in aqueous TiO_2 (2 g/L) was observed in the presence of 10^{-5} M concentrations of the transition metal ions Cu(II), Fe(III), and Mn(II) at pH 3. At higher concentrations of these metal ions and at higher pH, the oxidation rates decreased. The order of activity was Cu(II) > Fe(III) > Mn(II), with an approximately twofold increase in initial rate by Fe(III) + TiO_2 in comparison with TiO_2 alone. Ni(II) and Zn(II) ions had no influence on the oxidation rates. This order of activity had the same general trend as the standard reduction potentials of these metals. The proposed mechanism involved a reactive complex between the metal ion in solution, the organic compound, and an oxygen-containing species, which may be $O_2{\cdot}^-$ or H_2O_2. The decreased rates at higher metal ion concentrations may be due to absorption of UV light, thus competing with the excitation of the TiO_2 photocatalyst.[93]

9. Benzoquinone

Ultraviolet illumination of benzoquinone in deoxygenated aqueous suspensions of TiO_2 or ZnO resulted in reduction, with production of hydroquinone in 75% chemical yield. From methylbenzoquinone, the major product was methylhydroquinone.[94]

10. Furfuryl Alcohol

In order to determine the relative contributions of $\cdot OH$ radicals and $^1O_2$ molecules in the ZnO-promoted phototransformation of furfuryl alcohol, the reaction was compared with that observed by excitation of rose bengal, in which singlet oxygen was the reactive intermediate, and with that which occurred by excitation of nitrite ions, which produced hydroxyl radicals:[95]

$$NO_2^- + h\nu \rightarrow NO + \cdot O^- \tag{46}$$

$$\cdot O^- + H^+ \rightarrow \cdot OH \tag{47}$$

In all these reactions, the major product was 6-hydroxy-(2H)-pyran-3-(6H)-one (P), while furan-2-carboxaldehyde (F) was formed in small yields.[95] In the photooxygenation of furfuryl alcohol by singlet oxygen (with rose bengal), P was the only product, presumably formed by a mechanism of addition of singlet oxygen to the diene. Oxidation by hydroxyl radicals resulted in a mixture of P (92%) and F (8%). From a detailed kinetic study, it was concluded that the phototransformation over ZnO suspensions was predominantly due to $\cdot OH$ radicals, with only a minor contribution of $^1O_2$ involvement. Interestingly, this kinetic study also indicated that in the heterogeneous photocatalysis, the

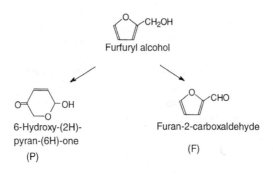

SCHEME III.7

hydroxylation of furfuryl alcohol partly occurred in the homogeneous aqueous phase. This was explained as due to the low value of K, the adsorption equilibrium constant of furfuryl alcohol on ZnO.[95] The absence of formation of $^1O_2$ molecules in illuminated ZnO suspensions (mainly at 365 nm) was further shown by a specific test: the lack of oxygenation of sodium 1,3-cyclohexadiene 1,4-diethanoate to its endoperoxide. In basic solutions (pH 11), the oxidation of this cyclohexadiene derivative was mainly due to the reaction of hydroxyl radicals, as indicated by the inhibitory effect of 2-propanol. In neutral solutions, kinetic data showed that in addition to hydroxyl radicals, positive holes may also be involved, by oxidizing substrate molecules adsorbed on the ZnO.[96]

11. Phthalic and Maleic Anhydrides

As noted in Chapter II, Section II.A.7, the direct photolysis of phthalic anhydride in oxygen-free aqueous solutions resulted in slow degradation, forming mainly polymeric products.[52] The photodegradation of both phthalic and maleic anhydrides in oxygenated aqueous solutions was enhanced considerably in the presence of TiO_2 and ZnO. No photocatalytic reaction occurred in N_2-purged solution. ZnO was the more effective photocatalyst. First-order rate constants for the disappearance of phthalic anhydride in O_2 saturated solutions without photocatalyst or in the presence of 1 g/L of TiO_2 or ZnO were 1.0×10^{-4}, 7.1×10^{-4}, and 8.7×10^{-4} s^{-1}, respectively. For the disappearance of maleic anhydride, the corresponding rate constants were 6.8×10^{-5}, 3.0×10^{-4}, and 4.9×10^{-4} s^{-1}, respectively.[53]

12. 2-Methylisoborneol

Substances responsible for the musty odor and off flavor in freshwater include the algal metabolites 2-methylisoborneol, released by *Phormidium tenue* and *Oscillatoria tenue,* and geosmin, released by *Anabena macrospora.*[97]

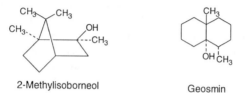

2-Methylisoborneol   Geosmin

SCHEME III.8

For removal of the off flavor in commercial catfish ponds, the expensive treatment with sodium carbonate peroxy hydrate had been proposed.[98] Almost complete photocatalytic degradation of 2-methylisoborneol was achieved by illuminating an aqueous solution of the compound (initially 1000 ppm) in the presence of TiO_2, either in suspended form (5 g/L) or supported on glass beads (2 mm diameter). Using a flow-through photoreactor packed with the TiO_2-coated glass beads, under illumination with blacklight, at flow rates of 20 to 50 mL/min, the concentration of 2-methylisoborneol decreased after a single pass to half its initial value of 1000 ppt within about 1 min. The process was developed to degrade the musty odor in Lake Biwa, Japan.[97]

13. Cyclic Acetals

2-Ethoxytetrahydropyran underwent photocatalytic oxidation by illumination in aqueous dispersions of TiO_2 in the presence of oxygen.[99]

C. Oil Spills

Photocatalytic oxidation by sunlight of films of crude oil or of *n*-octane on fresh or seawater was achieved by spreading n-TiO_2-coated oleophilic glass microbubbles on the water surface. These microbubbles, of 80 μm average diameter and 0.37 cm^{-3} density, stayed on the air–oil interface and promoted the photodegradation of the oil with air as the oxidant. In an initial very rapid process (<5 min), the microbubbles absorbed the oil, forming oil-containing aggregates, and leaving much of the water surface clean. The microbubbles were reported to provide for complete mineralization, with destruction of intermediates, at a weekly rate of about twice their weight of oil. The oxidation was presumably initiated by photogenerated holes or ·OH radicals, causing hydrogen abstraction from the hydrocarbons, leading to peroxy radicals, which underwent further decomposition, leading eventually to CO_2 and water. Photooxidation eliminated both aliphatic and aromatic compounds, including polycyclic aromatics. The process of oil cleanup was reported to be accelerated by waves, and thus to proceed even in high seas, where skimming would be impossible.[26,100–102]

Even just beach sand in contact with crude-oil residue and air under irradiation with a high-pressure Hg lamp caused photocatalytic oxidation of the hydrocarbons. This was ascribed to the catalytic properties of magnetite and ilmenite contained in the beach sand. Such results predict a natural abiotic "self-cleaning" process by sunshine on crude-oil-contaminated sandy beaches.[103]

See Chapter VII on polycyclic aromatic sulfur heterocyclic compounds as major components of most crude oils.

III. LONG-CHAIN COMPOUNDS

Industrial effluent waters often contain high concentrations of surfactants and other detergents, as well as pollutants specific to certain industries, such as from textile-dying plants or from paper and pulp processing. Many of these pollutants are resistant to biological degradation.

A. Surfactants

Detergents and other surfactants are important water pollutants that are difficult to degrade by the usual methods of wastewater treatment. Also, with some of these surfactants, biodegradation leads to toxic intermediates or products. Thus, alkylbenzene sulfonates are highly resistant to bacterial degradation. Alkylphenol polyethoxylated derivatives are converted by biodegradation to the very toxic 4-nonylphenol, C_9H_{19}-C_6H_4OH.[104] Cationic surfactants are particularly resistant to biodegradation because of their bactericidal properties.[105]

The photocatalytic oxidation of anionic surfactants such as dodecylbenzene sulfonate (DBS), cationic surfactants such as benzyl dodecyldimethylammonium chloride (BDDAC), and nonionic surfactants such as *p*-nonyl phenyl poly(oxyethylene) (NPE-n) was successfully achieved by sunlight illumination in the presence of suspended TiO_2 and oxygen.[106–108] The rates of photodegradation under similar experimental conditions were in general in the order anionic > nonionic > cationic surfactants.[109] Both DBS and sodium benzene sulfonate (BS) were rapidly degraded in aerated aqueous suspensions of TiO_2 under illumination ($\lambda > 330$ nm). Oxygen was required for the process. The reaction with DBS occurred in two stages. The more rapid oxidation of the aromatic ring was followed by a slower degradation of the aliphatic chain. The mechanism of oxidation of the aromatic ring presumably involves the action of hydroxyl radicals, but may also occur by direct action of holes on the semiconductor surface with adsorbed DBS molecules. The reactive aromatic intermediates may be hydroxycyclohexadienyl radicals, which in the presence of oxygen are oxidized to phenols.[106] The photodegradation of both the anionic DBS and the cationic BDDAC surfactants led to the intermediate formation of organic peroxides and aldehydes, which were, however, destroyed by longer irradiation times (>5 h).[109] Similarly, the photocatalytic degradation in aqueous dispersions of TiO_2 of the nonionic surfactants of the general formula C_9H_{19}-$C_6H_4O(CH_2CH_2O)_n$-H, with average number of ethoxy groups n = 2, 5, 6, 7, 12, 17, and 50 (trade name IGEPAL, Aldrich Chemicals), led to complete mineralization (>95%) within 1 to 2 h. In this photocatalytic process, the toxic 4-nonylphenol, which is resistant to bacterial degradation, was also completely destroyed.[104]

In a comparison of the photodegradation rates in aqueous dispersions of TiO_2 of the two anionic isomeric surfactants sodium dodecylbenzene sulfonate,

$$CH_3-(CH_2)_{11}-C_6H_4-SO_3^-Na^+ \text{ (DBS)}$$

and sodium 12-phenyldodecyl sulfonate,

$$C_6H_5-(CH_2)_{12}-SO_3^-Na^+ \text{ (PDS)}$$

the photooxidation of the aromatic moiety was faster in DBS than in PDS. The primary step in the photodecomposition presumably involves adsorption of the negatively charged sulfonate group to the positively charged TiO_2 surface. The difference in reactivity was explained to be due to the close proximity of the aromatic ring in DBS to the surface of the TiO_2 particles, where it is available to attack by surface adsorbed OH radicals. In PDS the aromatic ring at the end of the long aliphatic chain dangles freely in solution.[110]

The photodegradation of the cationic surfactant N-dodecylpyridinium chloride (N-DPCl), $C_{17}H_{30}NCl$, was measured in aqueous dispersions of TiO_2, TiO_2/Pt, and ZnO. The degradation followed first-order kinetics and could be fitted to a simple Langmuir–Hinshelwood model. TiO_2 and ZnO were equally effective as photocatalysts, while TiO_2/Pt was less effective. The reaction was observed by the disappearance of the 259 nm absorption peak (which is due to the $\pi \text{->} \pi^*$ transition of the pyridine ring). The destruction of the pyridine ring was accompanied by the photooxidation of the aliphatic chain, as observed by the increase in surface tension. After 8 h of irradiation with simulated sunlight (at 85 mW cm^{-2}), the surface tension of the medium was close to that of pure water (72.2 mN m^{-1} at 25°C). The presence of oxygen or air was essential in order to achieve complete mineralization, leading to the final products, CO_2, NH_4^+, and NO_3^-.[105] In another study of the TiO_2-photocatalyzed degradation of N-DPCl in oxygenated aqueous solutions, about 97% yields of NH_4^+ ions were reached within about 8 h of illumination, with concomitant evolution of CO_2 from the pyridinium ring. Complete mineralization and oxidation of ammonium ions to nitrate required much longer illumination times. The relatively slow photodegradation of this cationic surfactant was explained by the difficulty of adsorption of the pyridinium group on the positively charged illuminated TiO_2 surface.[111,112]

Photodegradation rates as a function of substrate concentrations in aqueous dispersions of TiO_2 were studied in a series of compounds. These included the anionic surfactant DBS, as well as the anionic reference compounds sodium benzene sulfonate and sodium dodecyl sulfate, and also the cationic surfactant BDDAC and the cationic reference compounds benzyl trimethylammonium chloride and hexadecyl trimethylammonium bromide. The initial rates could be fitted to the simple Langmuir–Hinshelwood equation, with linear relationships between the reciprocal rates 1/R and the reciprocal initial concentration of the surfactant 1/S,

$$1/R = 1/(kKS) + 1/S \qquad (48)$$

For all the preceding compounds, k had the same approximate value of 5 to 7 μM min^{-1}, irrespective of the structure of the surfactant, as indicated by the same intercept value for 1/k. This suggested that k was dependent only on the light source, catalyst activity, and the reaction medium. Differences in the slopes for different surfactants, and thus widely different values of K, indicated the effect of different structures and electrical charges of the substrates. The intermediate action of hydroxyl radicals ·OH was proven by spin trapping of these radicals and detection by electron spin resonance.[113] With the cationic surfactant benzyl tetradecyldimethylammonium chloride in aqueous dispersions of TiO_2, the benzyl moiety was photooxidized very rapidly, while the nitrogen atoms were released more slowly, mainly as ammonium ions, and only partly as nitrate.[112]

As noted in Chapter I, Section IV.E, illumination of aqueous suspensions of TiO_2 caused a positive shift in the ζ potential of the TiO_2 particles, due to the formation of positive charge carriers on the particle surfaces. Addition of the strongly charged anionic surfactants decreased the ζ potential of the TiO_2 particles. Under illumination at an acidic pH, the ζ potential gradually changed to more positive values with concomitant oxidation of the surfactant. The more rapid photooxidation of the anionic DBS surfactant relative to the cationic BDDAC surfactant was explained by the stronger adsorption of the anionic surfactant. The differences in adsorptivity are thus simple electrostatic effects. Cationic surfactants such as BDDAC were strongly adsorbed in TiO_2 dispersions only in alkaline media (pH 12), changing the ζ potential of the TiO_2 particles from negative to more positive values. Illumination caused a further shift of the ζ potential to even more positive values. With the nonionic surfactant p-nonylphenylpoly(oxyethylene) ether (NPE-9), the ζ potential of the TiO_2 particles was also shifted to more positive values upon illumination, both in acidic and alkaline pH.[109]

The anionic surfactant sodium dodecyl bis (oxyethylene) phosphate in an aqueous dispersion of TiO_2 (Degussa P25, 2 g/L) under UV illumination ($\lambda >$ 330 nm) released 85% of its phosphorus as orthophosphate within 2 h, while its carbon was initially oxidized to formic acid, and very much more slowly to CO_2. The rapid release of phosphate from this anionic surfactant was explained by the ease of adsorption of the anionic phosphate group of the surfactant to the positively charged illuminated TiO_2.[111]

B. Lignin Sulfonates and Kraft Wastewater

The photooxidation of lignin is the cause of the rapid yellowing of the mechanically produced wood pulps used in inexpensive paper, such as in newsprint. Such wood pulps, derived mainly from softwoods containing about 30% by weight of lignin, are bleached to whiteness by treatment with hydrogen peroxide. The major light-absorbing groups of lignin in the 300 to 400 nm region were proposed to be aromatic ketones such as α-phenoxylacetophenone, which participate in the three-dimensional cross-linking of the rigid and insoluble lignin macromolecule.[114]

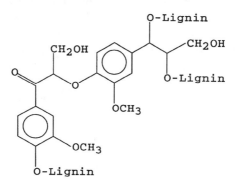

SCHEME III.9

The main degradation reactions of lignin include hydrogen abstraction from phenolic groups and β-cleavage of aryloxy ketones. Photochemical cleavage of the α-phenoxylacetophenones yielded phenoxyl and phenacyl radicals. Oxidation of phenoxyl radicals by molecular oxygen resulted in the formation of quinones, which cause the yellow color of light-exposed lignin.[114–116]

The primary steps of formation of short-lived intermediates in the photolysis of bleached thermomechanical pulp of black spruce were studied in diffuse reflectance laser flash experiments, using 8-ns laser pulses at 354 nm from a Nd:YAG laser. The transient absorption spectrum observed had a maximum at 440 to 450 nm. The intensity of the transient peak was decreased by the presence of oxygen or phenolic hydroxyl groups during the laser flash irradiation. Also, reduction of the bleached pulp with sodium borohydride considerably decreased the transient intensity, presumably by reducing the light-absorbing ketonic groups. The results indicated that the transient may be assigned to the lowest triplet state of the aromatic carbonyl chromophores in the bleached lignin. Several compounds are effective in inhibiting the yellowing of lignin under UV illumination. These include polyethylene glycol, ascorbic acid, tartaric acid, 3-mercapto-1,2-propanediol, and ethylene glycol bis(thioglycolate). Thermomechanical pulp containing these compounds gave under laser flash illumination either the same or even increased intensities of the transient absorption spectrum. Possibly these compounds acted as radical scavengers.[114–116]

A useful model compound for the study of the β-cleavage processes in lignin is α-guaiacoxyacetoveratrone. Laser excitation of deaerated solutions of this compound in mixtures of water with either acetonitrile, dioxane, or alcohols caused the appearance of a transient with $\lambda_{max} = 400$ nm, which was identified to be due to the triplet state of α-guaiacoxyacetoveratrone. The photochemical processes thus involve excitation of the model compound to its triplet state, followed by (1) scavenging of phenoxyl and phenacyl radicals by H-donors, (2) isomerization of the compound, (3) dimerization of phenacyl radicals, and (4) production of oligomers containing chromophoric groups such as quinones, causing the yellowing reaction. In protic solvents, such as in 40% water in dioxane or acetonitrile, the quantum yield of the photodegradation of α-guaiacoxyacetoveratrone was >20%, and was quenched by the presence of 0.1 *M* sorbic acid, a known triplet quencher. This result was attributed to the effect of hydrogen bonding on β-phenyl quenching of the triplet state. This β-phenyl quenching competes with the cleavage of the α-C-O bond of the substrate, which generates phenacyl and phenoxy radicals.[116,117]

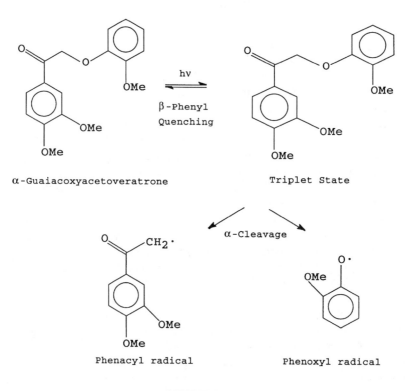

SCHEME III.10

Another mechanism for generating phenoxy radicals is the oxidation of arylglycerol β-*O*-aryl structures in bleached lignin by oxygen (or hydroxyl radicals) forming ketyl radicals, which then undergo homolytic cleavage to phenoxy radicals and enol derivatives, which rapidly tautomerize to acetophenone derivatives.[118]

For the photocatalytic treatment with TiO_2 of effluents from a pulp factory, an aerated recirculation reactor was used. The crude wastewater was at pH 4 and contained 60% of TOC (total organic carbon) of lignin sulfonates, 15% of TOC of oligosaccharides, low-molecular-weight organic acids, as well as SO_3^{2-}, SO_4^{2-}, and had an initial COD (chemical oxygen demand) of 93,000 mg L^{-1}. With the concentrated effluent, the photodegradation was extremely slow. Dilution by water considerably enhanced the rate of photodegradation. Thus, with 1:100 dilution, the disappearance of turbidity was practically complete within about 15 h, while the COD dropped from 800 mg L^{-1} to 360 mg L^{-1}.[119]

Kraft black liquor is the spent liquid from wood pulping operations, in which NaOH and sodium sulfide are used to remove lignin. This liquor contains lignin, polysaccharides and other polymers, and low-molecular-weight compounds. Photodegradation of kraft black liquor diluted with water (1:250) in the presence of ZnO (2%) was achieved by irradiating oxygenated mixtures with a Hg lamp through a quartz window ($\lambda > 254$ nm). After 60 min of irradiation, there occurred a 57% reduction in COD (chemical oxygen demand) and 80% bleaching of the color. Using Pt-impregnated ZnO (0.15 g Pt/1.0 g ZnO), the complete elimination of color was attained within 60 min, but the Pt/ZnO catalyst did not improve the reduction of the COD. Oxygen was required for the reaction, and near-UV light ($\lambda > 300$ nm) was ineffective.[120]

Ultraviolet illumination ($\lambda = 254$ nm) of kraft wastewater containing TiO_2 + O_3, followed or preceded by ultrafiltration membrane separation (to reject high-molecular-weight components), was found useful as a pretreatment, prior to biological treatment with activated sludge. Such a pretreatment significantly improved the quality of the final effluent, compared with only biological treatment. The order of the sequence photocatalysis/ultrafiltration had only minor effect on the final quality of the biological effluent. Improvements included substantial decreases in TOC, COD, color, and microtoxicity.[121]

Partial substitution (10 to 50%) of elemental chlorine by ClO_2 in the bleaching of softwood, followed by treatment of the kraft wastewater by membrane ultrafiltration, photocatalysis with TiO_2, and aerobic biotreatment, resulted in marked decreases in the TOC, COD, color, adsorbable organic halide (AOX), and toxicity levels.[122]

C. Phthalate Esters

Esters of phthalic acid are important plasticizers for many polymeric materials. When discarded as waste, such as in landfills, these esters are leached out and become serious water contaminants. Bis-2-ethylhexyl phthalate (BEHP, di-*iso*-octyl phthalate) has been used extensively as a plasticizer, especially for polyvinyl chloride (PVC). Several phthalate esters, such as bis-2-ethylhexyl phthalate, are known as suspect cancer agents and are priority pollutants.[123] In addition, several phthalate esters were found to be toxic to aquatic organisms.[124] These compounds are highly resistant to treatment with hypochlorite, chlorine, monochloramine, or chlorine dioxide. Ozonation caused an average total phthalate reduction of 11%, while filtration through sand or granulated activated carbon (GAC) resulted in phthalate ester removal of up to 20 and 50%, respectively.[125]

The direct photodegradation of a PVC sheet plasticized with 40 wt% BEHP by irradiation in air with a 300-W xenon lamp resulted in the release of CO_2, 2-ethylhexene-1,2-ethylhexanol, and phthalic acid (or phthalic anhydride).[126]

The homogeneous-phase photolysis of diethyl phthalate and of di-*n*-butyl phthalate at $\lambda > 290$ nm in the presence of hydrogen peroxide and oxygen resulted in half-lives of 290 and 188 d, respectively.[127] Effective photocatalytic degradation of diethyl phthalate was observed in aqueous dispersions of TiO_2 under illumination with UV light or sunlight.[128] Quite rapid oxidation of aqueous solutions of di-*n*-butyl phthalate was obtained by illumination with a 70-W high-pressure Hg lamp in the presence of TiO_2 (1 g/L). In the presence of both TiO_2 (1 g/L) and H_2O_2 (0.01 M or 0.1 M) the oxidation of the phthalate ester was slower than with TiO_2 alone. An alternative process for di-*n*-butyl phthalate photodegradation was by using a photo Fenton type reaction. In the presence of Fe^{3+} (1 mM) and H_2O_2 (50 mM), the half-time for degradation was 1.0 h. Even faster decomposition was achieved by illumination in the presence of TiO_2 (1 g/L) and sodium periodate (0.1 M), resulting in $t_{1/2} = 0.5$ h.[129]

D. Polymers

In the photooxidative degradation of polymers in the presence of H_2O_2, $FeCl_3$, and Fenton reagents, the initiation step mainly involves $\cdot OH$ and $\cdot HO_2$ radicals, produced by light absorption either by chromophoric groups in the polymer chain or by light-absorbing impurities. The presence of Fe^{3+} enables the formation of complexes with poly(ethylene oxide), poly(methylmethacrylate), and poly(vinylchloride) (PVC). Illumination with near-UV light causes electron transfer, with formation of Fe(II) coordinated to an oxidized polymer moiety, which then undergoes bond fission to oxidized products. $FeCl_3$ considerably accelerated the photodegradation of polyethylene, poly(methylmethacrylate), poly(vinylchloride), poly(vinyl alcohol), poly(ethylene oxide), poly(ethylene glycols), poly(caproamide), and cellulose. The outdoor lifetime of most polymers is considerably decreased by the incorporation of transition-metal ions. This may be advantageous particularly for the development of photodegradable plastics, for example, to be used in agriculture.[130]

IV. PHOTOINDUCED NITROSATION AND NITRATION

Mutagens known to be formed in the atmosphere include nitroaromatic compounds produced by the combustion of fossil fuels and by the photochemical reactions of polycyclic aromatic compounds with nitrogen oxides (NO_x). The studies of Suzuki et al.[131–139] showed that mutagens are also formed by UV or sunlight irradiation of aqueous solutions of alkylamines, aromatic amino acids, polycyclic aromatics such as biphenyl and pyrene, and naphthols and phenylphenols, in the presence of either nitrite or nitrate ions. Mutagens have been found in river sediments and may be produced from industrial and domestic sewage by chemical and biochemical transformations — with mutagen formation induced by photochemical nitration, nitrosation, and hydroxylation.

A. Dimethylamine

The highly carcinogenic nitrosodimethylamine has been detected in seawater, in municipal sewage, and in the effluent of sewage treatment plants. Dimethylamine (0.05 M) in aqueous solutions of sodium nitrite (0.05 M), under illumination with a 100-W high-pressure Hg lamp ($\lambda > 300$ nm) or with sunlight, was nitrosated to produce nitrosodimethylamine:

$$\left(CH_3\right)_2 NH + NO_2^- + h\nu \rightarrow \left(CH_3\right)_2 N - NO + OH^- \qquad (49)$$

The rate of formation of nitrosodimethylamine was very slow at acidic pH, but increased considerably above pH 8. At pH 8.2, under sunlight exposure for 3 d, the yield of nitrosodimethylamine was about 2%. Since both nitrite and secondary amines are constituents of natural waters, these results indicate that nitrosodialkylamines may be produced photochemically under natural conditions.[131]

B. Aromatic Amino Acids

Irradiation with a 100-W high-pressure Hg lamp ($\lambda > 300$ nm) of neutral aqueous solutions of the aromatic amino acids tryptophan, phenylalanine, and tyrosine in the presence of nitrite ions and in contact with the atmosphere caused the formation of some strongly mutagenic products, which could be extracted with ether. The mutagenicity decreased in the order tryptophan > phenylalanine > tyrosine.[134] The Ames test with the bacterial strains *Salmonella typhymurium* TA98 and TA100 was used to measure the mutagenicity.[140] The mutagenicity increased with irradiation time up to 3 h, and then remained constant. This indicated that the mutagens formed were stable to irradiation. When nitrate was used instead of nitrite, there was only weak mutagen formation with tryptophan, and none with the other amino acids. The nature of the mutagens produced was not identified. The same UV irradiation of nitrite with the aliphatic amino acids alanine, threonine, cysteine, glutamate, arginine, and proline did not result in mutagen formation. These results indicated that there may be production of mutagens in environmental waters containing nitrite ions and aromatic amino acids.[134]

C. Phenol

The photolysis of phenol in aqueous solutions of nitrite resulted in the formation of the nonmutagenic nitrosophenol, the yield of which increased with increasing initial nitrite concentration and with longer irradiation times. The presence of oxygen or NaSCN inhibited the photonitrosation of phenol, suggesting a radical mechanism. Presumably, photolysis of nitrite ion yielded ·OH and ·NO radicals. The hydroxyl radicals caused hydrogen abstraction at the *para* position of phenol, which was then attacked by the ·NO radicals.[132]

D. 2-Phenylphenol

2-Phenylphenol (2-hydroxybiphenyl, 2-biphenylol) is widely used as a fungicide for fruit preservation.[141] By photolysis of aqueous solutions of biphenyl or 2-phenylphenol containing nitrate or nitrite ions, nitration and hydroxylation products were formed, which included the mutagenic hydroxynitrobiphenyl derivatives.[136,139] In a mechanistic study of this reaction, aerated solutions at pH 5.7 of 2-phenylphenol (0.25 mM) and KNO$_3$ (0.1 M) were irradiated at $\lambda > 190$ nm. The main products identified were 2-hydroxydibenzofuran, phenylhydroquinone, phenylbenzoquinone, and 2-hydroxy-3-nitrobiphenyl. The cyclization product 2-hydroxydibenzofuran was also formed in the direct photolysis of 2-phenylphenol (in the absence of nitrate).[141] The primary step in the nitrate-promoted reaction was proposed to be excitation of the nitrate ion, leading to the formation of nitric oxide (or its dimer N$_2$O$_4$), which dissociated, producing hydroxyl radicals:[142]

$$NO_3^- + h\nu \rightarrow \cdot NO_2 + \cdot O^- \tag{50}$$

$$\cdot O^- + H^+ \rightarrow \cdot OH \tag{51}$$

Hydroxyl radicals were shown to be involved in the formation of phenylhydroquinone, phenylbenzoquinone, and 2-hydroxydibenzofuran, because in the presence of ethanol (2% v/v; a known hydroxyl radical quencher), the production of these three compounds was completely prevented. The presence of methanol did not appreciable affect the production of 2-hydroxy-3-nitrobiphenyl and of 2-hydroxy-5-nitrobiphenyl. These nitration reactions were thus proposed to be due to reaction of 2-phenylphenol with NO$_2$ or N$_2$O$_4$.[142]

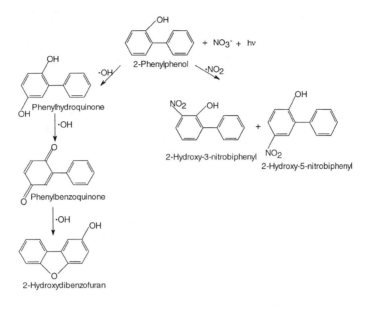

SCHEME III.11

E. 2-Naphthol

Ultraviolet irradiation of 2-naphthol in aqueous nitrite solutions in the presence of air caused the formation of mutagens. The products identified included the weakly mutagenic 1-nitro-2-naphthol, 5- or 8-nitro-2-naphthol, a dinitro-2-naphthol, and isocoumarin, as well as the toxic 1-nitroso-2-naphthol and 1,2-naphthoquinone, and the nonmutagenic monoquinoid dimer of 2-naphthol. The monoquinoid dimer of 2-naphthol upon further irradiation with aqueous nitrite produced a strong mutagen. The mutagen production from 2-naphthol and nitrite was faster under oxygen than under air, and did not occur under nitrogen. 1-Naphthol did not form a strong mutagen under UV irradiation in the presence of aqueous nitrite, but was instead mainly converted to the stable and weakly mutagenic isocoumarin. It was concluded that photochemical mutagen formation from 2-naphthol and nitrite requires oxygen and involves both oxidation to a quinone and nitration. The primary photochemical step in the UV irradiation of both 1- and 2-naphthol was considered to be formation of naphthoxy radicals. These radicals with oxygen presumably produced 1,2- and 1,4-naphthoquinones.[137,138]

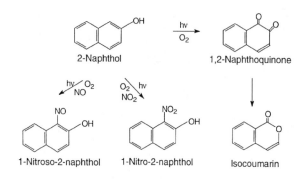

SCHEME III.12

F.   Biphenyl

Biphenyl in aqueous nitrate solutions, under UV illumination with a 100-W Hg lamp through a quartz tube (λ_{max} = 365 nm), underwent hydroxylation and nitration. The major photoproducts identified, 2-hydroxy-3-nitrobiphenyl and 4-hydroxy-3-nitrobiphenyl, were nonmutagenic.[133]

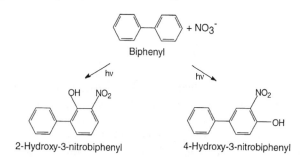

SCHEME III.13

However, four dinitro compounds were mutagenic: 4-hydroxy-3,5-dinitrobiphenyl, 4-hydroxy-3,3′-dinitrobiphenyl, a hydroxydinitrobiphenyl, and a dihydroxydinitrobiphenyl. Among these, the most strongly mutagenic compound was the dihydroxydinitrobiphenyl, the detailed structure of which was not determined. The reaction products were formed by both nitration and hydroxylation, presumably by ·NO$_2$ and ·OH radicals obtained in the UV photolysis of nitrate ions (λ_{max} ~ 300 nm).[133] If the illumination of the biphenyl solution in aqueous nitrate was through a Pyrex® glass tube (instead of a quartz tube), the hydroxylation and nitration reactions did not occur. On the other hand, irradiation of aqueous solutions of biphenyl with sodium nitrite (instead of nitrate), either by a Hg lamp through a glass tube (λ > 300 nm) or by sunlight, did cause the formation of the photoproducts, including the mutagenic hydroxynitrobiphenyls.[137]

G. Pyrene

Ultraviolet irradiation at $\lambda > 300$ nm of pyrene coated on silica gel and suspended in aqueous solutions of sodium nitrite resulted in formation of one strongly mutagenic product, 1-nitropyrene, which is a suspected cancer agent.[135]

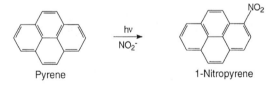

Pyrene   1-Nitropyrene

SCHEME III.14

Since wastewaters often contain high concentrations of nitrate and nitrite ions in addition to aliphatic and aromatic amines and polycyclic hydrocarbons, direct UV or sunlight irradiation may induce the formation of photostable mutagenic, carcinogenic, or toxic nitrosation and nitration products. Such products may, however, be degraded by heterogeneous photocatalytic oxidation.

REFERENCES

1. Fox, M. A., Organic heterogeneous photocatalysis: chemical conversions sensitized by irradiated semiconductors, *Acc. Chem. Res.,* 16, 314–321, 1983.
2. Fox, M. A., Charge injection into semiconductor particles — importance in photocatalysis, in: *Homogeneous and Heterogeneous Photocatalysis;* E. Pelizzetti and N. Serpone, (Eds.), *NATO ASI Ser., Ser. C,* 174, pp. 363–379. Reidel: Dordrecht, 1986.
3. Pichat, P., Partial or complete heterogeneous photocatalytic oxidation of organic compounds in liquid organic or aqueous phases, *Catal. Today,* 19, 313–334, 1994.
4. Papaconstantinou, E., Argitis, P., Dimoticali, D., Hiskia, A., and Ioannidis, A., Photocatalytic oxidation of organic compounds with heteropoly electrolytes. Aspects on photochemical utilization of solar energy, in: *Homogeneous and Heterogeneous Photocatalysis,* Pelizzetti, E. and Serpone, N. (Eds.), *NATO ASI Ser., Ser. C.* 174, 415–431, Reidel: Dordrecht, 1986.
5. Rudakov, E. S., Mitchenko, S. A., and Miroshnichenko, N. A., Oxidation of alkanes in aqueous solutions of mercury(II) salts and UV-radiation, *Kinet. Katal.,* 28, 187–191, 1987.
6. Green, P., Green, W. A., Harriman, A., Richoux, M. C., and Neta, P., The dehydrogenation of ethanol in dilute aqueous solution photosensitized by benzophenones, *J. Chem. Soc., Faraday Trans.* 1, 84, 2109–2127, 1988.
7. Kawaguchi, H., Photo-oxidation of formic acid in aqueous solution in the presence of hydrogen peroxide, *Chemosphere,* 26, 1965–1970, 1993.
8. Kawaguchi, H., Photooxidation of 2-propanol in aqueous solution in the presence of hydrogen peroxide, *Chemosphere,* 27, 577–584, 1993.

9. Bideau, M., Faure, L., Assemian, Y., and Claudel, B., Photooxidation of dissolved formic acid by gaseous oxygen in the presence of iron ions, *J. Mol. Catal.*, 43, 267–279, 1988.

10. Izumi, I., Dunn, W. W., Wilburn, K. O., Fan, F.-R. F., and Bard, A. J., Heterogeneous photocatalytic oxidation of hydrocarbons on platinized TiO_2 powders, *J. Phys. Chem.*, 84, 3207–3210, 1980.

11. Kraeutler, B. and Bard, A. J., Heterogeneous photocatalytic decomposition of saturated carboxylic acids on TiO_2 powder. Decarboxylative route to alkanes, *J. Am. Chem. Soc.*, 100, 5985–5992, 1978.

12. Chemseddine, A. and Boehm, H. P., A study of the primary step in the photochemical degradation of acetic acid and chloroacetic acids on a TiO_2 catalyst, *J. Mol. Catal.*, 60, 295–311, 1990.

13. Dillert, R. and Bahnemann, D., Photocatalytic degradation of organic pollutants: mechanisms and solar applications, *EPA Newslett.*, 52, 33–52, 1994.

14. Bideau, M., Claudel, B., Faure, L., and Rachimoellah, M., Homogeneous and heterogeneous photoreactions of decomposition and oxidation of carboxylic acids, *J. Photochem.*, 39, 107–128, 1987.

15. Bideau, M., Claudel, B., Faure, L., and Rachimoellah, M., Photooxidation of formic acid by oxygen in the presence of titanium dioxide and dissolved copper ions: oxygen transfer and reaction kinetics, *Chem. Eng. Commun.*, 93, 167–197, 1990.

16. Kisch, H. and Bücheler, J., Heterogeneous photocatalysis, VIII. Zinc sulfide catalyzed hydrogen formation from water in the presence of sodium formate, *Bull. Chem. Soc. Jpn.*, 63, 2378–2386, 1990.

17. Miyoshi, H., Mori, H., and Yoneyama, H., Light-induced decomposition of saturated carboxylic acids on iron oxide incorporated clay suspended in aqueous solutions, *Langmuir,* 7, 503–507, 1991.

18. Aguado, M. A. and Anderson, M. A., Degradation of formic acid over semiconducting membranes supported on glass: effects of structure and electronic doping, *Solar Mater. Solar Cells*, 28, 345–361, 1993.

19. Aguado, M. A., Anderson, M. A., and Hill, C. G., Influence of light intensity and membrane properties on the photocatalytic degradation of formic acid over TiO_2 ceramic membrane, *J. Mol. Catal.*, 89, 165–178, 1994.

20. Park, K. H. and Kim, J. H., Photocatalytic decompositions of carboxylic derivatives by semiconductors, *Bull. Kor. Chem. Soc.*, 12, 438–440, 1991.

21. Zeltner, W. A., Hill, C. G., and Anderson, M. A., Supported titania for photodegradation, *Chemtech,* 21, May 1993.

22. Kim, D. H. and Anderson, M. A., Photoelectrolytic degradation of formic acid using a porous TiO_2 thin-film electrode, *Environ. Sci. Technol.*, 28, 479–483, 1994.

23. Serpone, N., Lawless, D., Disdier, J., and Herrmann, J. M., Spectroscopic, photoconductivity, and photocatalytic studies of TiO_2 colloids. Naked and with the lattice doped with Cr^{3+}, Fe^{3+}, and V^{5+} cations, *Langmuir,* 10, 643–652, 1994.

24. Chen, L., Gu, W., Zhu, X., Wang, F., Song, Y., and Hu, J., Highly efficient hydrogen and ethylene photoproduction from aqueous methanol solution by ZnS and *in situ* spin trapping, *J. Photochem. Photobiol. A: Chem.*, 74, 85–89, 1993.

25. Chen, L. H., Zhu, X. W., Wang, F. D., and Gu, W. Z., Photoproduction of hydrogen and 1,2-propanediol from aqueous methanol and ethanol solution catalysed by ZnS, *J. Photochem. Photobiol. A: Chem.,* 73, 217–220, 1993.

26. Jackson, N. B., Wang, C. M., Luo, Z., Schwitzgebel, J., Ekerdt, J. F., Brock, J. R., and Heller, A., Attachment of TiO_2 powders to hollow glass microbeads — activity of the TiO_2-coated beads in the photoassisted oxidation of ethanol to acetaldehyde, *J. Electrochem. Soc.,* 138, 3660–3664, 1991.

27. Kaise, M., Nagai, H., Tokuhashi, K., Kondo, S., Nimura, S., and Kikuchi, O., Electron spin resonance study of photocatalytic interface reaction of suspended M/TiO_2 (M = Pt, Pd, Ir, Rh, Os, or Ru) with alcohol and acetic acid in aqueous medium, *Langmuir,* 10, 1345–1347, 1994.

28. Lepore, G. P., Pant, B. C., and Langford, C. H., Limiting quantum yield measurements for the disappearance of 1-propanol and propanal: an oxidation reaction study employing a TiO_2 based photoreactor, *Can. J. Chem.,* 71, 2051–2059, 1993.

29. Lepore, G. P., Vlcek, A., and Langford, C. H., The photocatalytic oxidation of propanol by TiO_2, in: *Photocatalytic Purification and Treatment of Water and Air,* Ollis, D. F. and Al-Ekabi, H. (Eds.), Elsevier Science Publishing, Amsterdam, 1993, pp. 95–109.

30. Moser, J. and Grätzel, M., Photoelectrochemistry with colloidal semiconductors: laser studies of halide oxidation in colloidal dispersions of TiO_2 and α-Fe_2O_3, *Helv. Chim. Acta,* 65, 1436–1444, 1982.

31. Rothenberger, R., Moser, J., Grätzel, M., Serpone, N., and Sharma, D. K., Charge carrier trapping and recombination dynamics in small semiconductor particles, *J. Am. Chem. Soc.,* 107, 8054–8059, 1985.

32. Arbour, C., Sharma, D. K., and Langford, C. H., Picosecond flash spectroscopy of TiO_2 colloids with adsorbed dyes, *J. Phys. Chem.,* 94, 331–335, 1990.

33. Muneer, M., Das, S., Manilal, V. B., and Haridas, A., Photocatalytic degradation of waste water pollutants: titanium dioxide-mediated oxidation of methyl vinyl ketone, *J. Photochem. Photobiol. A: Chem.,* 63, 107–114, 1992.

34. Kato, T., Butsugan, Y., Kato, K., Loo, B. H., and Fujishima, A., Decomposition of aqueous polyvinyl alcohol on photoexcited titanium dioxide, *Denki Kagaku,* 61, 876–877, 1993.

35. Lipczynska-Kochany, E., Degradation of aromatic pollutants by means of the advanced oxidation processes in a homogeneous phase: Photolysis in the presence of hydrogen peroxide versus the Fenton reaction, in: *Chemical Oxidation: Technologies for the Nineties,* Technomic Publishing, Lancaster, PA, Vol. 3, 1993, pp. 12–27.

36. Kawaguchi, H., Photooxidation of phenol in aqueous solution in the presence of hydrogen peroxide, *Chemosphere,* 24, 1707–1712, 1992.

37. Lipczynska-Kochany, E. and Bolton, J. R., Flash photolysis high-performance liquid chromatography method for studying the sequence of photochemical reactions. Direct photolysis of phenol, *Environ. Sci. Technol.,* 26, 2524–2527, 1992.

38. Augugliaro, V., Palmisano, L., Minero, C. and Pelizzetti, E., Photocatalytic degradation of phenol in aqueous titanium dioxide dispersions, *Toxicol. Environ. Chem.,* 16, 89–109, 1988.

39. Augugliaro, V., Davì, E., Palmisano, L., Schiavello, M., and Sclafani, A., Influence of hydrogen peroxide on the kinetics of phenol photodegradation in aqueous titanium dioxide dispersion, *Appl. Catal.,* 65, 101–116, 1990.

40. Kawaguchi, H., Rates of photosensitized photo-oxidation of 2,4,6-trimethylphenol by humic acid, *Chemosphere,* 27, 2177–2182, 1993.

41. Ruppert, G., Bauer, R., and Heisler, G., The photo-Fenton reaction. An effective photochemical wastewater treatment process, *J. Photochem. Photobiol. A: Chem.,* 73, 75–78, 1993.

42. Ruppert, G., Bauer, R., Heisler, G., and Novalic, S., Mineralization of cyclic organic water contaminants by the photo-Fenton reaction — influence of structure and substituents, *Chemosphere,* 27, 1339–1347, 1993.

43. Scully, F. E., Jr. and Hoigné, J., Rate constants for reactions of singlet oxygen with phenols and other compounds in water, *Chemosphere,* 16, 681–649, 1987.

44. Tratnyek, P. G. and Hoigné, J., Oxidation of substituted phenols in the environment: a QSAR analysis of rate constants for reaction with singlet oxygen, *Environ. Sci. Technol.,* 25, 1596–1604, 1991.

45. Tratnyek, P. G., Hoigné, J., Zeyer, J., and Schwarzenbach, R. P., QSAR analyses of oxidation and reduction rates of environmental organic pollutants in model systems, *Sci. Total Environ.,* 109/110, 327–341, 1991.

46. Tratnyek, P. G. and Hoigné, J., Photo-oxidation of 2,4,6-trimethylphenol in aqueous laboratory solutions and natural waters: kinetics of reaction with singlet oxygen, *J. Photochem. Photobiol. A: Chem.,* 84, 153–160, 1994.

47. Larson, R. A., Ellis, D. D., Ju, H. -L., and Marley, K. A., Flavin-sensitized photodecomposition of anilines and phenols, *Environ. Toxicol. Chem.,* 8, 1165–1170, 1989.

48. Foote, C. S., Photooxidation of biological model compounds, in: *Oxygen and Oxy-Radicals in Chemistry and Biology,* Rodgers, M. A. J. and Powers, E. L. (Eds.), Academic, Press, New York, 1981, pp. 425–440.

49. Okamoto, K., Hondo, F., and Itaya, A., Kinetics of dye-sensitized photodegradation of aqueous phenol, *J. Chem. Eng. Jpn.,* 15, 368–375, 1982.

50. Tatsumi, K., Ichikawa, I., and Wada, S., Flavin-sensitized photooxidation of substituted phenols in natural water, *J. Contam. Hydrol.,* 9, 207, 1992.

51. Ahel, M., Scully, F. E., Hoigné, J., and Giger, W., Photochemical degradation of nonylphenol and nonylphenol polyethoxylates in natural waters, *Chemosphere,* 28, 1361–1368, 1994.

52. Bajt, O., Sket, B., and Faganeli, J., Photochemical transformation of phthalic anhydride in natural waters, *Chemosphere,* 24, 673–679, 1992.

53. Bajt, O., Sket, B., and Faganeli, J., The effect of semiconductor oxides on the photochemical degradation of phthalic and maleic anhydrides in aqueous media, *Toxicol. Environ. Chem.,* 40, 267–273, 1993.

54. Goldberg, M. C., Cunningham, K. M., Aiken, G. R., and Weiner, E. R., The aqueous photolysis of α-pinene in solution with humic acid, *J. Contam. Hydrol.,* 9, 79–89, 1992.

55. Boule, P., Rossi, A., Pilichowski, J. F., and Grabner, G., Photoreactivity of hydroquinone in aqueous solution, *N. J. Chem.,* 16, 1053–1062, 1992.

56. Baykut, F., Benlioglu, G., and Baykut, G., Photocatalytic production of ascorbic acid. A secondary photosynthesis in plants, in: *Homogeneous and Heterogeneous Photocatalysis,* Pelizzetti E. and Serpone, N. (Eds.), *NATO ASI Ser. C,* Vol. 174, p. 163, Reidel: Dordrect, 1986.

57. Matthews, R. W., Kinetics of photocatalytic oxidation of organic solutes over titanium dioxide, *J. Catal.,* 111, 264–272, 1988.

58. Okamoto, K., Yamamoto, Y., Tanaka, H., Tanaka, M., and Itaya, A., Heterogeneous photocatalytic decomposition of phenol over anatase powder, *Bull. Chem. Soc. Jpn.,* 58, 2015–2022, 1985.

59. Okamoto, K., Yamamoto, Y., Tanaka, H., and Itaya, A., Kinetics of heterogeneous photocatalytic decomposition of phenol over anatase TiO_2 powder, *Bull. Chem. Soc. Jpn.,* 58, 2023–2028, 1985.

60. Richard, C. and Boule, P., Is the oxidation of salicylic acid to 2,5-dihydroxybenzoic acid a specific reaction of singlet oxygen? *J. Photochem. Photobiol. A: Chem.,* 84, 151–152, 1994.

61. Sclafani, A., Palmisano, L., and Davi, E., Photocatalytic degradation of phenol by TiO_2 aqueous dispersions: rutile and anatase activity, *N. J. Chem.,* 14, 265–268, 1990.

62. Wei, T.-Y., Wang, Y.-Y., and Wan, C.-C., Photocatalytic oxidation of phenol in the presence of hydrogen peroxide and titanium dioxide powders, *J. Photochem. Photobiol. A: Chem.,* 55, 115–126, 1990.

63. Wei, T.-Y. and Wan, C.-C., Kinetics of photocatalytic oxidation of phenol on TiO_2 surface, *J. Photochem. Photobiol. A: Chem.,* 69, 241–249, 1992.

64. Palmisano, L., Augugliaro, V., Schiavello, M., and Sclafani, A., Influence of acid-base properties on photocatalytic and photochemical processes, *J. Mol. Catal.,* 56, 284–295, 1989.

65. Al-Ekabi, H. and Serpone, N., Kinetic studies in heterogeneous photocatalysis. 1. Photocatalytic degradation of chlorinated phenols in aerated aqueous solutions over TiO_2 supported on a glass matrix, *J. Phys. Chem.,* 92, 5726–5731, 1988.

66. Matthews, R. W., Photooxidation of organic impurities in water using thin films of titanium oxide, *J. Phys. Chem.,* 91, 3328–3333, 1987.

67. Matthews, R. W., Carbon dioxide formation from organic solutes in aqueous suspensions of ultraviolet-irradiated TiO_2. Effect of solute concentration, *Aust. J. Chem.,* 40, 667–675, 1987.

68. Pichat, P., Guillard, C., Maillard, C., Amalric, L., and D'Oliveira, J.-C., TiO_2 photocatalytic destruction of water aromatic pollutants: intermediates; properties–degradability correlation; effects of inorganic ions and TiO_2 surface area; comparison with H_2O_2 processes, in: *Photocatalytic Purification and Treatment of Water and Air,* Ollis, D. F. and Al-Ekabi, H. (Eds.), Elsevier Science Publishing, Amsterdam, 1993, pp. 207–223.

69. Vidal, A., Herrero, J., Romero, M., Sanchez, B., and Sanchez, M., Heterogeneous photocatalysis: Degradation of ethylbenzene in TiO_2 aqueous suspension, *J. Photochem. Photobiol. A: Chem.,* 79, 213–219, 1994.

70. Kochany, J. and Maguire, R. J., Abiotic transformations of polynuclear aromatic hydrocarbons and polynuclear aromatic nitrogen heterocycles in aquatic environments, *Sci. Total Environ.,* 144, 17–31, 1994.

71. Das, S., Muneer, M., and Gopidas, K. R., Photocatalytic degradation of wastewater pollutants. Titanium dioxide-mediated oxidation of polynuclear aromatic hydrocarbons, *J. Photochem. Photobiol. A: Chem.,* 77, 83–88, 1994.

72. Matthews, R. W., Near UV-light induced competitive hydroxyl reactions in aqueous slurries of titanium oxide, *J. Chem. Soc., Chem. Communic.,* 177, 1983.

73. Matthews, R. W., Hydroxylation reactions induced by near-ultraviolet photolysis of aqueous titanium dioxide suspensions, *J. Chem. Soc., Faraday Trans. 1,* 80, 457–471, 1984.

74. Matthews, R. W., Photooxidation of organic material in aqueous suspensions of titanium dioxide, *Water Res.,* 20, 569, 1986.

75. Matthews, R. W. and Sangster, D. F., Measurement by benzoate radiolytic decarboxylation of relative rate constants for hydroxyl radical reactions, *J. Phys. Chem.,* 69, 1938–1946, 1965.

76. Richard, C. and Boule, P., Photocatalytic oxidation of phenolic derivatives. Influence of OH· and h+ on the distribution of products, *N. J. Chem.,* 18, 547–552, 1994.

77. Palmisano, L., Augugliaro, B., Sclafani, A., and Schiavello, M., Activity of chromium-doped titania for the dinitrogen photoreduction to ammonia and for the phenol photodegradation, *J. Phys. Chem.,* 92, 6710–6713, 1988.

78. Martin, C., Martin, I., Rives, V., Palmisano, L., and Schiavello, M., Structural and surface characterization of the polycrystalline system Cr_xO_y· TiO_2 employed for photoreduction of dinitrogen and photodegradation of phenol, *J. Catal.,* 134, 434–444, 1992.

79. Kawaguchi, H. and Uejima, T., Photochemical decomposition of phenol in the presence of photocatalyst, *Kagaku Kogaku Ronbunshu,* 9, 107–109, 1983. *Chem. Abstr.,* 98: 125172n.

80. Matthews, R. W. and McEvoy, S. R., Destruction of phenol in water with sun, sand and photocatalysis, *Solar Energy,* 49, 507–513, 1992.

81. Matthews, R. W. and McEvoy, R., Photocatalytic degradation of phenol in the presence of near-UV illuminated titanium dioxide, *J. Photochem. Photobiol. A: Chem.,* 64, 231–246, 1992.

82. Chatterjee, S., Sarkar, S., and Bhattacharyya, S. N., Photodegradation of phenol by visible light in the presence of colloidal Fe_2O_3, *J. Photochem. Photobiol. A: Chem.,* 81, 199–203, 1994.

83. Terzian, R., Serpone, N., Minero, C., Pelizzetti, E., and Hidaka, H., Kinetic studies in heterogeneous photocatalysis. 4. The photomineralization of a hydroquinone and a catechol, *J. Photochem. Photobiol. A: Chem.,* 55, 243–249, 1990.

84. Terzian, R., Serpone, N., Minero, C., and Pelizzetti, E., Photocatalyzed mineralization of cresols in aqueous media with irradiated titania, *J. Catal.,* 128, 352–365, 1991.

85. Howe, R. F. and Grätzel, M., EPR study of hydrated anatase under UV irradiation, *J. Phys. Chem.,* 91, 3906–3909, 1987.

86. Brezová, V., Brandsteterova, E., Ceppan, M., and Pies, J., Photocatalytic oxidation of *p*-cresol in aqueous titanium dioxide suspension, *Coll. Czech. Chem. Commun.,* 58, 1285–1293, 1993.

87. Brezová, V., Stasko, A., Ceppan, M., Mikula, M., Blecha, J., Vesely, M., Blazkova, A., Panak, J., and Lapcik, L., Photocatalytic activity and the formation of radical intermediates, in: *Photocatalytic Purification and Treatment of Water and Air,* Ollis, D. F. and Al-Ekabi, H. (Eds.), Elsevier, Amsterdam, 1993, pp. 659–664.

88. Brezová, V. and Stasko, A., Spin trap study of hydroxyl radicals formed in the photocatalytic system TiO$_2$–water–p-cresol–oxygen, *J. Catal.*, 147, 156–162, 1994.

89. Trillas, M., Peral, J., and Domènech, X., Photo-oxidation of phenoxyacetic acid by TiO$_2$-illuminated catalyst, *Appl. Catal. B: Environ.*, 3, 45–53, 1993.

90. Amalric, L., Guillard, C., Serpone, N., and Pichat, P., Water treatment: Degradation of dimethoxybenzenes by the TiO$_2$-UV combination, *J. Environ. Sci. Health, A*, 28, 1393–1408, 1993.

91. Amalric, L., Guillard, C., and Pichat, P., Use of catalase and superoxide dismutase to assess the roles of hydrogen peroxide and superoxide in the TiO$_2$ or ZnO photocatalytic destruction of 1,2-dimethoxybenzene in water, *Res. Chem. Intermed.*, 20, 579–594, 1994.

92. Fujihira, M., Satoh, Y., and Osa, T., Heterogeneous photocatalytic reactions on semiconductor materials. III. Effect of pH and Cu^{2+} ions on the photo-Fenton type reaction, *Bull. Chem. Soc. Jpn.*, 55, 666–671, 1982.

93. Butler, E. C. and Davis, A. P., Catalytic oxidation in aqueous titanium dioxide suspensions: the influence of dissolved transition metals, *J. Photochem. Photobiol. A: Chem.*, 70, 273–283, 1993.

94. Richard, C., Photocatalytic reduction of benzoquinone in aqueous ZnO or TiO$_2$ suspensions, *N. J. Chem.*, 18, 443–445, 1994.

95. Richard, C. and Lemaire, J., Analytical and kinetic study of the photo-transformation of furfuryl alcohol in aqueous ZnO suspensions, *J. Photochem. Photobiol. A: Chem.*, 55, 127–134, 1990.

96. Richard, C., Boule, P., and Aubry, J.-M., Oxidizing species involved in photocatalytic transformations on zinc oxide, *J. Photochem. Photobiol. A: Chem.*, 60, 235–243, 1991.

97. Ishibai, Y., Suita, T., Murakami, H., and Murasawa, S., Purification of water in Lake Biwa using TiO$_2$ photocatalyst, *Abstr. 2nd Int. Symp. on New Trends in Photoelectrochemistry,* University of Tokyo, March 1994, p. 57.

98. Martin, J. F., The use of sodium carbonate peroxyhydrate to treat off-flavor in commercial catfish ponds, *Water Sci. Technol.*, 25, 315–321, 1992.

99. Brezová, V., Breza, M., and Ceppan, M., The photocatalytic degradation of cyclic acetals in aqueous titanium dioxide suspensions, *Chem. Pap.*, 46, 359–363, 1992.

100. Heller, A., Nair, M., Davidson, L., Luo, Z., Schwitzgebel, J., Norrell, J., Brock, J. R., Lindquist, S.-E., and Ekerdt, J. G., Photoassisted oxidation of oil and organic spills on water, in: *Photocatalytic Purification and Treatment of Water and Air,* Ollis, D. F. and Al-Ekabi, H. (Eds.), Elsevier Science Publishers, Amsterdam, 1993, pp. 139–153.

101. Heller, A. and Brock, J. R., Accelerated photoassisted dissolution of oil spills, in: *Aquatic and Surface Photochemistry,* Crosby, D. G. and Helz, G. R. (Eds.), ACS Symposium Series, American Chemical Association, Washington, D.C., 1993, Chapter 29.

102. Heller, A., Schwitzgebel, J., and Ekerdt, J. G., Clean photocatalytic oxidation of organic films on water on n-TiO$_2$-coated buoyant glass microbubbles, *Abstr. 2nd Int. Symp. on New Trends in Photoelectrochemistry,* University of Tokyo, March 1994, p. 1.

103. Wise, H. and Sancier, M., Photocatalyzed oxidation of crude-oil residue by beach sand, *Catal. Lett.,* 11, 277–284, 1991.

104. Pelizzetti, E., Minero, C., Hidaka, H., and Serpone, N., Photocatalytic processes for surfactant degradation, in: *Photocatalytic Purification and Treatment of Water and Air,* Ollis, D. F. and Al-Ekabi, H. (Eds.), Elsevier Science Publishers, 1993, pp. 261–273.

105. Avranas, A., Poulios, I., Kypri, C., Jannakoudakis, D., and Kyriakou, G., Heterogeneous photocatalytic degradation of the cationic surfactant dodecylpyridinium chloride, *Appl. Catal. B: Environ.,* 2, 289–302, 1993.

106. Hidaka, H., Kubota, H., Grätzel, M., and Serpone, N., Photodegradation of surfactants. I. Degradation of sodium dodecyl benzene sulfonate in aqueous semiconductor dispersions, *Nouv. J. Chim.,* 9, 67–69, 1985.

107. Hidaka, H., Yamada, S., Suenaga, S., Kubota, H., Serpone, N., Pelizzetti, E., and Grätzel, M., Photodegradation of surfactants. V. Photocatalytic degradation of surfactants in the presence of semiconductor particles by solar exposure, *J. Photochem. Photobiol. A: Chem.,* 47, 103–113, 1989.

108. Hidaka, H., Yamada, S., Suenaga, S., Serpone, N., and Pelizetti, E., Photodegradation of surfactants. VI. Complete degradation of anionic, cationic and nonionic surfactants in aqueous semiconductor dispersions, *J. Mol. Catal.,* 59, 279–290, 1990.

109. Zhao, J., Hidaka, H., Takamura, A., Pelizzetti, E., and Serpone, N., Photodegradation of surfactants. 11. ζ-Potential measurements in the photocatalytic oxidation of surfactants in aqueous TiO_2 dispersions, *Langmuir,* 9, 1646–1650, 1993.

110. Zhao, J. C., Oota, H., Hidaka, H., Pelizzetti, E., and Serpone, N., Photodegradation of surfactants. X. Comparison of the photooxidation of the aromatic moieties in sodium dodecylbenzene sulphonate and in sodium phenyldodecyl sulphonate at TiO_2–H_2O interfaces, *J. Photochem. Photobiol. A: Chem.,* 69, 251–256, 1992.

111. Hidaka, H., Zhao, J., Nohara, K., Kitamura, K., Satoh, Y., Pelizzetti, E., and Serpone, N., Photocatalyzed mineralization of non-ionic, cationic, and anionic surfactants at TiO_2/H_2O interfaces, in: *Photocatalytic Purification and Treatment of Water and Air,* Ollis, D. F. and Al-Ekabi, H. (Eds.), Elsevier Science Publishers, Amsterdam, 1993, pp. 251–259.

112. Hidaka, H., Nohara, K., Zhao, J. C., Takashima, K., Pelizzetti, E., and Serpone, N., Photodegradation of surfactants. XIII. Photocatalytic mineralization of nitrogen-containing surfactants at the TiO_2/water interface, *N. J. Chem.,* 18, 541–545, 1994.

113. Hidaka, H., Zhao, J., Pelizzetti, E., and Serpone, N., Photodegradation of surfactants. 8. Comparison of photocatalytic processes between anionic sodium dodecyl benzene sulfonate and cationic benzyl dodecyl dimethylammonium chloride on the TiO_2 surface, *J. Phys. Chem.,* 96, 2226–2230, 1992.

114. Schmidt, J. A., Heitner, C., Kelly, G. P., Leicester, P. A., and Wilkinson, F., Diffuse-reflectance laser flash photolysis studies of the photochemistry of bleached thermomechanical pulp, *J. Photochem. Photobiol. A: Chem.,* 57, 111–125, 1991.

115. Schmidt, J. A., Heitner, C., Kelly, G. P., and Wilkinson, F., Diffuse reflectance laser-flash photolysis of mechanical pulp. 1. Detection and identification of transient species in the photolysis of thermomechanical pulp, *J. Pulp Paper,* 16, J111–J117, 1990.

116. Schmidt, J. A., Goldszmidt, E., Heitner, C., Scaino, J. C., Berinstain, A. B., and Johnston, L. J., Photodegradation of α-guaiacoxy-acetoveratrone: triplet state reactivity induced by protic solvents, *Am. Chem. Soc. Symp. Ser.,* Vol. 531, *Photochemistry of Lignocellulosic Materials,* pp. 122–128, American Chemical Society, Washington D.C., 1993.

117. Schmidt, J. A., Berinstain, A. B., De Rege, F., and Heitner, C., Photodegradation of the lignin model α-guaiacoxyacetoveratrone, unusual effects of solvent, oxygen, and singlet state participation, *Can. J. Chem.,* 69, 104–107, 1991.

118. Schmidt, J. A. and Heitner, C., Light-induced yellowing of mechanical and ultra-high yield pulps. Part 2. Radical-induced cleavage of etherified guaiacylglycerol-β-aryl ether groups is the main degradative pathway, *J. Wood Chem. Technol.,* 13, 309–325, 1993.

119. Tinucci, L., Borgarello, E., Minero, C., and Pelizzetti, E., Treatment of industrial wastewaters by photocatalytic oxidation on TiO$_2$, in: *Photocatalytic Purification and Treatment of Water and Air,* Ollis, D. F. and Al-Ekabi, H. (Eds.), Elsevier Science Publishers, Amsterdam, 1993, pp. 585–594.

120. Mansilla, H. D., Villaseñor, J., Maturana, G., Baeza, J., Freer, J., and Durán, N., ZnO-catalysed photodegradation of kraft black liquor, *J. Photochem. Photobiol. A: Chem.,* 78, 267–273, 1994.

121. Sierka, R. A. and Bryant, C. W., Biological treatment of kraft wastewater following pretreatment of the extraction waste stream by illuminating titanium dioxide and membranes, in: *Photocatalytic Purification and Treatment of Water and Air,* Ollis, D. F. and Al-Ekabi, H. (Eds.), Elsevier Science Publishers, Amsterdam, 1993, pp. 275–290.

122. Sierka, R. A. and Bryant, C. W., Enhancement of biotreatment effluent quality by illuminated titanium dioxide and membrane pretreatment of the kraft waste stream and by increased chlorine dioxide substitution, *Water Sci. Technol.,* 29, 209–218, 1994.

123. Bemis, A. G., Dindorf, J. A., Horwood, B., and Samans, C., in *Kirk-Othmer Encyclopedia of Chemical Technology,* Vol. 17, Wiley, New York, 3rd ed., 1978, p. 745.

124. DeFoe, D. L., Holcombe, G. W., Hammermeister, D. E., and Biesinger, K. E., Solubility and toxicity of eight phthalate esters to four aquatic organisms, *Environ. Toxicol. Chem.,* 9, 623–636, 1990.

125. Lykins, B. W. Jr., and Koffskey, W., Products identified at an alternative disinfection pilot plant, *Environ. Health Perspect.,* 69, 119–128, 1986.

126. Kawaguchi, H., Photodecomposition of bis-2-ethylhexyl phthalate, *Chemosphere,* 28, 1489–1493, 1994.

127. Mansour, M., Moza, P. N., Barlas, H., and Parlar, H., Photostability of environmental organic chemicals in the presence of hydrogen peroxide in an aqueous medium, *Chemosphere,* 14, 1469–1474, 1985.

128. Hustert, K., Kotzias, D., and Korte, F., Photocatalytic decomposition of organic compounds on titanium dioxide, *Chemosphere,* 12, 55–58, 1983.

129. Halmann, M., Photodegradation of di-*n*-butyl-*ortho*-phthalate in aqueous solution, *J. Photochem. Photobiol. A: Chem.,* 66, 215–223, 1992.

130. Lindén, L. A., Rabek, J. F., Kaczmarek, H., Kaminska, A., and Scoponi, M., Photooxidative degradation of polymers by HO· and HO$_2$· radicals generated during the photolysis of H$_2$O$_2$, FeCl$_3$ and Fenton reagents, *Coord. Chem. Rev.,* 125, 195–218, 1993.

131. Ohta, T., Suzuki, J., Iwano, Y., and Suzuki, S., Photochemical nitrosation of dimethylamine in aqueous solution containing nitrite, *Chemosphere,* 11, 797–801, 1982.

132. Suzuki, J., Yagi, N., and Suzuki, S., Photochemical nitrosation of phenol in aqueous nitrite solution, *Chem. Pharm. Bull.,* 32, 2803–2905, 1984.

133. Suzuki, J., Sato, T., and Suzuki, S., Hydroxynitrobiphenyls produced by photochemical reaction of biphenyl in aqueous nitrate solution and their mutagenicities, *Chem. Pharm. Bull.,* 33, 2507–2515, 1985.

134. Suzuki, J., Ueki, T., Shimizu, S., Uesugi, K., and Suzuki, S., Formation of mutagens by photolysis of amino acids in neutral aqueous solutions containing nitrite or nitrate ion, *Chemosphere,* 14, 493–500, 1985.

135. Suzuki, J., Hagino, T., and Suzuki, S., Formation of 1-nitropyrene by photolysis of pyrene in water containing nitrite ion, *Chemosphere,* 16, 859–867, 1987.

136. Suzuki, J., Sato, T., Ito, A., and Suzuki, S., Photochemical reaction of biphenyl in water containing nitrite or nitrate ion, *Chemosphere,* 16, 1289–1300, 1987.

137. Suzuki, J., Watanabe, T., and Suzuki, S., Formation of mutagens by photochemical reaction of 2-naphthol in aqueous nitrite solution, *Chem. Pharm. Bull.,* 36, 2204–2211, 1988.

138. Suzuki, J., Watanabe, T., Sato, K., and Suzuki, S., Roles of oxygen in photochemical reaction of naphthols in aqueous nitrite solution and mutagen formation, *Chem. Pharm. Bull.,* 36, 4567–4575, 1988.

139. Suzuki, J., Sato, T., Ito, A., and Suzuki, S., Mutagen formation and nitration by exposure of alkylphenols to sunlight containing nitrate or nitrite ion, *Bull. Environ. Contam. Toxicol.,* 45, 516–522, 1990.

140. Ames, B. N., Durston, W. E., Yamasaki, E., and Lee, F. D., Carcinogens are mutagens. Simple test system consisting of liver homogenates for activation and bacteria for detection, *Proc. Natl. Acad. Sci. USA,* 70, 2281–2285, 1973.

141. Seffar, A., Dauphin, G., and Boule, P., Oxidative photocyclization of 2-biphenylol in dilute aqueous solution, *Chemosphere,* 16, 1205–1214, 1987.

142. Sarakha, M., Boule, P., and Lenoir, D., Phototransformation of 2-phenylphenol induced in aqueous solution by excitation of nitrate ions, *J. Photochem. Photobiol. A: Chem.,* 75, 61–65, 1993.

Chapter IV

Halocarbons

These compounds include several ubiquitous water pollutants, which are widely used as degreasing agents, solvents, dry-cleaning materials, and in many other applications. Aromatic halocarbons include important pesticides and intermediates in chemical industry. Also, disinfection of water by chlorination converts many hydrocarbon compounds to halocarbons.[1] Chlorination of surface waters often results in the formation of trihalomethanes (THMs), such as chloroform. Very high THM concentrations were observed in the effluents of water treatment plants.[2] In lakewaters rich in bromide ions, such as in Lake Kinneret (Sea of Galilee, Israel), the predominant THM product was shown to be bromoform, which is lachrymatory and toxic. The overall reactions leading to these compounds are

$$HOCl + precursor \rightarrow CHCl_3 \tag{1}$$

$$HOBr + precursor \rightarrow CHBr_3 \tag{2}$$

The precursors presumably are humic and fulvic substances that contain active sites that undergo rapid haloform reactions.[3-5]

The production of THMs is drastically decreased by using chlorine dioxide (ClO_2), either liquid or gaseous, as the disinfectant. With this treatment, it has been possible to comply with the maximum contaminant level (MCL) of 0.10 mg/L for the total concentration of THMs in drinking water mandated by the U.S. Environmental Protection Agency (U.S. EPA). Chlorine dioxide treatment does, however, result in the formation of various halocarbon byproducts, as well as of several carboxylic acids such as maleic acid derivatives, but only in very low concentrations, of the order of 1 to 10 ng/L (ppt).[6-9]

Chloroform and carbon tetrachloride may be removed from wastewater by the combined action of activated carbon adsorption and coagulation precipitation with a mixture of $Ca(OH)_2$, $FeSO_4$, and a high-molecular-weight coagulant. These two halocarbon compounds were thus removed with average efficiencies of 82 and 85%, respectively.[10] THM compounds may be removed by ultrafiltration or by granular activated carbon (GAC) filtration. But the high-temperature reactivation of GAC may cause the formation of the very toxic polychlorinated dibenzo-*p*-dioxins[11,12] that are produced by high-temperature combustion in municipal incinerators.

Maximum recommended values for the total concentrations of trihalomethanes (TTHM) in drinking water are by the European Community, 1 µg/L; by the World Health Organization (WHO), 30 µg/L; and by the U.S. EPA, 100 µg/L.[13]

I. ALIPHATIC HALOCARBONS

The decomposition of various halocarbon compounds was achieved by illumination in the presence of suspended semiconductor particles.[14–28]

A. Chlorocarbon Photodegradation — Relative Rates

In a survey of the TiO_2-promoted photolysis of 12 organochlorine compounds that were di-, tri-, and tetrachloro derivatives of methane, ethane, and ethylene, the degradation rates were in the order $Cl_2 > Cl_3 > Cl_4$. Also, for those containing the same number of chlorine atoms, the order was ethylene > ethane. With TiO_2 as catalyst, the reaction involved release of Cl^- ions. With Pt-loaded TiO_2 (1.75 wt% Pt), the degradation rate was considerably enhanced. With this platinized catalyst the reaction changed in that the chloride concentration during photolysis only initially increased, passed through a maximum, and then decreased, with simultaneous formation of ClO_3^- ions. Thus, the reaction included photooxidation of chloride to chlorate. Addition of hydrogen peroxide to the TiO_2-catalyzed photolysis caused a marked rate enhancement for most compounds, but had no effect with other compounds.[22,23] In the photocatalyzed degradation of the toxic compound chloral hydrate, the presence of hydrogen peroxide considerably accelerated the rate of oxidation. Half-times for disappearance of the compound (1 mM) in the absence and presence of H_2O_2 (0.02 M) with TiO_2 as photocatalyst were 8 and 2 min, respectively. The stoichiometry for complete oxidation in the absence of H_2O_2 was[21]

$$CCl_3CH(OH)_2 + O_2 \rightarrow 2CO_2 + 3HCl \tag{3}$$

In the TiO_2-photocatalyzed degradation of chloral hydrate, di- and trichloroacetic acids were identified as intermediates. The formation of dichloroacetic acid was proposed to involve a chlorine bridged transition state.[29]

The relative rates of chlorocarbon mineralization at the 1 to 50 ppb levels in near-UV illuminated aqueous slurries of TiO_2 were dichloroacetaldehyde ≫ trichloroethylene > perchloroethylene > dichloroacetic acid ~ dichloromethane > trichloromethane ~ 1,2-dichloroethane ~ monochloroacetic acid > tetrachloromethane > trichloroacetic acid.[19,27] The sunlight-induced photooxidation of trichloroethylene and of trichloromethane (chloroform) was achieved in oxygen-saturated aqueous slurries of TiO_2. With initially 50 ppm of these chlorocarbon compounds, 98 to 99% mineralization to CO_2 and HCl was obtained within 3 to 5 h.[17]

The photocatalyzed degradation of 1,1,2-trichloroethane and trichloroethylene in aqueous solutions over TiO_2 coated on fiberglass was very considerably accelerated in the presence of the electron acceptor additives Oxone® (potassium peroxymonosulfate) and $(NH_4)_2S_2O_8$.[30]

In an effort to find the most effective photocatalyst for the oxidation of hazardous organic compounds under natural or simulated sunlight, several semiconductor materials were tested for their reactivity toward the degradation of trichloroethylene (TCE), toluene, methyl ethyl ketone (MEK), salicylic acid, and 2,4-dichlorophenol. The best photocatalyst on all these compounds was Pt (1% by weight) on TiO_2. Strontium titanate ($SrTiO_3$) and 1.5% $NiO-SrTiO_3$ catalysts were inactive toward the degradation of TCE. The optimal catalyst dosage was found to depend strongly on the concentration of the organic solute to be destroyed. With higher initial solute concentrations, larger dosages of the catalyst were required for maximal rates of photodegradation.[31]

Both dichloroacetic and trichloroacetic acids were formed in appreciable amounts as intermediate products during the TiO_2-mediated photodegradation of TCE and PCE. Dichloroacetic acid is toxic and known as an animal carcinogen. Thus, for the remediation of groundwater contaminated with these pollutants, it is essential to carry out the photodegradation until complete mineralization of the organochloro intermediates.[19,32–34] Two mechanisms of photodegradation were observed in the decomposition of TCE and PCE: a reductive pathway involving conduction-band electrons and leading to dichlorinated products, and an oxidative pathway leading both to trichlorinated products and to mineralization.[35]

B. Mixtures of Chlorocarbon Compounds

From a study of the relative rates of the TiO_2-catalyzed photodegradation of various saturated and unsaturated halocarbons, it was concluded that the relative reactivities may be predicted from the rate (k) and adsorption (K) parameters of the Langmuir–Hinshelwood equation for the individual compounds. Thus, in a mixture of dichloromethane, trichloromethane, 1,2-dichloroethane, 1,1,1-trichloroethane, trichloroethene, tetrachloroethene, 1,2-dichloropropane, and 1,3-dichloropropene, there occurred during the first 4 h a preferential oxidation of the unsaturated compounds. Only when the concentrations of the unsaturated halocarbons had decreased to below 10% of their initial value did the oxidation of the saturated dichloro compounds take over for the next 4 h. Finally, after consumption of most of the dichlorocarbons, the trichloro compounds were also oxidized. In a mixture of chlorobenzene, tetrachloroethene, and 1,1,1-trichloroethane, the order of degradation was chlorobenzene > tetrachloroethene > 1,1,1-trichloroethane. With the same halocarbons photodegraded separately, the order of oxidation was tetrachloroethene > chlorobenzene > 1,1,1-trichloroethane. The explanation given was that the higher electron density of the aromatic and unsaturated compounds led to stronger adsorption on electrophilic sites at the TiO_2 surface, while electron-withdrawing substituents led to weaker adsorption.[36]

C. Halomethanes

The photodegradation reactions of aqueous dichloromethane and chloroform in the presence of suspended TiO_2 were found to be described by the overall stoichiometry[14,19,37–40]

$$CH_2Cl_2 + O_2 = 2HCl + CO_2 \qquad (4)$$

$$CHCl_3 + H_2O + 1/2\,O_2 \rightarrow 3HCl + CO_2 \qquad (5)$$

In the photodegradation of chloroform in aqueous dispersions of TiO_2 at pH 7.2, the rate of mineralization was linearly related to the square root of the light intensity. At low light intensities, the quantum yield increased with the pH of the medium. This was explained by Bahnemann et al.[41] to be due to the effect of pH on the surface charge of the hydroxyl groups on the TiO_2 particles, which influenced the adsorption of chloroform molecules.[41] At acidic and alkaline pH, the TiO_2 surface was either positively or negatively charged. At alkaline pH, the negative charge on the semiconductor surface interacted with the positively polarized hydrogen atom of chloroform,

$$\equiv TiO^- + H\text{-}CCl_3 \longrightarrow \equiv TiO\cdots H\cdots CCl_3$$

SCHEME IV.1

leading to charge transfer and rapid release of chloride ions. In acidic media, such a mechanism cannot occur. The decrease in quantum yields at higher light intensities was explained by the competing recombination of hydroxyl groups,[41]

$$2 \cdot OH \rightarrow 2H_2O_2 \qquad (6)$$

The rate-determining step in the photocatalytic degradation of chloroform on TiO_2 was proposed to be the reaction of surface-bound hydroxyl groups with adsorbed $CHCl_3$ molecules.[42]

In the photocatalytic degradation of chloroform with dispersions of TiO_2, only the anatase form of the semiconductor was found to be active, while the rutile form of "naked" TiO_2 was reported to be without photocatalytic activity.[37] However, platinized TiO_2, even in the rutile form, was very effective in the photooxidation of chloroform. The rate of degradation was found to increase with the loading of platinum. Best results were reported with 10%Pt/TiO_2, and the chloroform concentration decreased under illumination with 20-W blacklight fluorescent lamps from initially 200 ppm to less than several ppm within 5 h. With "bare" rutile, only 12% had decomposed during this period.[25]

The photocatalytic activity of titanium dioxide as photocatalyst for the degradation of chloroform was enhanced by loading silver ions on the TiO_2. Using 1% (w/w) Ag/TiO_2 dispersed in an aqueous solution of chloroform (initially 200 mg/L), under illumination for 6 h with a 125-W medium-pressure Hg lamp, 44% of the chloroform had degraded. With the "bare" TiO_2, only 35% had decomposed under the same conditions.[43]

In order to overcome to problem of recovery of the suspended photocatalyst after the reaction, TiO_2 (anatase form) was coated on glass fiber cloth, by thermal decomposition of titanium isopropoxide followed by calcination at 500 to 550°C. This glass-supported catalyst was packed in quartz tubes, through which aqueous solutions of chloroform were circulated. Under irradiation with a low-pressure Hg lamp, the mineralization was essentially complete within about 6 h. Illumination with a blacklight fluorescent lamp was much less effective. The rate of photodegradation of chloroform per gram of TiO_2 was somewhat smaller with the catalyst coated on the glass cloth than with aqueous dispersions.[38-40]

In analogy with the mechanism proposed for the radiolysis of chlorinated methanes,[44] the photolytic degradation of chloromethanes in the presence of semiconductors such as TiO_2 may involve as primary steps hydrogen abstraction by hydroxyl groups, for example, for dichloromethane,

$$CH_2Cl_2 + OH\cdot \rightarrow \cdot CHCl_2 + H_2O \qquad (7)$$

In the absence of oxygen, dissociative electron capture may also occur:

$$CH_2Cl_2 + e^- \rightarrow \cdot CHCl_2 + Cl^- \qquad (8)$$

The time course of the photodegradation of chlorocarbon compounds may be followed by observing the release of chloride and hydronium ions. As an example, the results of illumination of dichloromethane in an aqueous suspension of titanium dioxide are shown in Figure IV.1.[45]

D. Chlorinated Ethanes

A comparison of the decomposition of chlorinated ethanes in aqueous solutions, either by photocatalytic oxidation over dispersed TiO_2, or by radiolysis, or by metabolic degradation, indicated that the same intermediate products were formed. Mao et al.[32] therefore proposed that in all these three methods the same mechanism prevailed. The common primary chemical step was interaction of an ·OH radical with a hydrogen atom of the chlorinated ethane, leading to C–H bond cleavage and yielding carbon-centered radicals. In the presence of oxygen, these carbon-centered radicals were converted at almost diffusion-

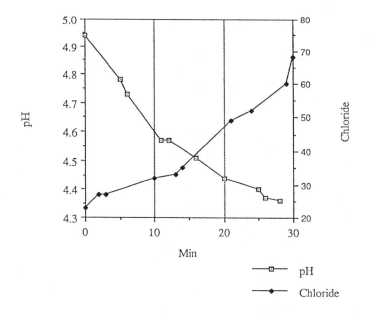

FIGURE IV.1.   Chloride ion release (m*M*) and pH decrease during photolysis of dichloromethane (initially 1 m*M*) in aqueous solution in the presence of suspended TiO$_2$ (0.3 g/L).

controlled rates into peroxy radicals, which then decomposed to the final products, usually chloro-substituted aldehydes and acetic acids. In the case of the photocatalytic oxidation, the hydroxyl radicals were probably surface-bound on the TiO$_2$ particles, and not free in solution. The main organic products from chlorinated ethanes (initially 3 m*M*) by TiO$_2$ (1.25 g/L) pro-moted photodegradation in air-saturated suspensions after 10 min of illumina-tion with a 450-W xenon lamp ($\lambda > 295$ nm) were chloroacetic acid and acetic acid from 1,1-dichloroethane (1,1-DCE); chloroacetic acid and chloroacetaldehyde from 1,2-dichloroethane (1,2-DCE); chloroacetic acid from 1,1,1-trichloroethane (1,1,1-TriCE); chloro- and dichloro acetic acids from 1,1,2-trichloroethane (1,1,2-TriCE); dichloroacetic acid from 1,1,2,2-tetrachloroethane (1,1,2,2-TetCE); dichloroacetic acid and trichloroacetaldehyde from 1,1,1,2-tetrachloroethane (1,1,1,2-TetCE); and trichloroacetic acid from pentachloroethane (PCE). The quantum yield for formation of chloroacetic acid from 1,1,1,-TriCE in aerated solutions at 360 nm in the presence of TiO$_2$ was 0.3%. No reaction occurred in homogeneous solution, in the absence of the photocatalyst, or in N$_2$-saturated solutions in the presence of the photocatalyst. Also, the photodegradation was suppressed in the presence of methanol, which is a known scavenger of valence band holes (h_{vb}^+) and of ·OH radicals. These effects indicate that the primary reaction step was the *oxidation* of the halocar-bon, and not a *reductive* process involving conduction band electrons (e_{cb}^-). In the photodegradation of these haloethanes, the oxygen molecules fulfill two functions: to scavenge the conduction band electrons and thus to inhibit the wasteful electron–hole recombination, and also to react with the organic radicals and form peroxy radicals. From a kinetic analysis of the competition between 1,1,1-TriCE and methanol for ·OH radicals (or valence-band holes), and taking the known value $k_2 = 4.1 \times 10^8$ M^{-1} s^{-1} for the rate constant of the oxidation of methanol by ·OH on TiO_2 particles,

$$CH_3OH + \cdot OH \rightarrow \cdot CH_2OH + H_2O \tag{9}$$

$$CH_3-CCl_3 + \cdot OH \rightarrow \cdot CH_2-CCl_3 + H_2O \tag{10}$$

a value of $k_2 = 4.9 \times 10^7$ M^{-1} s^{-1} was derived for the rate-determining step in the oxidation of 1,1,1,-TriCE.[32]

While the overall mechanism of the radiolytic and the photocatalytic oxidation of halogenated hydrocarbons is similar, as indicated by the qualitative identity of the products generated by both methods, such as organic acids and aldehydes, there exist definite differences between homogeneous solution γ-radiolysis and heterogeneous photocatalysis. These differences were revealed in the *1,2-chlorine shift* in β-chloroethyl radicals,

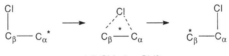

1,2-Chlorine Shift

SCHEME IV.2

An example was the radicals formed by hydrogen atom abstraction from 1,1,1-trichloroethane, in which the chlorine shift

$$CCl_3-CH_2 \cdot \rightarrow \cdot CCl_2-CH_2Cl \tag{11}$$

was much slower at the TiO_2 surface, with $k \sim 10^6$ s^{-1}, than in the homogeneous solution reaction (from γ-radiolysis), in which $k \geq 10^8$ s^{-1}. In homogeneous solution this rearrangement was faster than oxygen addition to the primary ·CH_2-CCl_3 radical. The slower heterogeneous reaction was ascribed to steric hindrance for the surface-adsorbed radical. The longer lifetime of the radical on the TiO_2 surface enabled it to react with oxygen, forming a peroxy radical, which hydrolyzed to monochloroacetic acid, which was the only organic acid formed by photocatalysis.

$$\cdot CCl_2-CH_2Cl + O_2 \rightarrow CH_2Cl-CCl_2-O_2 \cdot \tag{12}$$

$$CH_2Cl-CCl_2-O_2\cdot + H_2O \rightarrow CH_2Cl-COOH \qquad (13)$$

From 1,1,1-2-tetrachloroethane, after γ-radiolysis, the rearranged radical, after peroxidation, hydrolyzed to dichloroacetic acid as the only product, even at high oxygen concentrations.

$$\cdot CHCl-CCl_3 \rightarrow CHCl_2-CCl_2\cdot \rightarrow$$
$$CHCl_2-CCl_2-O_2\cdot \rightarrow CHCl_2-COOH \qquad (14)$$

The 1,2-chlorine shift occurred without transient liberation of chlorine atoms (as proven by absence of reaction with a Cl scavenger), indicating a chlorine-bridged intermediate in this rearrangement. On the other hand, in TiO_2-promoted heterogeneous photocatalysis, increasing amounts of trichloroacetic acid and decreasing amounts of dichloroacetic acid were formed with increasing oxygen concentrations.[29,32,46]

E.   Chlorinated Ethenes (Ethylenes)

During the TiO_2-photocatalyzed degradation of tetrachloroethene (tetrachloro-ethylene, perchloroethylene, PCE), intermediate chlorinated products observed were dichloroacetic acid and trichloroacetic acid.[46,47] The formation of trichlo-roacetic acid was accounted for by assuming a *1,2-chlorine shift* in the carbon-centered radical formed by attack of a hydroxyl radical:[29]

$$CCl_2=CCl_2 + \cdot OH_{ads} \rightarrow \cdot CCl_2-CCl_2(OH) \qquad (15)$$

$$\cdot CCl_2-CCl_2(OH) \rightarrow CCl_3-CCl(OH)\cdot \qquad (16)$$

$$CCl_3-CCl(OH)\cdot + O_2 + H_2O \rightarrow CCl_3-COOH \qquad (17)$$

The formation of dichloroacetaldehyde and dichloroacetic acid from the TiO_2-photocatalyzed degradation of both TCE and PCE was ascribed to one-electron reduction with conduction-band electrons as the primary steps. Thus, for the transformation of PCE the "electron capture" mechanism may be[47]

$$CCl_2=CCl_2 + e_{cb}^- \rightarrow \cdot CCl_2=CCl^- \qquad (18)$$

$$\cdot CCl_2=CCl^- + H^+ \rightarrow \cdot CCl_2-CHCl_2 \qquad (19)$$

$$\cdot CCl_2-CHCl_2 + O_2 \rightarrow \cdot O_2CCl_2-CHCl_2 \qquad (20)$$

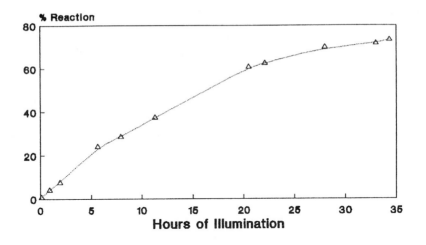

FIGURE IV.2. Chloride ion release as a function of reaction time during the photolysis of tetrachloroethylene (initially 1 m*M*) in the presence of titanium dioxide (0.1 g/L) and of H_2O_2 (initially 10 m*M*).

$$\cdot O_2CCl_2\text{--}CHCl_2 + H_2O \rightarrow CHCl_2COOH \qquad (21)$$

Another intermediate product from tetrachloroethene was oxalic acid. The rate of the photocatalyzed degradation of tetrachloroethene was considerably enhanced by addition of the electron scavenger sodium periodate, $NaIO_4$.[47]

In the presence of the hydroxyl radical scavengers bicarbonate and carbonate, the rates of the TiO_2-photocatalyzed degradation of both trichloroethylene (at pH 7) and of tetrachloroethene (at pH 8) were appreciably decreased. At higher concentrations of carbonate (to about 5 m*M*), tetrachloroethene formed higher yields of dichloroacetic acid and lower yields of trichloroacetic acid. This was suggested to indicate that trichloroacetic acid was produced by an oxidative mechanism. Carbonates are major natural groundwater components, and their presence affect not only the rates of disappearance of the chloroalkenes but also the composition of the degradation intermediates formed.[47]

With tetrachloroethylene in an aqueous dispersion of TiO_2 in the presence of H_2O_2, degradation was achieved by sunlight (Figure IV.2).[45]

In experiments in a continuously gas-sparged stirred-tank reactor, the times for elimination of 63% of dissolved tetrachloroethylene (TCE) in distilled water were compared by several methods. With only nitrogen purging, only ozonation, only UV photolysis, or UV photolysis together with ozonation, the times required were 100, 26, 20, and 7 min, respectively. Longer treatment times were needed with these treatments if the TCE was dissolved in filtered lake water.[48] In the photocatalytic oxidation of trichloroethylene in aqueous slurries of TiO_2, dichloroacetaldehyde was identified as an intermediate, which underwent further photodegradation, resulting in complete mineralization to CO_2 and HCl. The product chloride ion exerted an inhibitory effect on the photooxidation rate. The rates of disappearance of the trichloroethylene and of the intermediate as a function of the initial concentration of trichloroethylene and of the product Cl⁻ ion concentration could be fitted to a Langmuir–Hinshelwood type model.[16,27] Other intermediates identified during the TiO_2-induced photooxidation of TCE in the presence of air were dichloroacetic acid and formic acid.[47]

F.   Chloroacetic Acids

The homogeneous solution photolysis of monochloroacetic acid is often measured as a convenient actinometer solution. At $\lambda = 253.7$ nm, the quantum yields for Cl⁻ release in either argon- or oxygen-saturated solutions was 0.34 or 0.36, respectively. The primary process was proposed to be due to light absorption by an n–π* transition ($\lambda = 240$ nm) The main organic products were glycolic and acetic acids.[49]

In a comparison of the photooxidation of the chloroacetic acids in the presence of titanium dioxide and oxygen, the relative order of the initial rates of reaction was dichloroacetic acid > chloroacetic acid ≫ trichloroacetic acid.[19]

Chloro-, dichloro-, and trichloroacetic acids undergo photodegradation in illuminated suspensions of TiO_2, in reactions leading initially to intermediates such as formic and oxalic acids, and eventually to complete mineralization.[32,33] With chloro- and dichloroacetic acids, the oxidizing species was presumably the ·OH radical, acting by hydrogen abstraction. With trichloroacetic acid, no C–H bond exists, and the hydrogen abstraction mechanism is impossible. It was therefore proposed that with trichloroacetic acid, the primary step may be a photo Kolbe reaction with valence-band holes, followed by rapid hydrolysis of the trichloromethyl radical.[32]

$$CCl_3COO^- + h_{vb}^+ \rightarrow \left[CCl_3COO\cdot \right] \rightarrow \cdot CCl_3 + CO_2 \qquad (22)$$

As noted earlier (Chapter III, Section II.B.1), Chemseddine and Boehm[33] concluded from a detailed kinetic study of the photodegradation with TiO_2 of acetic acid and of the chloroacetic acids that the preceding photo Kolbe mechanism was of only secondary importance. The major contribution to the degradation was proposed to be attack at the α-C-H bonds, resulting in hydrogen abstraction. The reactive species carrying out this abstraction was the ·OH radical, and in the presence of oxygen also the superoxide ion, O_2^-.[33] This explained the very low reactivity of trichloroacetic acid in oxygenated solution and its high reactivity in nitrogen-saturated solutions. In a comparison of the photocatalytic degradation of acetic acid with that of the chloroacetic acids in oxygenated aqueous dispersions of TiO_2, the order of the initial rates of CO_2 release was dichloroacetic acid = chloroacetic acid > acetic acid ≫ trichloroacetic acid. A further relatively important mechanism, particularly for the destruction of trichloroacetic acid, was the capture of photogenerated electrons by the chlorine atoms, resulting in chloride ion release.[33]

In the photocatalytic degradation of dichloroacetic acid (DCA) in oxygen-saturated solutions in the presence of TiO_2 (Degussa P25, 0.5 g/L), the rates of reaction were linearly related to the light intensity (below 1.2×10^{-6} mol photons L^{-1} s^{-1}) when the reaction mixtures were at pH 2.6, 7, and 11. In these mixtures, the quantum yields were thus independent of the light intensity. However, surprisingly, when the reaction mixture was at pH 5, the degradation rate was proportional to the square root of the light intensity. From acid/base titrations of the surface-bound hydroxyl groups on the TiO_2 particles, it was concluded that only at pH 5 most of the surface hydroxyl groups exist in an uncharged form. A kinetic model was proposed to account for the square-root dependence of the rate on the light intensity.[34] Improved rates of photodestruction of DCA were achieved by platinization of the TiO_2 catalyst (with optimal loading at 1 wt% platinum), followed by calcining at 300 to 400°C for 1 h. The beneficial effect of Pt was ascribed to the inhibition of the $e_{cb}^- - h_{vb}^+$ recombination by the action of the Pt islands on the surface of TiO_2, acting as irreversible electron acceptors. In a comparison of several preparations of 1 wt% Pt/TiO_2, the quantum yields observed for the dechlorination of DCA in slurry reactors with TiO_2 obtained by a sol-gel process, with Degussa P25, and with a commercial Korean product (Hombikat) were about 7, 12, and 43%, respectively. The 1 wt% Pt/TiO_2 (Hombikat) was also coated as a thin film on the inside of a Pyrex® glass tube in the focus of a parabolic trough reflector, and was tested as a flow reactor. Under illumination with halogen lamps (1300 W/m^2) and a continuous supply of oxygen, with recirculation at a flow rate of 150 mL/min of 1 mM DCA, the decomposition of DCA observed pseudo first-order kinetics, with k = 0.016 min^{-1}.[50]

See also in Chapter VIII, Section V on the solar photodegradation of dichloroacetic acid over TiO_2 in parabolic trough and in thin film fixed bed reactors.

G. Chloroalkyl Ethers

1. 2-Chloroethyl Ether

2-Chloroethyl ether (bis-2-chloroethyl ether, $[ClCH_2CH_2]_2O$), is an important intermediate in the petrochemical, textile, and dyestuff industries. Since it is carcinogenic, its appearance in surface and groundwater is of serious environmental concern. Milano et al.[51] studied its hydrolytic degradation in pure water, as well as its direct photolysis, and its photodegradation in the presence of H_2O_2 and of TiO_2.[51] Rates of hydrolysis in pure water (pH 6.9) measured at 80 to 95°C observed first-order kinetics. From the Arrhenius equation, the activation energy of 108.3 kJ mol^{-1} was derived. By extrapolation, hydrolytic half-lives at 25 and 10°C were calculated to be 22 and 202 years, respectively. Thus, hydrolysis is inadequate to provide for the rapid degradation of this pollutant, which may accumulate in natural waters. The products of hydrolysis were 1,4-dioxane and HCl. Under illumination with a 100-W medium-pressure Hg lamp, through its quartz envelope, the decomposition was very much faster. In pure water (pH 6.9, containing 7 mg/L dissolved oxygen), with initially 2.6 mM 2-chloroethyl ether, the photoproducts after 50% conversion were 2-chloroethanol (46%), acetaldehyde (30%), ethanol (15%), methanol (6%), and 2-chloroethyl ethyl ether (3%), as well as traces of 2-chloroacetaldehyde, peracetic acid, and 2-chloro-epoxy-1′,2′-diethyl ether. Of these, 2-chloroethanol (ethylene chlorohydrin), the major photoproduct, is strongly mutagenic and highly toxic, and some of the other products are also toxic. Considerable enhancements in the rates of photodegradation of 2-chloroethyl ether were observed in the presence of either H_2O_2 (0.5 mM) or TiO_2 (5 mg/L). Most of the same photoproducts as in the direct photolysis were found, albeit in different proportions. Using TiO_2, no 2-chloroethyl ethyl ether was found, and instead, increased concentrations of 2-chloroepoxy-1′,2′-diethyl ether were produced. With H_2O_2 as oxidant, the primary step in the reaction was proposed to be formation of ·OH radicals, followed by hydrogen abstraction from the substrate,

$$·OH + (ClCH_2CH_2)_2O \rightarrow H_2O + ClCH_2CH_2-O-·CH-CH_2Cl$$

$$+ ·CHCl-CH_2-O-CH_2CH_2Cl$$

(23)

With TiO_2, the reaction of ·OH radicals was proposed to be with surface-adsorbed substrate molecules, followed by elimination of water to form an alkene, which underwent epoxidation to 2-chloroepoxy-1′,2′-diethyl ether:[51]

$$·OH_{ads} + (ClCH_2CH_2)_2O_{ads} \rightarrow ClCH_2CH_2-O-CH_2CH_2OH_{ads}$$

$$\rightarrow ClCH_2CH_2-O-CH=CH_{2(ads)}$$

(24)

2. 1,2-Bis(2-chloroethoxy)ethane

The biorefractory and toxic chloroether 1,2-bis(2-chloroethoxy)ethane (BCEE, triethylene glycol dichloride) was photodegraded in aqueous dispersions of TiO_2, using secondary wastewater as the medium. The photooxidation of BCEE was most rapid at pH 4. The rate of its photodegradation was inversely proportional to the concentration of soluble COD, while suspended solids had no effect on the rate of reaction. This study indicated that the photocatalytic destruction of biorefractory compounds was possible even in wastewater, as long as the soluble COD was sufficiently low.[52]

H. Reductive Mechanism

Evidence for a reductive mechanism in the photocatalytic degradation of CCl_4 in aqueous suspensions of TiO_2 was obtained by Hilgendorff et al.[53] In the presence of the hole scavenger *tert*-butanol (0.7 M, at pH 11, with 0.5 g/L TiO_2/Pt), the rate of photodegradation was double that in the absence of *tert*-butanol. The explanation given was that in the presence of the hole scavenger, the electron–hole recombination at the surface of the photocatalyst was partially inhibited, and thus more conduction-band electrons (e_{cb}^-) were available for the dissociative electron capture reaction,[53]

$$CCl_4 + e_{cb}^- \rightarrow \cdot CCl_3 + Cl^- \qquad (25)$$

With oxygen present, the consecutive degradation included formation of the $\cdot O_2CCl_3$ peroxy radical, which rapidly dissociated and hydrolyzed (possibly via phosgene) to CO_2 and HCl.[54]

Reductive photodehalogenation of bromoform ($CHBr_3$) was achieved under sunlight with a titanium dioxide–cobalt macrocycle hybrid catalyst.[55] The catalyst was prepared by silanizing TiO_2 (Degussa P25 anatase) with 3-(aminopropyl)triethoxysilane, which was then impregnated with cobalt tetrasulfophthalocyanine (CoTSP).[56] Under anaerobic conditions, the dehalogenation of bromoform with this hybrid catalyst, suspended in 50% (v/v) aqueous 2-propanol, was much faster than the direct (uncatalyzed homogeneous) photolysis of bromoform. The reaction products were dibromomethane (CH_2Br_2), bromomethane (CH_3Br), HBr, and traces of methane. For mixtures containing initially 34 mM $CHBr_3$ in 50% (v/v) aqueous 2-propanol, the initial rates of reaction for the direct photolysis, for the TiO_2-catalyzed photolysis, and for the TiO_2/CoTSP-catalyzed photolysis were 0.32, 0.28, and 1.4 mM min^{-1}, respectively. In these reactions, 2-propanol served as a sacrificial electron donor, which was oxidized to acetone, presumably by valence-band electrons. In the proposed mechanism, conduction-band electrons from the photoexcited TiO_2 cause the reduction of the cobalt(II) complex to a cobalt(I) complex,

$$TiO_2 - Co(II)TSP + e^- \rightarrow TiO_2 - Co(I)TSP \qquad (26)$$

This complex is a strong nucleophile, which interacts in a thermal (dark) reaction with the electrophilic carbon of bromoform, displacing the bromide ion:

$$TiO_2 - Co(I)TSP + CHBr_3 \rightarrow TiO_2 - Co(III)TSP - CHBr_2 + Br^- \qquad (27)$$

followed by photolytic fission of the Co–C bond, which may be rate-determining at high initial bromoform concentrations,

$$TiO_2 - Co(III)TSP - CHBr_2 + h\nu \rightarrow TiO_2 - Co(II)TSP + \cdot CHBr_2 \quad (28)$$

and proton abstraction from 2-propanol,[55]

$$\cdot CHBr_2 + (CH_3)_2 CHOH \rightarrow CH_2Br_2 + \cdot C(CH_3)_2 OH \quad (29)$$

As noted earlier, electron-capture reactions by halocarbon compounds are possible only in the absence of molecular oxygen. This has been studied mainly by pulse radiolysis. Thus, in 1,2-dibromoethane, electron capture occurs at diffusion-limited rate,[57]

$$k_2 = 1.2 \times 10^{10} \text{ mol}^{-1} \text{ dm}^3 \text{ s}^{-1}$$

The primary radical formed by electron capture from 1,2-dibromoethane was found to dissociate in a fast unimolecular process,

$$BrCH_2CH_2 \cdot \rightarrow C_2H_4 + Br \cdot \quad k = 2.8 \times 10^6 \text{ s}^{-1} \quad (30)$$

resulting in the strongly oxidizing bromine atom, which reacts with Br^-, producing the Br_2^- radical anion.[57]

Hydroxyl radicals also react rapidly in a hydrogen abstraction reaction,

$$CH_2Br - CH_2Br + OH \cdot \rightarrow CH_2Br - CHBr \cdot + H_2O \quad (31)$$

$$k_2 = 2.1 \times 10^8 \text{ mol}^{-1} \text{ dm}^3 \text{ s}^{-1}$$

followed by decay of the radical to vinyl bromide and Br atoms,[44]

$$CH_2Br - CHBr \cdot \rightarrow Br \cdot + CH_2 = CHBr \quad (32)$$

I.   Halogenated Peroxy Radicals

In the presence of oxygen, chlorocarbon radicals were shown by pulse radiolysis experiments to be rapidly converted to halogenated peroxy radicals $(RO_2 \cdot)$:[44,58,59]

$$\cdot CHCl_2 + O_2 \rightarrow \cdot O_2CHCl_2 \quad (33)$$

$$\cdot CH_2Cl + O_2 \rightarrow \cdot O_2CH_2Cl \qquad (34)$$

$$\cdot CCl_3 + O_2 \rightarrow \cdot O_2CCl_3 \qquad (35)$$

The toxic action of halocarbons, such as of carbon tetrachloride as a liver toxin, has been linked to the oxidizing properties of the halogenated peroxyl radicals, causing hydrogen abstraction from unsaturated fatty acids, thus initiating the peroxidation of lipids.[44,59–61]

J. Bromocarbons

1. 1,1- and 1,2-Dibromoethane

1,2-Dibromoethane (ethylene dibromide, EDB, CH_2BrCH_2Br) has been used in leaded gasoline, and also as a soil and grain fumigant. Since it is carcinogenic and mutagenic, its accumulation as a groundwater contaminant is of considerable concern. Aqueous solutions of both 1,2-dibromoethane and its isomer, 1,1-dibromoethane, are quite resistant to direct photolysis by near-UV light. In the presence of suspended TiO_2, both isomers readily underwent photocatalyzed degradation, leading to complete mineralization to CO_2 and HBr. The initial reaction rates could be described by simple Langmuir–Hinshelwood kinetics. The relative order of reactivity of the bromoethanes and of the bromomethanes was $CH_2Br_2 \sim CHBr_3 \sim CHBr_2CH_3 > CH_2BrCH_2Br$. In the photocatalytic degradation of 1,2-dibromoethane, an observed intermediate was vinyl bromide, $CH_2=CHBr$. The proposed mechanism of its formation was by hydroxyl radical attack on TiO_2-surface-adsorbed 1,2-dibromoethane, with displacement of a bromine atom, leading to 2-bromoethanol, which then underwent dehydration to vinyl bromide. This intermediate was rapidly oxidized to CO_2 and HBr.[18]

2. 1,2-Dibromopropane

During UV-photolysis in pure water in the presence of oxygen, 1,2-dibromopropane yielded the following photoproducts: 1-bromo-2-propanol, 1-bromo-2-propanone, 3-bromopropene, allylic alcohol, acetone, methanol, and acetic acid. The mechanism proposed involved intermediate formation of the halogenated radical $BrCH_2\cdot CH\text{-}CH_3$.[62,63]

3. 1,2-Dibromo-3-chloropropane (DBCP)

This very toxic halocarbon compound had been widely used as nematicide (nemagon) from 1955 to 1977. It has since been banned because of carcinogenicity and because of causing male sterility. This pesticide is very persistent in groundwater. It is a serious water-quality problem in California, requiring the closure of several wells.[64]

The direct photolysis in aqueous solution was studied using its UV absorption ($\lambda_{max} = 205$ nm). Under illumination with a medium-pressure Hg lamp, the photohydrolysis of DBCP occurred even without a sensitizer:

$$CH_2Br-CHBr-CH_2Cl + h\nu \rightarrow$$

$$CH_2OH-CHBr-CH_2Cl + CH_2Br-CHOH-CH_2Cl \tag{36}$$

Further intermediates included the highly toxic epichlorohydrin, while the final products were glycerol and acrolein.[65] In a further study of the UV photolysis of DBCP in dilute aqueous solutions (initially 400 ppm) to 50% conversion, the main photoproducts were 2-bromo-3-chloropropanol (30%), 1-bromo-3-chloro-2-propanol (10%), 1-bromo-3-chloro-2-propanone (3%), 3-bromo-1-chloropropene Z and E (5%), 1-chloro-2,3-epoxypropane (15%), acetone (20%), and methanol (10%). Thus, the direct photolysis of DBCP produces several intermediates, which themselves are very toxic. The rate of photodecomposition of DBCP was found to be accelerated in the presence of low concentrations of hydrogen peroxide, which, however, did not change the distribution of products.[63]

The nonphotochemical degradation of DBCP by the superoxide ion (O_2^-) was studied in DMF solutions. The initial reactions formed a complex mixture of dehydrohalogenation products, including the mutagenic 2-bromoacrolein. The proposed mechanism involved nucleophilic attack of O_2^- on the bromocarbon bond, with Br^- ion displacement and formation of an $R_2CHOO\cdot$ radical. Subsequent steps included displacement of the second bromine atom with intermediate formation of a dioxetane ring,

$$\begin{array}{c} CH_2 - CH - CH_2Cl \\ | \quad\quad | \\ O \; - \; O \end{array} \tag{37}$$

which then broke down to the final products, formaldehyde and formic acid.[66,67]

K. Groundwater Remediation

In photocatalytic experiments on water that was highly contaminated with DBCP (from a well in California), in the presence of suspended titanium dioxide, illumination caused a decrease in the concentration of DBCP from the initial high value of 2.9 ppb to 0.4 ppb.[26,45]

The photodegradation of halocarbon compounds in well water heavily polluted with pesticides from a nearby chemical plant in Israel was tested by three photocatalytic treatments: (1) $TiO_2 + H_2O_2$, (2) $Fe^{3+} + H_2O_2$, and (3) $TiO_2 + H_2O_2 + Fe^{3+}$. The experiments were performed in batch mode in round-bottomed glass-stoppered flasks exposed to 75 h of actual sunlight, with 1 g/L

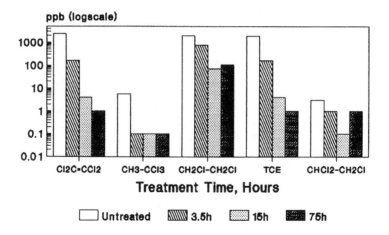

FIGURE IV.3. Solar photodegradation of aliphatic halocarbons in polluted well water in the presence of Fe^{3+} (0.3 mM) + TiO_2 (1 g/L) + H_2O_2 (0.1 M). Effect of treatment time.

TiO_2 and initially 0.1 M H_2O_2 and 0.3 mM $Fe_2(SO_4)_3$. Best results were obtained by treatment 2, which presumably occurred by a photo Fenton mechanism. The degradation with $TiO_2 + H_2O_2 + Fe^{3+}$ was also determined after 3.5 and 15 h of sunlight exposure. As shown in Figure IV.3 for several common halocarbon solvents, decreases of an order of magnitude were achieved by method 3 after 3.5 h, but reduction to the limit of detection (0.1 ppb) was not reached for tetrachloroethylene and TCE even after 75 h.[68]

L. Fluorocarbons

Direct photolysis in the vapor phase by the 185-nm emission from a 50-W low-pressure Hg lamp caused the degradation of $CHCl_3$, CCl_4, and of a chlorofluorocarbon (CFC-113). The presence of oxygen and water vapor was required for efficient destruction of these halocarbon compounds, which was performed in a flow system. The effluent from the photolysis cell was passed through a column packed with a metal, for example, 5 wt% Fe/SiO_2, which trapped the Cl and F species released. The carbon atom of the halocarbon compounds was quantitatively converted to CO_2.[69]

Fluorocarbon compounds generally reacted more slowly than chloro- and bromocarbons toward photodegradation in the presence of aqueous suspended TiO_2. Sabin et al.[70] observed that for 1,1-difluoro-1,2,2-trichloroethane (FC 122) and for 1,1-difluoro-1,2-dichloroethane (FC 132b) the half-times for decomposition under standardized conditions were 24.7 and 17.3 min. Fluorotrichloromethane (FC 11 or Freon 11) and 1,1,1-trifluoro-2,2,2-trichloroethane (FC 113) were not degraded at all.

1. Fluoroalkenes

The volatile two- and three-carbon fluoroalkenes were photooxidized in the presence of air in a gas–solid reaction over moist TiO_2 and were compared with that of ethylene. The reaction was performed under illumination with an 8-W low-pressure Hg lamp placed inside a Pyrex® glass photoreactor. The TiO_2 was coated on the inside of a glass tube between the lamp and the reactor, and was moistened by spraying water on the photocatalyst. No reaction occurred in the absence of either moist TiO_2, oxygen, or illumination. Irradiation was carried out at atmospheric pressure. The order of reactivity between the fluoroalkenes was hexafluoropropene > tetrafluoroethylene ≫ 1,1-difluoroethylene > ethylene > 3,3,3-trifluoropropene = fluoroethylene. In the photodecomposition of hexafluoropropene, an intermediate product identified was trifluoroacetic acid, in addition to the total mineralization products, CO_2 and HF. The products agreed with the stoichiometries,

$$CF_3CF=CF_2 + 2H_2O_2 + O_2 \rightarrow CF_3COOH + CO_2 + 6HF \qquad (38)$$

$$CF_3CF=CF_2 + 3H_2O + O_2 + 3/2\,O_2 \rightarrow 3CO_2 + 6HF \qquad (39)$$

In the proposed mechanism, the primary chemical step is nucleophilic attack on the double bond, either by the superoxide anion radical (formed by electron capture by surface adsorbed oxygen), or by injection of an electron from the TiO_2 into the alkene, followed by addition of oxygen to the carbanion produced. In either case, the intermediate product may be a peroxy organic compound, which then may decompose to the final products.[71]

M. Vacuum-UV Photolysis of Halocarbons

A xenon-excimer light source was developed as a powerful tool for various photochemical applications (see description in Chapter I, Section V.B).[72]

For groundwater contaminated with volatile organic compounds, a common treatment is air stripping, which transfers the pollutants to the gaseous phase. Halocarbon contaminants may be scrubbed out with granulated activated carbon (GAC). However, the heat reactivation of the activated carbon may cause the production and release of the very toxic polychlorinated dioxins. The photolysis of gaseous chloro- and fluorocarbons was achieved in humid air, by illumination with a xenon–xenon excimer (Xe_2^*) lamp.[73,74] At its emission peak of 172 nm, the lamp output was 0.8 W. The removal of the halocarbons followed pseudo first-order kinetics. For chloroform, carbon tetrachloride, and 1,1,2-trichlorotrifluoroethane ($Cl_2FC_2F_2Cl$, CFC-113), the rate constants were 3.1×10^{-3}, 2.0×10^{-3}, and 0.2×10^{-3} s^{-1}, respectively. The times t_{95} (for 95% decomposition) were 16, 25, and 240 min, respectively. Thus, the vacuum UV photolysis at 172 nm enabled the destruction even of the trichlorotrifluoroethane, which was resistant to photodegradation in aqueous dispersions of TiO_2.[70] The vapor-phase decomposition of carbon tetrachloride was markedly enhanced by the presence of oxygen and was only slightly affected by the concentration of water vapor. These results suggested that the photolysis was due mainly to oxidation of CCl_4 by atomic oxygen $O(^1D)$, formed by the photolysis of O_2. Carbon tetrachloride in *aqueous solution* under illumination with the (Xe_2^*) excimer lamp underwent first-order degradation, with $k = 2.2 \times 10^{-4}$ s^{-1}. At 172 nm, water is strongly absorbing ($\varepsilon = 1970$ L/m-cm), and the degradation of CCl_4 was predominantly due to reaction with $\cdot H$ and $\cdot OH$ radicals formed by the photolysis of water.[73,74]

II. AROMATIC AND OTHER CYCLIC COMPOUNDS

Haloaromatic compounds are among the more serious pollutants, derived from widely used pesticides, herbicides, and electrical insulators. These compounds undergo partial degradation by homogeneous photooxidation with UV light. However, in many cases these irradiation products are even more toxic than the starting compounds. Among these compounds, the chlorophenols are priority pollutants, due to their toxicity to aquatic life and poor biotreatability, and because they cause unpleasant taste and odor in water. Chlorophenols are widely used in the manufacture of dyes, herbicides, fungicides, and drugs. They are also formed by the degradation of phenoxy herbicides and during the chlorination of drinking water containing aromatic impurities.[75] Their photocatalytic degradation is usually quite rapid, leading to complete mineralization. Chlorinated dioxins are very toxic pollutants and are quite recalcitrant to many methods of degradation.

s-Triazine herbicides, such as atrazine, are widely used agrochemicals and are persistent contaminants of groundwater. The photodegradation of these chloro-substituted aromatic nitrogen compounds is described in Chapter V, Section I.

A. Haloaromatics

1. Reaction Rates with ·OH Radicals

Since the advanced oxidation techniques for water purification, such as treatment with ozone, ozone–hydrogen peroxide, UV–hydrogen peroxide, UV–TiO_2, or UV–Fe(II)–hydrogen peroxide, all depend primarily on the action of hydroxyl radicals, it was important to determine the rate constants of these ·OH radicals with the pollutant molecules in aqueous solution. For a few such molecules, such as benzene and chlorobenzene, the rate constants had been previously known from pulse radiolysis experiments. Kochany and Bolton[76] developed a competition kinetics method to determine the second-order rate constants k_2 for the reaction of ·OH radicals with some halobenzenes, using an electron paramagnetic resonance (EPR) spin trapping technique with the spin trap 5,5′-dimethylpyrroline *N*-oxide (DMPO) (see also Chapter I, Section II.A, on this method). Values thus obtained at pH 7.0 of k_2 for benzene, chloroben-zene, bromobenzene, iodobenzene, *o*-dichlorobenzene, *m*-dichlorobenzene, and *p*-dichlorobenzene were 7.6×10^9, 4.3×10^9, 4.8×10^9, 5.3×10^9, 4.0×10^9, 5.7×10^9, and 5.4×10^9 M^{-1} s^{-1}, respectively.[76]

A quantitative structure–activity relationship (QSAR) was developed for the photohydrolysis in water of meta-substituted halobenzene derivatives.[77] In these direct photolysis reactions, irradiation resulted in fission of the carbon–halogen bond with the lowest bond strength, with formation of the correspond-ing hydroxyl derivative. The observed quantum yields for the photolysis of these compounds, in dilute aqueous solutions under irradiation at 250 to 350 nm, could be correlated with the bond strength BS, the sum of the Hammett σ constants and of the inductive constants σ_I of additional substituents to the benzene ring, and the sum of the steric factors E_S. The following three-parameter equation provided a good correlation (to the 95% confidence level) for most compounds,

$$\log \Phi = -1.02\sigma_I + 0.30E_s - 0.005BS - 0.4 \tag{40}$$

The correlation failed for 3-chlorobenzonitrile, 3,5-dichlorobenzonitrile, 3-fluorobenzonitrile, and 3-fluorobenzoic acid, which did not undergo photohydrolysis.[77] For the photohydrolysis of various structurally related 1,3-disubstituted and 1,3,5-trisubstituted halobenzene derivatives, quantum yields could be correlated with the carbon-halogen bond strength BS, and the summa-tion E_S of the steric factors of all substituents,[78]

$$\log \Phi = 0.665E_s - 0.01BS + 1.031 \tag{41}$$

2. Semiconductor Photocatalysis

The photodegradation of many haloaromatic compounds in the presence of semiconductors has been reported.[14,19,79–85]

Among several catalysts tested on the photodegradation of haloaromatic compounds, the order of activity was $TiO_2 > ZnO > CdS > WO_3 > SiO_2$. For practical application, the TiO_2 was immobilized in a thin film on glass surfaces, thus obtaining an effluent free of the catalyst.[28,81,82] See Chapter I, Section V.A.2, for more detail on immobilized photocatalysts.

3. Chlorobenzenes

Chlorobenzene in dilute aqueous solution was practically resistant to direct photolysis by UV illumination (with 15-W blacklight fluorescent lamps). In the presence of suspended TiO_2 (Fisher certified grade, BET area 7 m^2/g), Ollis et al.[19] found that chlorobenzene (initially 70 ppm) underwent slow dechlorination, with pseudo first-order kinetics. The dependence of reaction rates on the initial concentration of chlorobenzene followed the Langmuir–Hinshelwood relationship. Primary intermediates detected were *ortho-* and *para-* chlorophenol and the corresponding hydroquinones, which after more prolonged irradiation were dechlorinated to *ortho-* and *para*-benzoquinones. Starting with a higher initial concentration of chlorobenzene (130 ppm), the same primary intermediates observed were *o-* and *p*-chlorophenol, but these after continued irradiation dimerized to the biphenyl products 4,4′-dichloro-1,1′-biphenyl and 4,5′-dichloro-1,1′-biphenyl. *o*-Dichlorobenzene by photocatalytic oxidation with TiO_2 was initially converted into *o*-chlorophenol, 2,3-dichlorophenol, and 3,4-dichlorophenol. In this study, the formation of CO_2 was not observed. The mechanism of photoassisted oxidation of the chlorobenzenes can be understood as electrophilic hydroxyl radical substitution at *ortho* and *para* positions on the benzene ring.[19] In a further study of the photocatalyzed degradation of aerated aqueous solutions of chlorobenzene, using a different source of TiO_2 (Degussa P25, BET surface area 50 m^2/g), and with illumination by a 100-W medium-pressure Hg lamp, Matthews[79] obtained complete mineralization to CO_2 and HCl. The improved photooxidation of chlorobenzene, by comparison with the incomplete oxidation reported by Ollis et al.,[19] was explained by the higher surface area and hence better reactivity of the Degussa P25 catalyst.

The three isomers of dichlorobenzene are important intermediates in chemical industry. Due to their toxicity, their occurrence in wastewater is a serious environmental problem. In a careful study of the photochemical transformation of the dichlorobenzenes in aqueous suspensions of ZnO under illumination with a low-pressure Hg lamp (mainly at 360 nm), the primary step involved hydroxylation at the *ortho* or *para* positions with respect to the chlorine atoms. With 1,2- and 1,4-dichlorobenzenes, in which one of these positions is occupied by chlorine atoms, substitution of chlorine by hydroxyl also occurred. The initial rates increased with the concentration of ZnO, reaching a plateau at 2 g/L. Oxygen was required for these phototransformations. From 1,2-dichlorobenzene (0.46 mM), 1,3-dichlorobenzene (0.3 mM), and 1,4-dichlorobenzene (0.15 mM), the main intermediates identified in solution after 10 min of irradiation and 23 to 24% conversion were[84] as shown in Scheme IV.3. In addition, several water-insoluble tetrachloro- and polychloropolyhydroxybiphenyls were identified after dissolution of the ZnO. The initial quantum yields, assuming total absorption of the available light by the ZnO, were evaluated to be 0.046 ± 0.008 for 1,2- and 1,4-dichlorobenzene, and 0.038 ± 0.008 for 1,3-dichlorobenzene. Hydrogen peroxide was produced, along with the photoconversion of the dichlorobenzenes, at initial rates similar to those of the photodegradation. In order to determine to what extent the observed hydroxylation reactions were due either to direct interaction of positive holes on the ZnO surface with the adsorbed aromatic compounds, or to the attack by hydroxyl radicals, the photoconversion was measured in the presence of the

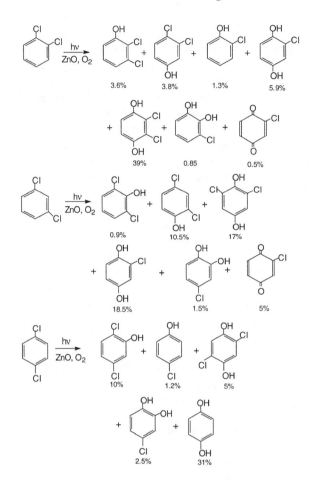

SCHEME IV.3

known ·OH scavenger ethanol. In the presence of ethanol (0.4% v/v, 68 m*M*), the hydroxylation of the dichlorobenzenes was 95% inhibited, and with 1% ethanol, the inhibition was complete. This was suggested to be evidence that hydroxyl radicals were exclusively involved in the primary step of the oxidation.[84]

Improved photocatalysts for the degradation of the toxic 1,4-dichlorobenzene were prepared by chemical precipitation of molybdic and tungstic acids on TiO_2, which after drying at 150°C and calcination at 600°C for 12 h produced MoO_3/TiO_2 and WoO_3/TiO_2 (5 w/o) catalysts. In oxygenated solutions of 2,4-dichlorobenzene containing 0.5 g/L of the photocatalyst, under illumination with a 400-W Hg lamp, the order of dearomatization was $MoO_3/TiO_2 > WoO_3/TiO_2 > TiO_2$. About 93% mineralization was reached after 7 h.[86]

Considerable enhancement in the TiO_2-photocatalyzed degradation of chlorobenzene and of 2,4-dichlorophenol was achieved by adding electron acceptors, such as hydrogen peroxide, ammonium persulfate [$(NH_4)_2S_2O_8$], potassium bromate, and potassium peroxymonosulfate ($2KHSO_5 \cdot KHSO_4 \cdot K_2SO_4$, Oxone®). These additives were considered to act both by scavenging conduction-band electrons, thus preventing the wasteful electron–hole recombination, and also by forming further oxidizing species, such as $\cdot OH$ and $\cdot SO_4^-$ radicals. The reactions were performed in a flow-through reactor in which TiO_2 was coated on fiberglass mesh, under UV illumination (330 to 400 nm), and without provision for aeration or oxygenation. At a flow rate of 3 L/min, in a single-pass mode, with initially 40 ppm chlorobenzene, the percent degradation without additive, or in the presence of 2.8 mM of either H_2O_2, $(NH_4)_2S_2O_8$, $KBrO_3$, or Oxone® was 30, 37, 43, 45, and 85%, respectively. Here Oxone® was clearly the most effective additive. A completely different order of reactivity was observed in the photodegradation of 2,4-dichlorophenol, which in the absence of additive, or in the presence of either $(NH_4)_2S_2O_8$, H_2O_2, Oxone®, or $KBrO_3$ was 23, 35, 42, 80, and 86%, respectively. Here, both the persulfate and the bromate ions were very effective. An even higher degree of degradation in a single-pass mode, to 96%, was attained by using a mixture of $(NH_4)_2S_2O_8$, $KBrO_3$, and Oxone® as additives.[30]

Ceramic membranes of TiO_2 supported on glass were prepared by a sol-gel technique, and were applied in a photoreactor enabling continuous flow-through of gaseous and liquid feeds. Using 3-chlorosalicylic acid as a model of haloorganic compounds, its rate r of mineralization to CO_2 and HCl as a function of the initial concentrations of oxygen [O_2] and 3-chlorosalicylic acid [S] was found to be compatible with a Langmuir–Hinshelwood type equation,[87]

$$r = kK_1[O_2]K_2[S]\big/\big\{\big(1 + K_1[O_2]\big)\big(1 + K_2[S]\big)\big\} \qquad (42)$$

In an attempt to enhance the photocatalytic efficiency of these ceramic membranes with visible light, the TiO_2 was doped with niobium. However, niobium doping did not increase the rate of mineralization of 3-chlorosalicylic acid.[88]

4. Polychlorobenzenes in Surfactant Micelle Solutions

Solubilization of hydrophobic organic pollutants by surfactants followed by photolytic destruction was proposed as a soil remediation strategy. This approach was tested for the photodegradation of polychlorobenzene congeners. Polychlorobenzenes undergo very slow direct photolysis in water, giving rise to the more toxic dibenzofuran compounds. Thus, the photolysis of hexachlorobenzene in distilled water under irradiation at 253.7 nm occurred with an initial lag stage, in which the quantum yield was only $\Phi = 0.0043$, followed by a first-order decay, with $\Phi = 0.058$. The photodechlorination was very much enhanced in surfactant micellar solutions. With micellar solutions of the anionic surfactant sodium dodecyl sulfate, and of the nonionic surfactants Brij35, Brij58, Tween20, and Tween80, the initial quantum yields were 0.29, 0.32, 0.37, 0.50, and 0.51, respectively, without any lag phase. The observed photolysis products included pentachlorobenzene, all possible tetra-, tri-, and dichlorobenzenes, monochlorobenzene, benzene, phenol, and HCl. The mechanism presumably included reductive dechlorination, as well as photohydrolysis and isomerization. The rate of dechlorination was markedly accelerated by high concentrations (20 mM) of the strong reductant sodium borohydride.[89]

5.   Photoinduced Electron Transfer by Organic Anions

One approach to the degradation of chloroaromatic compounds has been by photodechlorination, using electron transfer from anionic sensitizers. Photodechlorination of 2-chloronaphthalene and of 4-chlorobiphenyl was achieved in methanolic solutions of NaOH (0.04 M) containing β-naphthol (2 mM) as sensitizer, by illumination with a 12-W blacklight lamp (mainly at 350 nm). The products identified after 4 h were naphthalene (73%) and biphenyl (66%), respectively, and the turnover numbers were 10 and 3.3%, respectively. The primary excitation step was proposed to be the formation of the naphthoxide singlet state, which was quenched by electron transfer to the haloaromatic compound, which also quenched the fluorescence of the naphthoxide. Both the aromatic radical formed and the naphthoxyl radical react with solvent molecules SH, releasing the hydrocarbon product, and reforming the naphthol sensitizer.[90,91]

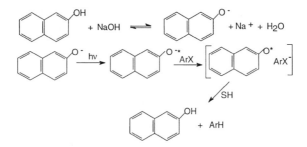

SCHEME IV.4

6.   Chlorobenzoic Acids

The chlorobenzoic acids underwent efficient photodegradation in aerated aqueous suspensions of TiO$_2$ (Degussa P25), leading to complete mineralization, according to

$$Cl-C_6H_4-COOH + 7O_2 \rightarrow 7CO_2 + HCl + 2H_2O \qquad (43)$$

The relative rates of disappearance of the three isomers of chlorobenzoic acid were *para* > *ortho* > *meta.* Identified intermediates included chlorophenols, suggesting that the primary photodegradation steps were decarboxylation and hydroxylation of the aromatic rings.[92]

7.   Haloaromatic Ethers

Haloaromatic ethers have important uses as solvents, pesticides, intermediates in chemical synthesis, and dyestuff production. Many of these ethers are very toxic and are priority pollutants. The direct photolysis of 2-bromoethyl phenyl ether, 2-bromoethyl 4-bromophenyl ether, and 4-bromophenyl ethyl ether (4-bromophenetole) in aqueous solutions or in a water–acetonitrile mixture (1:1) was studied under illumination with a 100-W medium-pressure Hg lamp, in the absence or presence of hydrogen peroxide. The presence of H_2O_2 (2.4 mM) considerably enhanced the rates of degradation, but did not substantially change the distribution of the reaction products. Very many intermediate degradation products were identified, some of which are extremely toxic. These included phenolic compounds, carboxylic acids, and polymers. For the photodecomposition of 2-bromoethyl phenyl ether, the primary excitation step was proposed to result in fission of the carbon–bromine bond, leading to a carbon-centered radical and a bromine atom,[93]

$$C_6H_5-O-CH_2CH_2-Br + h\nu \rightarrow \left[C_6H_5-O-CH_2CH_2-Br\right]^*$$
$$\rightarrow C_6H_5-O-CH_2CH_2\cdot + \cdot Br \tag{44}$$

The bromine atom may abstract a hydrogen atom from a solvent molecule or from another substrate molecule, while the organic radical may abstract a hydrogen atom to form products such as phenetole (phenyl ethyl ether),

$$C_6H_5-O-CH_2CH_2\cdot + RH \rightarrow C_6H_5-O-CH_2CH_3 + \cdot R \tag{45}$$

or it may undergo ring closure to form 2,3-dihydrobenzofuran,

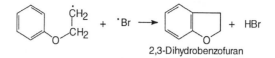

2,3-Dihydrobenzofuran

SCHEME IV.5

In the presence of oxygen, the same radical may be converted to a peroxy radical, which then may undergo hydrolysis to form the observed phenoxy acetaldehyde and the highly toxic phenoxyethanol:

$$C_6H_5-O-CH_2CH_2\cdot + O_2 \rightarrow C_6H_5-O-CH_2CH_2-O_2\cdot \tag{46}$$

$$C_6H_5-O-CH_2CH_2-O_2\cdot + H_2O \rightarrow$$
$$C_6H_5-O-CH_2CHO + C_6H_5-O-CH_2CH_2-OH \tag{47}$$

With 2-bromoethyl 4-bromophenyl ether and with 4-bromophenyl ethyl ether, the preferred initial reaction step was shown to be dehalogenation of the aromatic ring:[93]

$$Br-C_6H_4-O-CH_2CH_2-Br + h\nu \rightarrow \cdot C_6H_4-O-CH_2CH_2-Br$$
$$\rightarrow C_6H_5-O-CH_2CH_2-Br \tag{48}$$

$$Br-C_6H_4-O-CH_2CH_3 + h\nu \rightarrow \cdot C_6H_4-O-CH_2CH_3$$
$$\rightarrow C_6H_5-O-CH_2CH_3 \tag{49}$$

4-Bromodiphenyl ether (4-bromophenyl phenyl ether) in aqueous solution under irradiation with a 100-W medium-pressure Hg lamp ($\lambda \geq 254$ nm) both in the absence and in the presence of hydrogen peroxide underwent direct photolysis. Among the many intermediate products identified, the major compounds were diphenyl ether, phenol, benzene, bromobenzene, a dibromobenzene, dibenzofuran, muconic acid, oxalic acid, malic acid, and glycolic acid, as well as several hydroxylated diphenyl derivatives. In the proposed mechanism, the first step was assumed to be dehalogenation, forming diphenyl ether and *p*-hydroxydiphenyl ether. In the second step, diphenyl ether decomposed to phenol and benzene, while in the third step opening of the aromatic ring led to carboxylic acids and finally to mineralization.[94]

B. Halogenophenols

Chlorophenols are toxic, and since they are resistant to biodegradation, considerable effort has been exerted to develop alternative treatment methods for the destruction of chlorophenol wastes.

1.   Homogeneous Photolysis

The direct photolysis of chlorophenols in aqueous solutions resulted primarily in the replacement of the chlorine atom by a hydroxyl group.[95] The phototransformation of 4-halogenophenols in deoxygenated alkaline aqueous media under illumination with a low-pressure Hg lamp resulted in the formation of hydroquinone, phenol, 2,4′-dihydroxy-4-halogeno-biphenyl and 2,4′-dihydroxy-biphenyl. Among the 4-halogenophenols, the yields of hydroquinone decreased in the order Cl > Br > I. The mechanism of photolysis of the 4-halogenophenols was proposed to proceed by homolytic fission of the carbon–

halogen bond. With 3-chlorophenol, photolysis in alkaline solutions caused the formation of resorcinol in high yield, and the suggested mechanism involved an electrophilic nature for the excited 3-chlorophenol. In aqueous CN⁻ solutions, irradiation of 4-chlorophenol resulted in 4-cyanophenol, while 3-chlorophenol yielded mainly resorcinol, with small amounts of 3-cyanophenol.[96]

The photohydrolysis of 3-chlorophenol to resorcinol under illumination at 254 nm was strongly sensitized by phenol. On the other hand, the photohydrolysis of 4-chlorophenol in aerated aqueous solution to hydroquinone and 5-chloro-2,4′-dihydroxybiphenyl was unaffected by the presence of phenol. By direct excitation 2-chlorophenol was oxidized to a cyclopentadienic acid, in addition to the formation of catechol.[97]

A flash photolysis/high performance liquid chromatography (HPLC) method was applied to study the direct photolysis of halogenophenols. Irradiation with an intense brief light flash resulted in a significant conversion of the pollutant, with only minor secondary reactions, while HPLC enabled separation and identification of the relatively stable reaction products. In aerated aqueous solutions of 4-chlorophenol or of 4-bromophenol, after a single flash with a xenon lamp, the only observed product was *p*-benzoquinone. After multiple flashes, or with much more intense single flashes, hydroxy-*p*-benzoquinone and hydroquinone were also formed:[98,99]

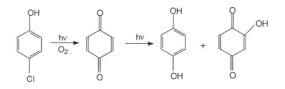

SCHEME IV.6

The presence of oxygen was required for these oxidation reactions, but the source of the new oxygen atom in *p*-benzoquinone was shown to be water and not molecular oxygen.[100]

In another study, unbuffered air-saturated dilute aqueous solutions of 4-chlorophenol (initially 0.2 mM) were irradiated at 296 nm. The initial photoproducts included hydroquinone (about 30%) and benzoquinone, resulting from the direct photoexcitation of 4-chlorophenol. The kinetics of formation of hydroquinone indicated that it was a primary photoproduct. In more concentrated media (2 mM), benzoquinone was the major photoproduct, and the formation of 5-chloro-2,4′-dihydroxybiphenyl was also observed. Prolonged irradiation resulted in the formation of hydroxybenzoquinone. In deoxygenated dilute solutions of 4-chlorophenol (0.2 mM), hydroquinone was the major photoproduct, while in concentrated solutions (2 mM), the main product was 5-chloro-2,4′-dihydroxybiphenyl as well as 2,5,4′-trihydroxybiphenyl. Benzoquinone was not produced in oxygen-free solutions. One of the possible mechanisms to account for the phototransformations of 4-chlorophenol involves its conversion to a triplet state, which then reacts with another molecule of 4-chlorophenol, forming an exciplex. The exciplex in deoxygenated solutions rearranges to form 5-chloro-2,4'-dihydroxybiphenyl, and in oxygenated solutions leads mainly to benzoquinone.[101]

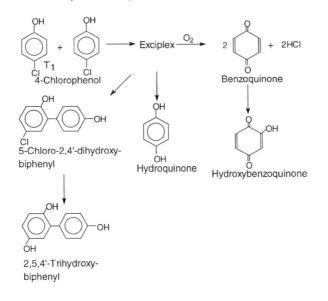

SCHEME IV.7

The minor product 2,5,4'-trihydroxybiphenyl may result from the photohydrolysis of 5-chloro-2,4'-dihydroxybiphenyl. Its production was enhanced by the addition of hydroquinone.[101]

In this direct photolysis of halogenophenols in aerated aqueous solutions, the primary free radicals were identified by EPR (electron paramagnetic resonance), using the spin-trapping technique, with DMPO (5,5'-dimethylpyrroline *N*-oxide) as the spin trap. The only primary radicals detected during the photolysis of chloro- and bromophenol were an aryl radical, presumably the 4-hydroxyphenyl radical, a hydrated electron, which underwent protonation forming the H atom, the hydroxyl radical, and the *p*-benzosemiquinone anion. *p*-Benzoquinone was the only detectable stable primary photoproduct, and its photolysis probably generated the ·OH radical.[98,99,102–105]

a. UV/Ozone on 4-Chlorophenol

While UV light alone had almost no effect on 4-chlorophenol in aqueous solution, and ozone alone caused rather slow oxidation, the combination of ozonation with UV irradiation had a marked synergistic effect, resulting in rapid degradation. Mineralization of the photolysis intermediates from 4-

chlorophenol was slower, and the decrease of TOC (total organic carbon) was first order with respect to the TOC concentration.[106]

b. Reaction Rates with ·OH Radicals

In a kinetic study, the EPR spin-trapping technique was applied to derive rate constants for the reaction of hydroxyl radicals with chlorophenols, by applying competition kinetics, using the known rates of hydroxylation of phenol and of formate. Thus, the secondary rate constants for the reaction of ·OH radicals with phenol, 2-chlorophenol, 3-chlorophenol, 4-chlorophenol, 2,4-dichlorophenol and 3,5-dichlorophenol were 1.4×10^{10}, 1.7×10^{10}, 0.9×10^{10}, 2.5×10^{10}, 2.7×10^{10}, and 0.8×10^{10} M^{-1} s^{-1}, respectively. 3-Chloro-substituted phenols were less reactive than 4-chloro-substituted phenols. These extremely high rate constants exceeded the diffusion-controlled rate constant, 0.7×10^{10} M^{-1} s^{-1}. The explanation given was that ·OH radicals, in analogy to H^+ and OH^- ions, may be propagated through water without actual movement of atoms.[107]

c. Effect of H_2O_2

The rate of photolysis of 2-chlorophenol in acidic aqueous solutions was very much enhanced by the presence of hydrogen peroxide. Under illumination with a 10-W low-pressure Hg lamp ($\lambda > 220$ nm) at pH 3, with initially 7.8 μM 2-chlorophenol and 0.1 mM H_2O_2, the oxidation efficiency of the hydroxyl radicals for 2-chlorophenol was almost unity. On the other hand, in alkaline solutions, at pH 13, the presence of hydrogen peroxide had only a minor effect on the rate of photolysis of 2-chlorophenol. Hence, it was concluded that the photodegradation in alkaline solutions was predominantly due to the direct photolysis, and the oxidation efficiency of the hydroxyl radicals was only 0.05.[108]

In the photodegradation of 4-chlorophenol in the presence of hydrogen peroxide, the primary photochemical reaction was dissociation of the hydrogen peroxide to hydroxy radicals. With a large excess of hydrogen peroxide (20 times that of the chlorophenol), the oxidation products included p-benzoquinone, 4-chlorocatechol, hydroquinone, and 1,2,4-trihydroxybenzene.[102]

d. Sensitization by Hydroquinone

The phototransformation of 2-halogenophenols in acidic aqueous solutions was sensitized by the presence of hydroquinone. When irradiating solutions of 2-fluorophenol, 2-chlorophenol, or 2-bromophenol (0.2 mM) containing hydroquinone (0.2 mM) at 302 nm, the hydroquinone was selectively excited (its λ_{max} is 289 nm, while the λ_{max} of the halogenophenols is about 270 nm). The main photoproduct was catechol. Additional products were cyclopentadiene carboxylic acids (two isomers) and 2,5,2'-trihydroxybiphenyl.[109]

These products were the same as those observed in the direct photolysis at lower wavelength (280 nm), which occurred by a mechanism of heterolytic hydrolysis. In the absence of hydroquinone there was almost no

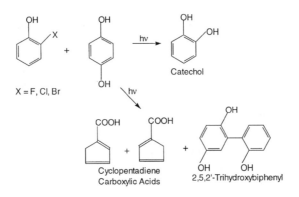

SCHEME IV.8

phototransformation at 302 nm. Quantum yields for the hydroquinone sensitized transformation at pH 4.0 and 302 nm for 2-fluorophenol, 2-chlorophenol, and 2-bromophenol were 0.35, 0.10, and 0.09, respectively. The sensitization by hydroquinone is possible because the energy level of its triplet state is higher than that of the halogenophenols.[109] Similarly, the photolysis of 3-chlorophenol in aqueous solutions at 296 nm was sensitized by hydroquinone, the photoproduct being resorcinol. This reaction was proposed to be due to energy transfer from the excited triplet state of hydroquinone to 3-chlorophenol.[110]

e. Dye-Sensitized Photooxidation of Chlorophenols

The photooxidation of mono- and polychlorophenols in aerated or oxygenated alkaline aqueous solutions was tested in the presence of various dye sensitizers. Riboflavin was found to be inactive,[111] while rose bengal, methylene blue, and eosin were highly effective.[112] The photooxidation of these chlorophenols by visible light ($\lambda > 550$ nm) was shown to be due to a singlet oxygen ($O_2-{}^1\Delta_g$) mechanism, as proven by the suppression of oxidation in the presence of the singlet oxygen inhibitors DABCO (50 mM) or NaN$_3$ (5 mM). Using kinetic competition experiments with known singlet oxygen quenchers, quantum yields (Φ) were derived for the photooxidation of 2-chlorophenol (0.02), 3-chlorophenol (0.0063), 4-chlorophenol (0.027), and 2-chloroanisol (<10^{-4}). The efficiency of photooxidation was thus highly dependent on the structure of the phenolic compound. This suggested that singlet molecular oxygen quenching by chlorophenols involved an intermediate with partial charge transfer,

$$P + {}^1O_2 \leftrightarrow \left[P-{}^1O_2\right] \rightarrow \text{Products} \tag{50}$$

The nature of the final products was not identified.[112] The same mechanism was observed in the dye-sensitized photooxidation of 2,4-dichlorophenol, 2,6-dichlorophenol, and 2,4,6-trichlorophenol, which occurred with a total rate constant of 10^8 to 10^9 mol s^{-1}, and with an estimated quantum yield of 0.01. The high rates suggested the possibility of using singlet oxygen for the decontamination of industrial wastes containing polychlorophenols.[113,114] The dye-sensitized photooxidation of polychlorophenolic pesticides was strongly affected by the presence of surfactants, such as cetyltrimethylammonium chloride (CTAC). This was explained to be due to comicellization of the phenolic compounds.[115]

Using methylene blue as a dye sensitizer, the photodegradation of chlorophenols, from monochlorophenol to pentachlorophenol, was achieved with visible light in a continuous-flow reactor system. The amount of air required for the photooxygenation amounted to about 50 to 80 L/g of chlorophenol. Within 1 h, 90 to 99% of the chlorophenols were degraded, and the chemical oxygen demand (COD) had decreased by 30 to 60%. The excess of unused methylene blue could be removed by adsorption in a column of a natural zeolite.[116]

f. Vacuum-UV Photolysis of 4-Chlorophenol

An Xe excimer lamp emitting at 172 nm was used by Jakob et al.[117] in the oxidative degradation of 4-chlorophenol in aqueous solution. Due to the very high absorbance of water at 172 nm, total absorption of the incident light was reached within less than 1 mm. When pure water was illuminated, hydrogen peroxide was produced, in a concentration which within about 30 min reached a stationary concentration of 0.1 mM. The primary step in the photoexcitation of water in the vacuum ultraviolet is dissociation to hydrogen atoms and hydroxyl radicals.

$$H_2O + h\nu \rightarrow H\cdot + \cdot OH \qquad (51)$$

These radicals can combine in a variety of reactions, some of which lead to hydrogen peroxide, for example,

$$\cdot OH + \cdot OH \rightarrow H_2O_2 \qquad (52)$$

Illumination of aerated aqueous solutions of 4-chlorophenol (0.5 mM) with the excimer light source caused rapid degradation of this compound, 50% of which had disappeared after only 10 min, and practically 100% after 45 min. The rate of this reaction, which observed first-order kinetics, was independent of the presence of oxygen. The release of chlorine atoms was also rapid, and was quantitative within 80 min. The decay of total organic carbon (TOC) was slower and was complete after 180 min, indicating that some intermediate products were formed and destroyed. In the absence of oxygen, these secondary (possibly thermal) reactions were much slower. Since more than 99% of the incident light in such dilute solutions of 4-chlorophenol was absorbed by water, the primary reaction must be the dissociation of water, involving the radical species formed in pure water. The reactions of hydroxyl radicals with 4-

chlorophenol (or with other organic substrates) may then be similar to the reactions of hydroxyl radicals generated from the photolysis of hydrogen peroxide, or during photocatalysis with TiO_2, such as formation of a chloro-cyclohexadienyl radical intermediate. There was no interference in this excimer light source degradation of 4-chlorophenol by the presence of bicarbonate or nitrate ions, which often are components of environmental waters.[117]

g. Fenton's Reagent

Chlorinated phenols may be oxidized even in the dark using Fenton's reagent, an aqueous solution of hydrogen peroxide and ferrous iron salts. The oxidation of tartaric acid by this catalyst system had been discovered by Fenton a century ago,[118] and its mechanism has been shown by Haber and Weiss[119] to involve as a primary step the reduction of hydrogen peroxide, with formation of the hydroxyl radical.

$$Fe^{2+} + H_2O_2 \rightarrow Fe^{3+} + OH^- + \cdot OH \qquad (53)$$

In a kinetic study of the dark reaction of the three monochlorophenol isomers and of five of the six dichlorophenol isomers with the Fenton reagent, both dechlorination and partial mineralization were observed. Monochlorophenols reacted much more rapidly than dichlorophenols, and were also mineralized to a larger extent than dichlorophenols. The rate of oxidation was found to be highly dependent on the initial ferrous iron concentration. Highest rates of mineralization were obtained with a semibatch reactor, in which hydrogen peroxide was injected continuously in order to maintain a steady concentration of the oxidant. However, complete mineralization was not achieved in the dark reaction with Fenton's reagent.[120]

In contrast with the *dark* Fenton reaction, which ceases after the consumption of either the hydrogen peroxide or the ferrous ion, the photo Fenton causes regeneration of the consumed Fe^{2+} ions, with production of additional hydroxyl radicals:

$$Fe^{3+} + H_2O + h\nu \rightarrow Fe^{2+} + \cdot OH + H^+ \qquad (54)$$

The photodegradation of 4-chlorophenol by the photo Fenton reaction was carried out under illumination at $\lambda > 320$ nm, and was found to be accelerated with increasing light intensity. Since the Fe^{2+} ions function as catalysts, their concentration may be small, which is advantageous for practical wastewater treatment. Under illumination with a 400-W Hg lamp, a 1.0 mM solution of 4-chlorophenol containing 10 mM H_2O_2 and 0.25 mM Fe^{2+} was completely mineralized within about 30 min.[121]

Several advanced oxidation processes (AOPs) were compared for the degradation of 4-chlorophenol in aqueous solutions, UV alone, UV/TiO_2, UV/

$\cdot H_2O_2$, $UV/TiO_2/H_2O_2$, and the dark Fenton reaction. Illuminations were made with a 400-W high-pressure Hg lamp ($\lambda > 310$ nm), and with initially 1 mM 4-chlorophenol. The disappearance of 4-chlorophenol by the dark Fenton reaction was complete within a few minutes, but TOC removal reached only 42% after 24 h. The most effective method for TOC removal was $UV/TiO_2/$ H_2O_2, with 2 g/L TiO_2 and 10:1 molar excess H_2O_2 relative to the 4-chlorophenol. This method combined homogeneous and heterogeneous reactions, and resulted in TOC degradation of 80% after 24 h. Direct photolysis (UV alone) caused slow degradation of 4-chlorophenol, while the TOC had scarcely decreased during 24 h.[121–123]

In a further study of the mineralization of 4-chlorophenol, the photo Fenton reaction using Fe^{2+} (initially 0.25 µM) and H_2O_2 (initially 10 mM) was most effective. The order of rates of TOC degradation was $UV/H_2O_2/Fe^{2+} > UV/O_3/$ $Fe^{2+} > UV/O_3 > UV/H_2O_2 = UV/TiO_2$.[124] A direct comparison was also made of the rate of photodegradation of 4-chlorophenol by the photo Fenton reaction with $Fe^{2+}/H_2O_2/UV$, with $Fe^{3+}/O_2/UV$, and with TiO_2 (slurry)/UV. The most rapid degradation occurred by this photo Fenton reaction, in which the 4-chlorophenol was completely mineralized within 30 min. With $Fe^{3+}/O_2/UV$ and with TiO_2/UV, the 4-chlorophenol had mineralized by only 3 and 5% within this period. A drawback of the photo Fenton reaction is that solutions should be acidified to pH < 4, to prevent precipitation of iron hydroxides.[125]

The photochemical generation of hydroxyl radicals from ferric ions at 313 nm was used to determine the rate and quantum yield of hydroxyl radical production in acidic aqueous solutions.[126] The method involved competition kinetics, with 2-chlorophenol as a test compound and 1-octanol as a radical scavenger. The primary step of hydroxyl radical formation depended on the intense light absorption by the Fe(III)–hydroxy complex, $Fe(OH)^{2+}$, at 313 nm ($\varepsilon = 1760$ M^{-1} cm^{-1}):

$$Fe^{3+} + H_2O = Fe(OH)^{2+} + H^+ \tag{55}$$

$$Fe(OH)^{2+} + h\nu \rightarrow Fe^{2+} + \cdot OH \tag{56}$$

Using the known values for the second-order rate constants of reactions of $\cdot OH$ with 2-chlorophenol (1.2×10^{10} M^{-1} s^{-1}) and with 1-octanol (7.5×10^7 M^{-1} s^{-1}), the average rate of hydroxyl radical generation in the pH range 2 to 3 was 0.5×10^{-6} M^{-1} min^{-1}, and the average quantum yield was $\Phi = 0.019$. This quantum yield of hydroxyl radical production from ferric ions is much smaller than the quantum yield of production of $\cdot OH$ by the photodissociation of hydrogen peroxide at 254 nm, in which $\Phi = 1.0$. However, the absorptivity of H_2O_2 at 254 nm is only 20 M^{-1} cm^{-1}, and thus the rate of hydroxyl radical generation is higher in the ferric ion system. In the presence of oxygen, the ferric ion is regenerated from ferrous ion by the "dark" Fenton reaction.[126]

$$Fe^{2+} + O_2 \rightarrow Fe^{3+} + O_2^- \tag{57}$$

$$2O_2^- + 2H^+ \rightarrow H_2O_2 + O_2 \tag{58}$$

$$Fe^{2+} + H_2O_2 \rightarrow Fe^{3+} + \cdot OH + OH^- \tag{59}$$

The rate of photodecomposition of 2-chlorophenol depended on its concentration. Above 0.05 mM 2-chlorophenol, its concentration decreased linearly with irradiation time, indicating zero-order kinetics. Below 0.03 mM 2-chlorophenol, the initial rates of photolysis decreased with lowered concentration. The results could be accounted for by a mechanism in which the primary step is the steady-state formation of hydroxyl radicals by the photodissociation of the Fe(III)–hydroxy complex.[127]

See also Chapter VIII, Section I, on the photolysis of 2-chlorophenol in natural waters.

h.   Chlorophenoxy Herbicides

The chlorophenoxy herbicides 2,4-D (2,4-dichlorophenoxyacetic acid) and 2,4,5-T (2,4,5-trichlorophenoxyacetic acid)

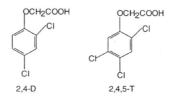

2,4-D   2,4,5-T

SCHEME IV.9

are very widely used, but are a serious environmental problem due to their toxicity, particularly for fish. 2,4-D in aqueous solution was practically stable to direct photolysis at $\lambda > 340$ nm but was slowly photolysed at $\lambda > 220$ nm (26.5 mW/cm^2). The homogeneous photolysis rate was considerably enhanced in the presence of H$_2$O$_2$. The degradation observed pseudo first-order kinetics, and the main observed intermediate was chlorohydroquinone. With initially 0.36 mM 2,4-D and 2.9 mM H$_2$O$_2$, the mineralization was essentially complete within 3 h.[128]

Both the dark and the photo Fenton reactions were applied to the degradation of these chlorophenoxy herbicides. In these reactions, Fe(III) chelates of either picolinic acid, gallic acid, or rhodizonic acid together with hydrogen peroxide were applied to aqueous solutions of the herbicides (at pH 6). The thermal (dark) transformation of the herbicides with these Fe(III) chelates was rapid, with complete chlorine release within several hours. However, mineralization, with loss of TOC, was much slower, and was incomplete even after 24 h. Under UV illumination with blacklight lamps (mainly in the 300 to 400 nm region), mineralization was much accelerated, and using the Fe(III)–picolinate complex was 90% complete within 2 h. With Fe(III) chelates of oxalate and citrate, in the presence of H_2O_2, the thermal reaction on these herbicides was very slow, but under UV illumination the mineralization of 2,4-D was quite rapid. The oxidation and mineralization of these herbicides was thus proposed to occur in two stages. Hydroxyl radicals, both dark- and photogenerated, are effective in dechlorination and in the conversion of the first 40% of the ring and carboxy carbon atoms of 2,4-D to carbon dioxide. In the second stage, which seemed not to involve hydroxyl radicals, the mineralization of the remaining 60% of the carbon atoms occurred by photolysis of Fe(III) complexes.[129-131] The explanation for the enhancement of these reactions by UV light was photodissociation of the Fe(III) chelate by ligand-to-metal charge-transfer excitation:[132]

$$L\text{-}Fe(III) + h\nu \rightarrow Fe^{2+} + L^* \tag{60}$$

followed by the "dark" Fenton reaction of Fe(II) with hydrogen peroxide,

$$Fe^{2+} + H_2O_2 \rightarrow Fe^{3+} + OH^- + \cdot OH \tag{61}$$

See also Section II.B.2 on the heterogeneous photodegradation of 2,4-D in slurries of TiO_2.

2. Heterogeneous Photodegradation

The heterogeneous photocatalysis enabled the degradation of halocarbon derivatives even by illumination with near-UV light, which is ineffective for direct photolysis. Thus, 3-chlorophenol in dilute aqueous solution was stable to irradiation at 365 nm, but was 70% destroyed when irradiated in the presence of ZnO (2 g/L).[83]

a. *1,2,4-Trichlorobenzene*

Irradiation of the toxic 1,2,4-trichlorobenzene, adsorbed on TiO_2 (0.25 g/L) and suspended in water, with simulated AM1 sunlight ($\lambda > 340$ nm), resulted after 6 min in extensive decomposition of the trichlorobenzene, with the intermediate formation of three isomeric trichlorophenols, 2,3,6-trichlorohydroquinone, dichlorophenols, and di- and trichloroquinones, as well as traces of the potentially very toxic polyhydroxypolychlorobiphenyls. Dibenzodioxine derivatives were not found. More prolonged irradiation, for 2 h, resulted in complete mineralization, with disappearance of all these intermediates.[133]

b. Chlorophenols

While the direct photolysis of 2-chlorophenol and 3-chlorophenol in acidic aqueous solution was essentially a photohydrolysis with C–Cl fission, leading to catechol (1,2-dihydroxybenzene) and to resorcinol (1,3-dihydroxybenzene), respectively,[83] the TiO_2-photocatalyzed degradation of both chlorophenols led initially to chlorohydroquinone and hydroxyhydroquinone and in the case of 2-chlorophenol also to traces of catechol.[134]

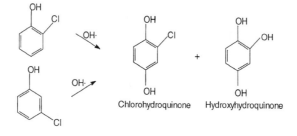

Chlorohydroquinone   Hydroxyhydroquinone

SCHEME IV.10

In the photocatalytic oxidation mediated by TiO_2, electrophilic reactions of ·OH radicals with the aromatic ring resulted in *ortho*- and *para*-hydroxylation, leading to 3-chlorocatechol and chlorohydroquinone. The photoexcitation of 2-, 3-, and 4-chlorophenol in dilute aqueous solutions with suspensions of ZnO (2 g/L) by near-UV light (mainly 365 nm) also led to *ortho* and *para* hydroxylation as the major reactions. If the *ortho* and *para* positions were occupied by chlorine atoms, dechlorination occurred, with replacement by hydroxyl groups. This preferential orientation with respect to the phenol function occurred because these sites present the highest electron density. From 2-chlorophenol (initially 0.55 mM), the intermediate products identified after 10 min of irradiation and 14.5% conversion were chlorohydroquinone (49%), chlorobenzoquinone (6.5%), catechol (1%), and 3-chlorocatechol (6%). From 3-chlorophenol after similar exposure, resulting in 30% conversion, the main photoproducts were chlorohydroquinone (32%), 4-chlorocatechol (10%), 3-chlorocatechol (9%), and chlorobenzoquinone (8%). 4-Chlorophenol after 30% conversion formed hydroquinone (9%), 4-chlorocatechol (16%), and benzoquinone (1.4%). In the presence of ethanol (0.08% v/v, 14 mM), the ZnO-photocatalyzed transformation of the chlorophenols was about 60 to 70% inhibited. Ethanol is known as an effective scavenger of OH radicals. It was therefore concluded that in the photocatalyzed transformation of the chlorophenols there participate two pathways, one involving hydroxyl radicals and the other possibly due to direct action of positive holes on the ZnO surface.[84,85]

Prolonged direct photolysis of 2-chlorophenol yielded cyclopentadienic acid. On the other hand, the semiconductor photocatalyzed reaction eventually achieved complete mineralization. The presence of oxygen was required for the photocatalytic reaction, for which the main pathway was *para*-hydroxylation. The dependence of the rate of the photocatalytic reaction on light intensity was linear below a radiant flux of about 20 mW cm^{-2}, but became proportional to the square root of the radiant flux at higher light intensity. Thus, the quantum yield obviously decreased with higher light flux, presumably due to recombination of the photoproduced charge carriers in the TiO$_2$.[134]

See also Chapter I, Section IV.B, on the effect of light intensity on the photodegradation of 4-chlorophenol.[135]

c. 4-Chlorophenol

In a comparison of TiO$_2$ (Degussa P25, mainly anatase), ZrO$_2$, and MoO$_3$ as photocatalysts for the degradation of 4-chlorophenol, TiO$_2$ was found to be the most effective. A rough estimate for the initial quantum yield in the TiO$_2$-photocatalyzed destruction of 4-chlorophenol (initially 0.15 m*M* aqueous solution, pH 3 to 6) was 0.01. In contrast to the direct photolysis of 4-chlorophenol, which was slow and incomplete and produced a large variety of relatively stable intermediates, including the toxic dihydroxy-, trihydroxy-, and chlorodihydroxybiphenyl derivatives, the photocatalytic degradation was much faster and led to complete mineralization. The photocatalytic process with TiO$_2$ was studied by illumination at $\lambda \geq 340$ nm in order to identify the photocatalytic reactions without interference by the direct homogeneous photolysis of 4-chlorophenol. Doping TiO$_2$ with Cr^{3+}, or surface garnishing with Pt particles (0.5 wt% Pt/TiO$_2$), caused decreased photocatalytic activity compared with undoped and "bare" TiO$_2$. The detrimental influence of Cr^{3+} was explained by an "inner filter" effect of the chromium ions. The inhibition by Pt could be due to electron transfer from TiO$_2$ to Pt, thus competing with electron transfer to oxygen, and hence causing decreased production of ·OH radicals. The initial rates r_0 of disappearance of 2-chlorophenol versus its initial concentrations c_0 fitted the Langmuir–Hinshelwood relationship, $r_0 = kKc_0/(1 + Kc_0)$, yielding values of the reactivity constant k = 6.2 µmol/h and the adsorption constant K = 1.66×10^4 L mol^{-1}.[125] This adsorption constant was an order of magnitude larger than the "dark" adsorption constant of 4-chlorophenol (see also Chapter I, Section IV.B).[137] The discrepancy was explained to be due possibly to a significant effect of photoadsorption, or due to occurrence of some of the reaction in the double layer. In the range of light fluxes of 2 to 5 mW cm^{-2}, the initial rate of photodegradation of 4-chlorophenol was linearly related to the square root of the radiant flux. This effect shows that the recombination of the electron–hole pairs significantly interfered with the oxidation of adsorbed 4-chlorophenol.[136]

The photomineralization of 4-chlorophenol sensitized by Degussa P25 TiO$_2$ in oxygen saturated aqueous solution was described earlier (Chapter I, Section IV.H) as a proposed standard test system to facilitate comparison with the photodegradation reactions of other pollutants.[138,139] The intermediates in this mineralization were found by Mills et al.[140] to be 4-chlorocatechol (4-CC),

hydroquinone (HQ), benzoquinone and 4-chlororesorcinol. These were produced by at least three parallel routes, leading initially to either 4-CC, HQ, or
an unidentified unstable (presumably short-lived) intermediate, with relative
probabilities of 48, 10, and 42%.

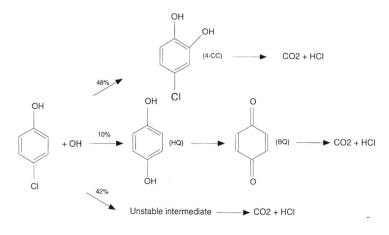

SCHEME IV.11

The unidentified species that eluted in HPLC analysis at the solvent front
was considered to be some aliphatic carboxylic acid. The ring opening was
proposed to occur via chloro-dihydroxycyclodienyl radicals formed by the
initial oxidation of the chlorophenol by hydroxyl radicals. The decrease in the
concentration of 4-CP, and the rise and fall in the concentrations of the
intermediates CC and HQ, as well as the appearance of CO_2, could be fitted to
a model using a Langmuir–Hinshelwood type equation.[140]

Stafford et al.[141,142] compared the oxidation of 4-chlorophenol in aqueous
solution by TiO_2-promoted photocatalysis with the γ-radiolytic degradation.
Water radiolysis yields as main transient intermediates ·OH, e^-_{aq} and ·H radicals. In order to identify the radiolytic oxidation by ·OH radicals, radiolysis
was made in N_2O saturated solutions, in which the hydrated electron was
converted to the hydroxyl radical:

$$e^-_{aq} + N_2O + H_2O \rightarrow N_2 + \cdot OH + OH^- \tag{62}$$

The main products thus obtained by ·OH attack on 4-chlorophenol were 4-
chlorocatechol and hydroquinone. γ-Radiolysis could also be carried out under
reducing conditions, in the presence of *tert*-butanol, which scavenges ·OH
radicals according to

$$\cdot OH + \left(CH_3\right)_3 OH \rightarrow \cdot CH_2\left(CH_3\right)_2 COH + H_2O \tag{63}$$

The only major product under these reducing conditions was phenol. Direct electron-transfer oxidation of 4-chlorophenol was carried out by radiolysis in the presence of sodium azide. Azide ions scavenge ·OH radicals, forming azide radicals, which then abstract electrons from 4-chlorophenol (4-CP):

$$N_3^- + \cdot OH \rightarrow \cdot N_3 + OH^- \tag{64}$$

$$\cdot N_3 + 4\text{-CP} \rightarrow N_3^- + (4\text{-CP})^+ \tag{65}$$

The result of this direct electron-transfer oxidation was an unidentified mixture of many nonaromatic fragments. Pulse radiolysis indicated the formation of a transient species assigned as a 4-chlorophenoxy radical. In the photocatalytic oxidation of 4-chlorophenol over TiO_2, the main intermediates were also 4-chlorocatechol and hydroquinone, indicating a common ·OH-mediated pathway, as in γ-radiolysis. However, in the photocatalyzed degradation, the concentrations of the intermediates were lower, and the yield of hydroquinone relative to that of 4-chlorocatechol, particularly at high TiO_2 loading, was higher than in γ-radiolysis. This suggested that in the photocatalysis, the oxidation products were due not only to the solution-phase ·OH reactions, but also to reactions on the semiconductor surface. The direct oxidation of substrates by photogenerated holes on the TiO_2 surface is a possibility that may therefore not be discounted. The analogy in γ-radiolysis is the direct oxidation of 4-chlorophenol by azide radicals, shown to occur by electron transfer. It was thus proposed that the degradation of 4-chlorophenol in aqueous dispersions of TiO_2 occurred by both hydroxyl radical and direct hole oxidation processes. At increased TiO_2 loading, enabling increased adsorption of 4-chlorophenol, the contribution of the direct reaction with surface holes increased.[141]

In order to enhance the photocatalytic activity for the photodegradation of haloorganic compounds, the effect of metal doping on thin-film catalysts of TiO_2 was tested. The TiO_2 thin films containing Cu, Ni, Co, and Mn were prepared by a sol-gel process and were coated on glass plates. Using a flow system under illumination by a solar simulator or with sunlight, with 4-chlorophenol as a model pollutant, the highest rates of photodegradation were observed with Cu- or Mn-doped TiO_2.[143] An aerated aqueous suspension of Ag-loaded TiO_2 (optimally 0.25 g/L of Ag–TiO_2) in the presence of H_2O_2 (0.2 mM) was used in an open upflow reactor, with irradiation by three 125-W high-pressure Hg lamps, to decompose phenol, 2,4-dichlorophenol, 2,3,5-trichlorophenol, and pentachlorophenol. At a flow rate of 145 mL min^{-1} and initial concentrations of 0.1 mM for these phenols, the half-lives for their destruction were 14, 16, 18.5, and 19.5 min, respectively.[144]

Enhanced efficiency in the TiO_2/UV reaction was achieved by illumination through fused-silica fibers coated with immobilized TiO_2. The quantum efficiencies estimated for the photodegradation of 4-chlorophenol with this reactor and with a slurry of TiO_2 were 2×10^{-4} and 5×10^{-5}, respectively.[125]

See also Chapter I, Section V.A, on the photomineralization of 4-chlorophenol on TiO_2 immobilized on fused-silica glass fibers,[145] and Chapter V, Section III.B, on synergism in the photodegradation of mixtures of 4-chlorophenol and 4-nitrophenol in aqueous slurries of TiO_2.[146]

d. Fluorophenols

The three monofluorophenol isomers and difluorophenols were photodegraded by irradiation in aqueous suspensions of TiO_2, resulting in quite rapid mineralization to CO_2 and HF. Identified intermediates included fluorohydroquinone, fluorocatechol, and catechol.[147]

e. Hammett Correlation in p-Substituted Phenols

In a kinetic study of the TiO_2-photocatalyzed degradation of a variety of *para*-substituted phenols in oxygen-saturated aqueous solutions, the initial rates were consistent with the Langmuir–Hinshelwood model. For phenol itself and for 4-fluoro-, 4-chloro-, 4-bromo-, and 4-iodophenol, both the apparent reaction rate constants k and the adsorption equilibrium constants derived from the Langmuir–Hinshelwood equation were linearly related to the Hammett substituent constants σ. This suggested that the photodegradation of the monohalogenophenols proceeded by a similar mechanism, possibly through a positively charged reaction center. The correlation with the Hammett substituent constant failed for the nonhalogenated phenols studied: 4-methoxyphenol, *p*-cresol, 4-hydroxyacetophenone, α,α,α-trifluoro-*p*-cresol, and 4-cyanophenol.[148]

f. 2,4-Dichlorophenoxyacetic Acid (2,4-D)

In the presence of TiO_2 (2.5 g/L) in aerated aqueous suspensions, under illumination with a 125-W high-pressure Hg lamp ($\lambda > 340$ nm), 2,4-D (initially 80 mg/L) photodecomposed with a lower limit of the quantum yield of 0.03. The half-life of 2,4-D irradiated at $\lambda = 290$ nm in the presence of TiO_2 was 60 times shorter than by homogeneous photolysis.[149] The 2,4-D disappeared within about an hour, but complete mineralization required several hours. The aliphatic moiety of the molecule was readily split from the aromatic ring, presumably by a photo Kolbe reaction, releasing 2,4-dichlorophenol. Other intermediates identified included 2,4-dichloromethoxybenzene, chlorohydroquinone, chloro-*p*-benzoquinone, 4-chlorocatechol, hydroxyquinone, 2,4-dichlorophenyl formate, 3,5-dichlorocatechol, and formate.[128,150]

In an effort to enhance the rate of the photocatalytic degradation of 2,4-D, the activity of bare TiO_2 was compared with TiO_2-supported MoO_3 and WoO_3 (5 wt%). The bare TiO_2 was far superior to the molybdenum- or tungsten-coated photocatalysts. The rates of photodegradation were much higher in oxygen-saturated than in aerated suspensions. With TiO_2 (0.5 g/L) in oxygen-ated solutions of 2,4-D (initially 200 ppm) under illumination with a 400-W Hg lamp, mineralization was 93% complete after 7 h.[86]

g. Effects of Peroxydisulfate and Periodate

A considerable enhancement of the photodegradation of several haloaromatic compounds in aqueous dispersions of TiO_2 was observed in the presence of the inorganic oxidants peroxydisulfate and periodate.[151] In the photooxidation of 2-chlorophenol, the presence of the peroxydisulfate or periodate ions caused very rapid mineralization. The initial step was proposed to be formation of the cyclohexadienyl radical, followed by further hydroxylation, ring opening, and decarboxylation of organic acid intermediates. Peroxydisulfate may react either by direct one-electron oxidation of organic substrates or by its photolytic or thermal dissociation to the sulfate radical anion, which may oxidize water to the hydroxyl radical:

$$S_2O_8^{2-} \rightarrow 2SO_4 \cdot \tag{66}$$

$$SO_4 \cdot + H_2O \leftrightarrow \cdot OH + SO_4^{2-} \tag{67}$$

The effect of periodate in accelerating the TiO_2-photocatalyzed degradation of 2-chlorophenol may be due to the scavenging of the conduction-band electrons:[151]

$$IOH_4^- + 8e^- + 8H^+ \rightarrow 4H_2O + I^- \tag{68}$$

h. Di- and Tri-Chlorophenols

Using time-resolved diffuse reflectance spectroscopy, the radical intermediate formed upon band-gap irradiation of TiO_2 (Degussa P25) in aqueous solutions of 2,4,6-trichlorophenol was identified as the phenoxyl radical. This was in contrast to the radical species formed by reaction of the freely solvated hydroxyl radical in homogeneous solution by pulse radiolysis, which reacted with 2,4,6-trichlorophenol to form the cyclohexadienyl radical. It was thus proposed that in photocatalyzed reactions on TiO_2, oxidation often occurred by direct electron transfer between the substrate and positive holes.[152]

Very effective decomposition of 2,4-dichlorophenol in aqueous suspensions of TiO_2 was obtained by illumination with a low-pressure Hg lamp, at 254 nm. The rate of removal of 2,4- dichlorophenol increased with increasing TiO_2 concentration, reaching a limiting value at a catalyst loading of 1.4 g/L. Almost total disappearance of 2,4-dichlorophenol was obtained within 4 to 5 h of illumination, but for complete mineralization of the reaction intermediates, about 10 h was required.[153]

In a kinetic study of the photodegradation of six dichlorophenols and three trichlorophenols in aqueous suspensions of TiO_2, D'Oliveira et al.[154] related the degradability of the various isomeric chlorophenols to their molecular structure, as expressed in their Hammett substituent coefficients σ and their 1-octanol/water partition coefficient K_{ow}.[154] The partition coefficient K_{ow} was taken as a parameter related to the distribution of the organic compounds between water and a nonaqueous phase (such as TiO_2). For the three monochlorophenols, five dichlorophenols, and two trichlorophenols, the apparent first-order rate constants of the disappearance k (h^{-1}) were found to be represented by[155]

$$k = -10\sigma + 5.2 \log K_{ow} - 7.5 \qquad (69)$$

The main photodegradation pathways were elucidated by identification of the aromatic intermediates, and indicate preferable substitution of hydroxyl radicals on sites of higher electron density on the aromatic ring. Thus, these phenols were primarily substituted at the *ortho* and *para* positions, replacing either chlorine or hydrogen atoms at these positions.[155]

As an example of the pathways observed during the photolysis of di- and trichlorophenols, the photodegradation of 2,6-dichlorophenol in the presence of suspended TiO_2 and oxygen is shown in Scheme IV.12 which shows the percentages of the intermediates formed after 63% of the initial 2,6-dichlorophenol had disappeared,

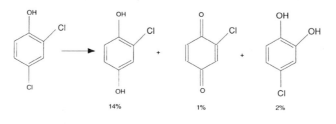

14%   1%   2%

SCHEME IV.12

The lifetime of these intermediates was much shorter than that of the initial chlorophenols, and after 2.5 to 8 h all the compounds had decomposed completely to carbon dioxide and chloride ions.[154]

i.   CdS as Photocatalyst

Cadmium sulfide is a small-band-gap semiconductor absorbing strongly in the visible region. As a photocatalyst for water purification, it has the advantage of excellent overlap with the visible part of the solar spectrum. The severe drawbacks are the toxicity of CdS (a suspect cancer agent), and its sensitivity to corrosion and dissolution to the very toxic cadmium salts. A study of the photodegradation of several chlorophenol derivatives in aqueous solution in the presence of CdS indicated that the Cl removal was faster at higher chlorine substitution. The rates of oxidation decreased in the order pentachlorophenol > 2,4,6-trichlorophenol > 2,4-dichlorophenol, 2-chlorophenol, phenol. Chlorine removal was enhanced at lower pH.[156,157] The photooxidation rates of substituted phenols do not correlate with those expected for electrophilic substitution. The important effect of adsorption was shown by a correlation with the octanol–water partition coefficient, as well as with the dependence of the surface charge on pH.[158]

j. Pentachlorophenol

The highly toxic pentachlorophenol is widely used as a wood preservative, and it appears also in the wastewaters of the paper pulp industry. Pentachlorophenol in aqueous solution (0.45 μM) was only very slowly degraded by illumination at 350 nm, but underwent quite rapid direct photolysis in the presence of oxygen at 254 nm, with complete mineralization after 45 min. Even more rapid photodegradation was observed at 350 nm in the presence of TiO_2 (3 g/L) and O_2. The oxygen could be replaced by hydrogen peroxide (1% by volume). The observed intermediates of the photocatalytic degradation, tetrachlorocatechol and tetrachlororesorcinol, were also rapidly photooxidized in the presence of TiO_2. This direct comparison of UV photodegradation of pentachlorophenol indicated an advantage for the TiO_2–O_2 system over the H_2O_2–O_2 system if the oxygen was continuously fed through the reaction mixture.[159]

Another study reported that under irradiation for 5 h at 254 nm (from a 15-W low-pressure Hg lamp), aqueous solutions of pentachlorophenol at pH 8 were condensed and partly dechlorinated to even more toxic products, including octachlorodibenzo-*p*-dioxin (OCDD), 1,2,3,4,6,7,8-heptachlorodibenzo-*p*-dioxin (HpCDD), octachlorodibenzofuran (OCDF), and 1,2,3,4,6,7,8-heptachlorodibenzofuran (HpCDF).[160]

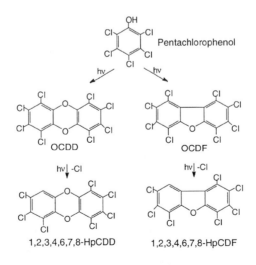

SCHEME IV.13

The photocatalytic degradation of pentachlorophenol was studied in aqueous slurries of several semiconductor materials: TiO_2, ZnO, CdS, WO_3, and SnO_2. Of these, TiO_2 was most effective, followed by ZnO and CdS. WO_3 and SnO_2 were quite ineffective. With 2 g/L TiO_2, in the presence of oxygen, at initially 12 ppm pentachlorophenol, under simulated sunlight exposure ($\lambda >$ 310 nm), the half-lives were 20 and 15 min when the initial pH was 3.0 and 10.5, respectively. Complete mineralization into CO_2 and HCl was achieved by long-term irradiation, 10 to 15 h. Under natural sunlight illumination, with initially 45 mM pentachlorophenol, the half-life was about 8 min.[161]

In a detailed study of the photocatalytic degradation of pentachlorophenol in air-saturated aqueous solutions containing suspended TiO_2 (0.2 g/L) at pH 5, illumination was carried out with filtered light (330 to 370 nm) from a 450-W Xe lamp. The initial quantum efficiency (rate of degradation/incident light flux) was 1.3%. The photodegradation during the first half-life followed zero-order kinetics. The disappearance of pentachlorophenol was complete within 3 h of illumination. The primary detected intermediates were *p*-chloranil (tetrachloro-1,4-benzoquinone), tetrachlorohydroquinone, *o*-chloranil, and hydrogen peroxide, while formate and acetate were observed after more prolonged irradiation. The quantum efficiencies for chloride ion release from pentachlorophenol were independent of the incident light intensity at low light intensities, but decreased at higher light intensities. The mechanism proposed involved mainly oxidation of pentachlorophenol by surface-bound hydroxyl radicals, and in addition direct electron transfer to a surface-trapped hole, forming the pentachlorophenoxyl radical (which had previously been identified by pulse radiolysis). The formation of *p*-chloranil and tetrachlorohydroquinone was explained by preferred orientation of ·OH attack on the *para* position of pentachlorophenol.[162]

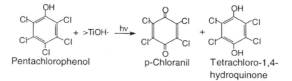

Pentachlorophenol   p-Chloranil   Tetrachloro-1,4-hydroquinone

SCHEME IV.14

See also Section II.C on the formation of polychlorinated dibenzo-*p*-dioxins from pentachlorophenol on fly ash and Chapter VIII, Section V, on the photodegradation of pentachlorophenol in a large-scale solar plant.

k.   Rose Bengal

The photodegradation of the haloaromatic dye rose bengal in aerated aqueous dispersions of TiO_2 or ZnO resulted in bleaching of the dye. This was ascribed to charge transfer between the surface-adsorbed dye molecules and the semi-conductors. Electron injection from the photoexcited dye molecules into the conduction band of these n-type semiconductors caused formation of surface-trapped electrons, which then were scavenged by oxygen molecules, while the semioxidized dye molecules underwent further degradations.[163]

$$D + h\nu \rightarrow D^* \tag{70}$$

$$D^* \rightarrow D^+ + e^- \tag{71}$$

$$e^- + 2O_2 \rightarrow O_2^- \tag{72}$$

3. TiO₂ on Photoelectrodes

As noted earlier (Chapter I, Section E.5), TiO_2 particulate electrodes, prepared by sintering TiO_2 particles on a conducting glass plate, were found to be very effective for the photocatalytic degradation of 4-chlorophenol. An aqueous suspension of TiO_2 was coated on a conducting glass plate and fired at 400°C for 1 h, producing an optically transparent electrode (OTE), on which the semiconductor thin film adhered strongly. With such an electrode, at a bias potential of +0.6 V (vs. SCE) under illumination at >300 nm, the photodegradation of 4-chlorophenol in aqueous solution was greatly acceler-ated relative to the rate of decomposition with the same electrode in the absence of an external bias, or relative to that in aqueous suspensions of TiO_2.[164] In another study, a high-surface-area TiO_2 anode was used, prepared by coating and baking 14 layers of $TiCl_4$ dissolved in ethanol on conducting glass. The photoelectrochemical oxidation of 4-chlorophenol (2.6 µM, in O_2-saturated aqueous solution, in a single-compartment cell) was achieved at an anodic bias of 0.8 V (vs. SCE), under illumination with a 500-W Xe lamp. Within 7.5 h, the chlorophenol was completely mineralized to CO_2 and HCl. Since the photocurrents observed were independent of the concentration of 4-chlorophenol at the micromolar level but increased linearly with the light power, it was proposed that the primary reaction step involved hydroxyl radicals formed by excitation of the TiO_2 electrode. These radicals then at-tacked the chlorophenol.[165] For separation of the anodic and cathodic processes in the photoelectrochemical degradation of 4-chlorophenol, a two-compart-ment cell was used, with a fine glass frit as separator, an optically transparent electrode (OTE) of nanocrystalline TiO_2 on conducting glass as photoelectrode, and a platinum gauze counterelectrode. With oxygen saturated solutions in both compartments, 4-chlorocatechol was degraded even in the absence of an external bias (open circuit conditions), but the rate of degradation was mark-edly enhanced by an anodic bias of 0.6 V on the photoelectrode. In oxygen-saturated solutions, 4-chlorocatechol was the predominant intermediate, which also degraded rapidly in oxygenated solutions under anodic bias. With nitrogen-saturated solutions in the anodic compartment, hydroquinone was the primary intermediate. For further rapid oxidation of the hydroquinone produced, the presence of oxygen was required.[166]

The kinetics of the photoelectrochemical oxidation of chloroform was studied at an n-type single-crystal rutile TiO_2 (001 face) rotating-disc electrode (RDE), together with oxygen reduction at a Pt counterelectrode. It was concluded that interfacial charge transfer, and not the production of electron–hole pairs, was the rate-limiting process at the TiO_2 surface. Therefore, it may in principle be possible to increase the efficiency of photocatalytic degradation of toxic wastes by catalyzing either the electron transfer (using more efficient electron acceptors than oxygen) or the hole transfer.[167]

C. Polychlorinated Dioxins, Dibenzofurans, and Biphenyls

Polychlorinated dibenzo-*p*-dioxins (PCDDs), polychlorinated dibenzofurans (PCDFs), and polychlorinated biphenyls (PCBs) are released in considerable amounts into the atmosphere from car exhausts and by incineration of industrial and municipal wastes.[168] These very toxic compounds are found in high levels in fly ash, as well as in sewage sludge.[169] Experiments on the production of PCDDs were made under conditions of simulated municipal solid waste incineration. Very rapid formation of PCDD compound formation occurred on fly ash from pentachlorophenol as a precursor, at a rate that was four orders of magnitude higher than the *de novo* synthesis from activated charcoal, with air, inorganic chloride, and Cu(II) as catalyst. Both types of reactions were performed at 250 to 350°C.[170]

One of the sources of PCDDs in the environment is the photochemical conversion of chlorophenols. Pentachlorophenol (PCP), which is widely used as a fungicide and wood preservative, undergoes photoinduced condensation to octachlorodibenzo-*p*-dioxin (OCDD). The reaction was studied under natural or simulated sunlight with PCP-treated wood.[171]

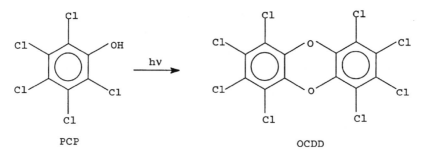

SCHEME IV.15

See also Section II.B on the formation of PCDDs and PCDFs during the direct photolysis of pentachlorophenol-containing water.

Polychlorinated biphenyls (with the trade name Aroclor) are highly toxic ubiquitous contaminants of water, soils, and sediments and are quite recalcitrant to biodegradation and direct solar photodegradation. Incineration, unless very carefully controlled, leads to their partial oxidation to the even more toxic polychlorinated benzofurans and dioxins, while landfill disposal leads to the pollution of groundwater. An example of the danger of PCBs is an accident in a warehouse in Canada in 1988, in which 1500 barrels of PCB-containing oil caught fire, releasing toxic fumes — costing several million dollars for cleanup of the site and the surrounding soil.[172]

The toxicity of these lipophilic dioxin-like compounds, such as PCDDs, PCDFs, and PCBs, is considered to be due to their interaction with the intercellular Ah (aryl hydrocarbon) receptor. The symptoms of dioxin poisoning include a severe wasting syndrome, chloracne, reproductive abnormalities, immune system suppression, and carcinogenicity. The toxicity varies considerably among the different congeners of the dioxin-like compounds.[173] Among the PCDDs, 2,3,7,8-tetrachlorodibenzo-*p*-dioxin is considered to be the most toxic.[174]

1. Homogeneous Photolysis

In a study of 22 isomeric tetrachloro-dibenzo-*p*-dioxins (TCDD), the photolytic half-lives were determined for their degradation in isooctane solutions and on soft glass surfaces. Half-lives varied from 57 min for the 2,3,7,8-TCDD to ~8400 min for the 1,4,6,9-TCDD isomer. The identification of the molecular structure of the different isomers was facilitated by a pattern recognition technique. The nomenclature used was the *Chemical Abstracts* numbering system for polychlorinated dibenzo-*p*-dioxins.[175]

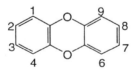

SCHEME IV.16

Octachlorodibenzo-*p*-dioxin (OCDD) adsorbed on soil underwent stepwise photolytic dechlorination, leading to 2,3,7,8-tetrachlorodibenzo-*p*-dioxin (TCDD) and to 1,2,3,7,8-pentachlorodibenzo-*p*-dioxin, as well as to three isomeric hexachlorodibenzo-*p*-dioxins.[176] A process for the sunlight photodegradation of soil containing 2,3,7,8-dibenzo-*p*-dioxin and spiked with the ^{14}C-labeled dioxin was demonstrated. These experiments were carried out in standard 100-ml glass beakers filled with loamy sand, providing 6-cm-deep soil columns. A solvent mixture of tetradecane/1-butanol (2:1) was added, so that the starting solvent on the soil was water (4%), tetradecane (8%), and 1-butanol (4%). During sunlight exposure, evaporation on the top surface of the soil resulted in a continued upward movement of the dissolved dioxin, and partial degradation. During 60 d of sunlight exposure, the disappearance of the starting dioxin reached up to 85%. The ^{14}C-activity measurements indicated partial conversion to ethanol-soluble and ethanol-insoluble products.[177]

In aqueous solutions, PCDDs undergo only very slow photodegradation. Quantum yields for the direct homogeneous phase phototransformation at 313 nm in water–acetonitrile (2:3 v/v) for the four PCCCs 1,2,3,7-T$_4$CDD (1,2,3,7-tetrachlorodibenzo-*p*-dioxin), 1,3,6,8-T$_4$CDD, 1,2,3,4,6,7,8-H$_7$CDD, and 1,2,3,4,6,7,8,9-O$_8$CDD were 5.4×10^{-4}, 2.2×10^{-3}, 1.5×10^{-5}, and 2.3×10^{-5}, respectively. Estimated sunlight midsummer half-lives in surface natural waters were 1.8, 0.31, 47, and 18 d, respectively.[178–180]

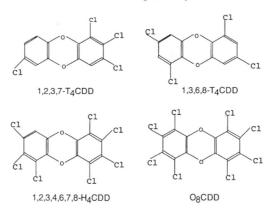

1,2,3,7-T$_4$CDD   1,3,6,8-T$_4$CDD

1,2,3,4,6,7,8-H$_4$CDD   O$_8$CDD

SCHEME IV.17

The photolysis of 2,3,7,8-tetrachlorodibenzo-*p*-dioxin in isooctane solution yielded in addition to 2,3,7-trichlorodibenzo-*p*-dioxin (2,3,7-TrCDD) also a rearrangement product, 4,4′,5,5′-tetrachloro-2,2′-dihydroxybiphenyl.[174]

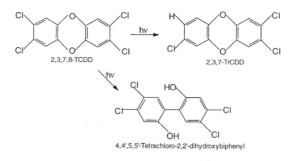

2,3,7,8-TCDD   2,3,7-TrCDD

4,4′,5,5′-Tetrachloro-2,2′-dihydroxybiphenyl

SCHEME IV.18

The direct photolysis of 2-chlorodibenzo-*p*-dioxin and 1,2,4-trichlorodibenzo-*p*-dioxin was studied in acetonitrile–water (1:1, v/v) and methanol–water (1:1, v/v) solutions.[181] The organic cosolvents were necessary because of the poor solubility of these compounds in water. Methanol serves as a good hydrogen atom donor, while acetonitrile is quite resistant to hydrogen abstraction. Under illumination either at 308 nm (monochromatic light from an Hg–Xe lamp) or at 222 nm (from a XeCl excimer lamp), the photolysis of 2-chlorodibenzo-*p*-dioxin resulted in a complex mixture of products, with dibenzodioxin as the major product. In addition, 2-hydroxydibenzodioxin, two isomers of trihydroxybiphenyl, two isomers of dihydroxychlorobiphenyl, and two isomers of trihydroxychlorobiphenyl were also identified. The formation of dibenzodioxin was explained to be due to a homolytic C–Cl bond fission, creating a Cl atom and a dibenzodioxanyl radical.

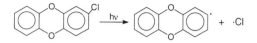

SCHEME IV.19

Evidence for this came from the kinetics of photolysis at 308 nm in the methanol–water solution, in which the rates of degradation under argon or oxygen atmospheres were equal. Since methanol is a good hydrogen donor, it competed effectively with O_2 for reaction with the carbon-centered radical, thus forming dibenzodioxin. In the acetonitrile–water mixture, the rate of dibenzodioxin formation was much slower under oxygen than under argon. Since acetonitrile is a very poor hydrogen donor, in this case O_2 could attack the carbon-centered radical, forming a peroxyl adduct, which decomposed with cleavage of the aromatic ring, resulting in formaldehyde as one of the observed products. The chloro- and hydroxybiphenyl products were proposed to be formed by a complex series of steps, initiated by homolytic cleavage of the C–O bond of the dioxin ring, followed by rearrangement to the biphenyl derivatives.[181]

Since chlorinated dibenzo-*p*-dioxins are produced during the chlorine bleaching of wood pulp, experiments were made on the photodegradation of 1,2,4-trichloro-*p*-dioxin adsorbed on pulp sludge (10% aqueous slurry) under irradiation at 222 nm. The photodecomposition of the dioxin followed first-order kinetics. Due to the opaque nature of the material, about 10-fold higher fluences were required to achieve rates of degradation similar to that in the homogeneous acetonitrile–water solutions.[181]

Homogeneous photolysis of PCDDs and PCBs in aqueous solutions or hydrocarbon solvents involves replacement of chlorine atoms by hydrogen atoms or hydroxyl groups, or condensation reactions with ring closure, as shown for the reactions of tetrachlorobiphenyl.[182]

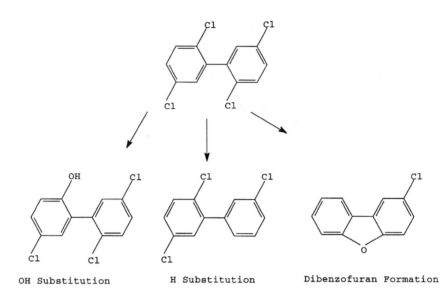

OH Substitution   H Substitution   Dibenzofuran Formation

SCHEME IV.20

Rapid and complete photodechlorination of a commercial PCB (Aroclor 1254) was achieved in alkaline 2-propanol solution. The dechlorination occurred by a chain reaction, thus leading to extremely high quantum yields. In 2-propanol solution (with 0.1 M NaOH) containing 1 g/L of the PCB in solution illuminated at 254 nm, the PCB was completely dechlorinated within 30 min, and the quantum yield was 35. The reaction could be sensitized to occur even with visible light, by adding phenothiazine (5 mM) as photosensitizer. With this sensitizer, under illumination at 350 nm, the quantum yield was 2.33, and complete dechlorination to biphenyl as the only organic product was achieved within 1 h. With exposure to sunlight, the sensitized dechlorination of the solution occurred within 4 h. The photosensitized reaction was applied to the dechlorination of PCB-contaminated soil, requiring only 2 h at 350 nm, and 20 h under sunlight. No interference was observed by the presence of small amounts of either water or methanol (<5% v/v). The reaction was quenched in the presence of pure oxygen gas, indicating a radical mechanism. However, the presence of air did not interfere in the photodechlorination, presumably because of the low solubility of oxygen in the solvent mixture.[172]

The proposed mechanism for the direct photolysis of the PCB involves light absorption by the ArCl molecule, forming the excited singlet state ^{1}ArCl. This undergoes intersystem crossing (ISC) to produce the triplet state ^{3}ArCl, which dissociates to aryl radicals and chlorine atoms.[172,183]

$$ArCl + h\nu \rightarrow {}^1ArCl \qquad (73)$$

$$^1ArCl \rightarrow {}^3ArCl \quad (ISC) \tag{74}$$

$$^3ArCl \rightarrow Ar\cdot + Cl\cdot \tag{75}$$

The chain reaction was propagated by hydrogen abstraction reactions of the aryl radicals and chlorine atoms with 2-propanol, producing the ketyl radical, $(CH_3)_2C(OH)\cdot$. This radical underwent proton transfer in the alkaline medium to its conjugate base, the radical-anion $(CH_3)_2CO\cdot^-$, which reacted with ArCl ground-state molecules by electron transfer to create an additional aryl radical and a chloride anion.

$$\left(CH_3\right)_2 CO\cdot^- + ArCl \leftrightarrow \left(CH_3\right)_2 CO + Ar\cdot^- Cl$$

$$Ar\cdot^- Cl \rightarrow Ar\cdot + Cl\cdot \tag{76}$$

In the phenothiazine-sensitized photodechlorination, the primary step is visible light absorption by the sensitizer, which produces its triplet state. This triplet-state molecule donates an electron to the chloroaromatic compound ArCl, producing the radical anion $Ar\cdot^-Cl$, which dissociates as just shown for the nonsensitized reaction.[172,183]

The rates of direct photolysis of polychlorinated dibenzofurans at 300 nm in aqueous solutions were compared with those in organic solvents. In order to overcome the difficulty of preparing aqueous solutions of such very weakly soluble compounds, the aqueous solutions were obtained by a generator-column technique (in which water was pumped through a column packed with 0.2-mm glass beads coated with such compounds). Photolysis half-lives for 2,7-dichlorodibenzofuran in water, hexane, 60% acetonitrile/water, and methanol were 0.5, 7.2, 11, and 8.2 h, respectively. Thus, the photolysis in water was much faster than in the organic solvents. In aqueous solutions, the first-order rate constants for photolysis of 2,7-dichlorodibenzofuran, 2,3,7,8-tetrachlorodibenzofuran, and 1,2,7,8-tetrachlorodibenzofuran were 1.3, 0.45, and 0.36 h^{-1}, respectively. In natural waters, the sunlight photolysis of poly-chlorinated dibenzofurans was enhanced by sensitized reactions due to natural compounds in the water.[184]

The rate of the direct photolysis under sunlight of 2,3,4,7,8-pentachlorodibenzofuran in lake water was 240 times faster than in distilled water–acetonitrile (25:10) solutions. This was suggested to indicate the action of sensitizers such as humic material in natural waters. The only product tentatively identified was a tetrachlorinated dibenzofuran.[185]

Nitropolychlorodibenzo-*p*-dioxins in dilute aqueous NaOH underwent rapid photolysis by UV light (>300 nm). A bioassay was applied to measure the disappearance of the starting compound, by using its affinity for the murine hepatic Ah receptor.[186]

2. Heterogeneous Photodegradation

A successful and complete degradation of many polychlorinated aromatic compounds was obtained by illumination with natural or simulated sunlight in aqueous slurries of various semiconductors.

2,3,7,8-Tetrachlorodibenzo-*p*-dioxin (TCDD) is a major environmental problem, due to its very high toxicity, low water solubility and low volatility, and considerable resistance to thermal and microbial degradation. With this compound, photocatalytic degradation seems currently to be the most effective method of disposal.[187]

Several semiconductor materials in aqueous dispersions were tested as photocatalysts in the degradation of PCDDs and PCBs in simulated sunlight. With aeration complete mineralization of 2-chloro-dibenzo-*p*-dioxin (CDD), 2,7-dichloro-dibenzo-*p*-dioxin (DCDD), and 3,3′-dichlorobiphenyl (DCB) was achieved. The TiO_2-promoted photodegradation of 2-chlorodibenzo-*p*-dioxin was found to be more rapid in alkaline media (pH 11) than in acidic media (pH 3). 2,7-Dichlorodibenzo-*p*-dioxin, which was relatively inert to photocatalytic degradation by aqueous TiO_2 alone, was degraded to more than 99% within 30 and 60 min in the presence of 0.1 M periodate or peroxydisulfate, respectively. Hydrogen peroxide only slightly enhanced the rate of degradation of the dioxin, while the chlorate ion was completely inactive. For DCB, the photocatalytic activity was in the order $TiO_2 \sim TiO_2/Pt$ (5% by weight) > WO_3 > ZnO, while Fe_2O_3 and CdS were practically inactive.[151,182,188]

Zhang et al.[189,190] studied the sunlight decomposition of PCBs in aqueous solutions, clay suspensions, and sediments in the presence of suspended TiO_2. After 4 h of sunlight exposure, the TiO_2-treated aqueous solutions and clay suspensions contained only 13 and 19%, respectively, of the original PCB concentration. With sediment suspensions, the TiO_2-promoted photodegradation destroyed 50% of the PCBs within 6 h. PCB congeners with low chlorine content, such as mono-and dichlorinated biphenyls, underwent more rapid photodegradation than highly chlorinated biphenyls.[189,190]

D. Halocarbon Pesticides

Considerable efforts have been directed toward the study of the UV photolysis of chlorinated hydrocarbon pesticides, such as DDT, 1,1,1-trichloro-2,2′-bis(*p*-chlorophenyl) ethane. DDT and other chlorocarbon insecticides when sprayed in the open were found to degrade rapidly under sunlight exposure.[191]

1. Permethrin and DDT

Permethrin is a pyrethroid-like insecticide and insect repellent, and is a suspected carcinogen and a priority pollutant. The toxic DDT is a persistent insecticide with contact and stomach action.[192]

"Dark" reactions of the superoxide ion O_2^- with the chlorinated pesticides permethrin and DDT were carried out, using KO_2 in DMF solutions at 20°C for 1 h. These reactions may provide a model for understanding the metabolic and environmental degradations of these compounds. *cis*-Permethrin reacted with O_2^-, yielding a chloroacetylenic product with 65% conversion.[67]

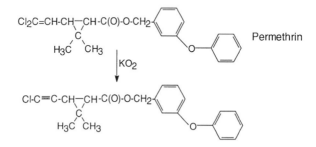

SCHEME IV.21

DDT reacted with the superoxide ion to form DDE in 95% conversion.[67]

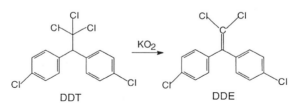

SCHEME IV.22

Since permethrin is hydrophobic and very insoluble in water, it tends to float on the surface and is quite resistant to biodegradation. Photodegradation of immiscible hydrophobic compounds may be facilitated by the use of surfactants. The photocatalytic degradation of permethrin was accomplished in an aqueous dispersion of TiO_2 with a nonphotodegradable fluoro surfactant. When using unmodified "naked" TiO_2 (P25) dispersed together with permethrin and either an anionic fluoro surfactant A (perfluorooctyl sulfonate), a cationic surfactant C (perfluorooctyl sulfoamido propyltrimethyl ammonium iodide), a nonionic surfactant N (perfluorooctyl sulfoamidoethyl tetraethoxyethanol), or a surfactant-free suspension W, the rates of aromatic cleavage and of mineralization to CO_2 decreased in the order $N > A > W > C$. In the presence of the surfactants, the hydrophobic permethrin dissolved inside the micellae of the surfactants. Under illumination, the TiO_2 dispersion became acidic and positively charged, thus attracting and strongly adsorbing the nonanionic and anionic surfactants. On the other hand, the cationic surfactant was less readily adsorbed on the positively charged TiO_2, due to electrostatic repulsion. Hence, the permethrin decomposed more readily when dissolved in the nonanionic and anionic surfactants than in the cationic surfactant.[193]

Permethrin was also efficiently photodegraded in TiO_2 dispersed in a hexane–water mixture. With illumination with UV light ($\lambda > 330$ nm), degradation occurred with pseudo first-order kinetics, with $k = 1.73 \times 10^{-3}$ min^{-1} and $t_{1/2} = 400$ min. Under sunshine on a clear day, the degradation was more than 90% complete within 8 h. The rates of aromatic ring opening and of dechlorination were practically equal, while the total mineralization to CO_2 was slower. Even better photoactivity than with "bare" TiO_2 was achieved with a hydrophobic catalyst, prepared by treating TiO_2 with octyltrimethoxylsilane.[194]

2. Chlordane

The insecticide chlordane is used for control of termites, ants, wasps and cockroaches. It is of serious mammalian toxicity, both chronic and cumulative, due to accumulation in lipid-containing organs. It causes damage to liver and kidneys. It is very toxic to fish and to bees.[192]

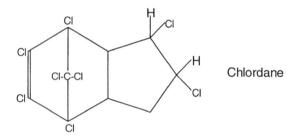

Chlordane

SCHEME IV.23

Effective decomposition of chlordane was achieved by photocatalytic degradation in aqueous dispersions of TiO_2.[195]

3. Mirex

Mirex, $C_{10}Cl_{12}$, is a perchlorinated compound with a "caged" structure containing a total of six aromatic rings.[196] It has been used as an insecticide against the imported fire ant[197] and has been banned since 1977 in the United States by the Environmental Protection Agency.[198] It is hydrophobic, of low chemical reactivity, very resistant to biodegradation, and toxic. The major photoconversion

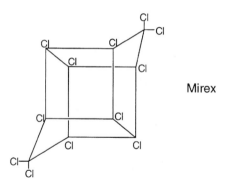

Mirex

SCHEME IV.24

product of its aqueous solution by illumination in sunlight is "photomirex" (8-monohydromirex), which also is chemically unreactive and toxic.[199] The rate of photoreduction of mirex to photomirex was enhanced by the presence of humic acids. In natural waters, a large fraction of mirex was found to be bound to the dissolved organic matter (DOM). Thus the rate of photoconversion of mirex in natural waters may be sensitized by its association with humic acids and other DOM.[200]

REFERENCES

1. Lykins, B. W., Jr. and Koffskey, W., Products identified at an alternative disinfection pilot plant, *Environ. Health Perspect.*, 69, 119–128, 1986.
2. El-Dib, M. A. and Ali, R. K., Trihalomethanes and halogenated organic formation in water treatment plant, *Bull. Environ. Contam. Toxicol.*, 49, 381–387, 1992.
3. Luong, T. V., Peters, C. J., and Perry, R., Influence of bromide and ammonia upon the formation of trihalomethanes under water-treatment conditions, *Environ. Sci. Technol.*, 16, 473–479, 1982.
4. Rebhun, M., Manka, J., and Zilberman, A., Trihalomethane formation in high-bromide Lake Galilee water, *J. Am. Water Works Assoc.*, 80, 84–89, 1988.
5. Rebhun, M., Heller-Grosman, L., Manka, J. Kimel, D., and Limoni, B., Trihalomethane formation and distribution in bromide-rich and ammonia containing lake water, *Water Chlorination*, 6, 665–680, 1990; *Chem. Abstr.* 114, 192121g.
6. Symons, J. M., Stevens, A. A., Clark, R. M., Geldreich, E., Love, O. T., Jr., and DeMarco, J., *Treatment Techniques for Controlling Trihalomethanes in Drinking Water*, U.S. Environmental Protection Agency, Cincinnati, OH; EPA-600/2–81–156; September 1981.
7. Symons, J. M., Stevens, A. A., Clark, R. M., Geldreich, E., Love, O. T., Jr., and DeMarco, J., Removing trihalomethanes from drinking water. An overview of treatment techniques, *Water Eng. Manage.*, 128, 65–76, 1981. *Chem. Abstr.*, 96, 24447j.

8. Richardson, S., Scoping the chemicals in your drinking water, *Today's Chemist at Work,* 3, 29–32, 1994.

9. Richardson, S. D., Thruston, A. D., Jr., Collette, T. W., Patterson, K. S., Lykins, B. W., Jr., Majetich, G., and Zhang, Y., Multispectral identification of chlorine dioxide disinfection byproducts in drinking water, *Environ. Sci. Technol.,* 28, 592–599, 1994.

10. Adachi, A. and Kobayashi, T., Simple method of removing chloroform and carbon tetrachloride from laboratory waste water, *Jpn. J. Toxicol. Environ. Health,* 39, 63–67, 1993.

11. Lykins, B. W. Jr., Clark, R. M., and Adams, J. Q., Granular activated carbon for controlling THMs, *J. Am. Water Works Assoc.,* 80, 85–92, 1988.

12. Clark, R. M., Fronk, C. A., and Lykins, B. W., Jr., Removing organic contaminants from groundwater. A cost and performance evaluation, *Environ. Sci. Technol.,* 22, 1126–1130, 1988.

13. Sayre, I. M., International standards for drinking water, *J. Am. Water Works Assoc.,* 80, 53–60, 1988.

14. Matthews, R. W., Kinetics of photocatalytic oxidation of organic solutes over titanium dioxide, *J. Catal.,* 111, 264–272, 1988.

15. Hsiao, C.-Y., Lee, C.-L., and Ollis, D. F., Heterogeneous photocatalysis: degradation of dilute solutions of dichloromethane (CH_2Cl_2), chloroform ($CHCl_3$), and carbon tetrachloride (CCl_4) with illuminated TiO_2 photocatalyst, *J. Catal.,* 82, 418–423, 1983.

16. Pruden, A. L. and Ollis, D. F., Photoassisted heterogeneous catalysis: the degradation of trichloroethylene in water, *J. Catal.,* 82, 404–417, 1983.

17. Ahmed, S. and Ollis, D. F., Solar photoassisted catalytic decomposition of the chlorinated hydrocarbons trichloroethylene and trichloromethane, *Solar Energy,* 32, 597–601, 1984.

18. Nguyen, T. and Ollis, D. F., Complete heterogeneously photocatalyzed transformation of 1,1- and 1,2-dibromoethane to CO_2 and HBr, *J. Phys. Chem.,* 88, 3386–3388, 1984.

19. Ollis, D. F., Hsiao, C. -Y., Budiman, L., and Lee, C.-L., Heterogeneous photoassisted photocatalysis: conversion of perchloroethylene, dichloroethane, chloroacetic acids, and chlorobenzenes, *J. Catal.,* 88, 89–96, 1984.

20. Tanguay, J. F., Suib, S. L., and Coughlin, R. W., Dichloromethane photodegradation using titanium catalysts, *J. Catal.,* 117, 335–347, 1989.

21. Tanaka, K., Hisanaga, T., and Harada, K., Efficient photocatalytic degradation of chloral hydrate in aqueous semiconductor suspension, *J. Photochem. Photobiol. A: Chemistry,* 48, 155–159, 1989.

22. Tanaka, K., Hisanaga, T., and Harada, K., Photocatalytic degradation of organohalide compounds in semiconductor suspensions with added hydrogen peroxide, *N. J. Chem.,* 13, 5–7, 1989.

23. Hisanaga, T., Harada, K., and Tanaka, K., Photocatalytic degradation of organochlorine compounds in suspended TiO_2, *J. Photochem. Photobiol. A: Chem.,* 54, 113–118, 1990.

24. Dibble, L. A. and Raupp, G. B., Kinetics of the gas-solid heterogeneous photocatalytic oxidation of trichloroethylene by near UV illuminated titanium dioxide, *Catal. Lett.,* 4, 345–354, 1990.

25. Murabayashi, M., Itoh, K., Ohta, Y., and Kamata, K., Photocatalytic degradation of chloroform on platinized TiO$_2$ powder, *Denki Kagaku,* 57, 1221–1222, 1989.

26. Halmann, M., Hunt, A. J., and Spath, D., Photoassisted degradation of halocarbon pollutants in water with TiO$_2$, *Abstr. Int. Conf. Photochem. Conv. Storage Solar Energy,* Palermo, Italy, 1990, p. 240.

27. Ollis, D. F., Contaminant degradation in water, *Environ. Sci. Tech.,* 19, 480–484, 1985.

28. Matthews, R. W., Photooxidation of organic impurities in water using thin films of titanium oxide, *J. Phys. Chem.,* 91, 3328–3333, 1987.

29. Mao, Y., Schöneich, C., and Asmus, K.-D., Radical mediated degradation mechanisms of halogenated organic compounds as studied by photocatalysis at TiO$_2$ and by radiation chemistry, in: *Photocatalytic Purification and Treatment of Water and Air,* Ollis, D. F. and Al-Ekabi, H. (Eds.), Elsevier Science Publishers, Amsterdam, 1993, pp. 49–65.

30. Al-Ekabi, H., Butters, B., Delaney, D., Ireland, J., Lewis, N., Powell, T., and Story, J., TiO$_2$ advanced photo-oxidation technology: effect of electron acceptors, in: *Photocatalytic Purification and Treatment of Water and Air,* Ollis, D. F., and Al-Ekabi, H. (Eds.), Elsevier Science Publishers, Amsterdam, 1993, pp. 321–335.

31. Suri, R. P. S., Liu, J., Hand, D. W., Crittenden, J. C., Perram, D. L., and Mullins, M. E., Heterogeneous photocatalytic oxidation of hazardous organic contaminants in water, *Water Environ. Res.,* 65, 665–673, 1993.

32. Mao, Y., Schöneich, C., and Asmus, K.-D., Identification of organic acids and other intermediates in oxidative degradation of chlorinated ethanes on TiO$_2$ surfaces en route to mineralization. A combined photocatalytic and radiation chemical study, *J. Phys. Chem.,* 95, 10080–10089, 1991.

33. Chemseddine, A. and Boehm, H. P., A study of the primary step in the photochemical degradation of acetic acid and chloroacetic acids on a TiO$_2$ catalyst, *J. Mol. Catal.,* 60, 295–311, 1990.

34. Bahnemann, D. W., Bockelmann, D., Goslich, R., Hilgendorff, M., and Weichgrebe, D., Photocatalytic detoxification: Novel catalysts, mechanisms and solar applications, in: *Photocatalytic Purification and Treatment of Water and Air,* Ollis, D. F. and Al-Ekabi, H. (Eds.), Elsevier Science Publishers, Amsterdam, 1993, pp. 301–319.

35. Glaze, W. H., Kenneke, J. F., and Ferry, J. L., Chlorinated by-products from the TiO$_2$-mediated photodegradation of trichloroethylene and tetrachloroethylene in water, *Environ. Sci. Technol.,* 27, 177–184, 1993.

36. Koster, T. P. M., Assink, J. W., Slaager, J. M., and van der Veen, C., Photocatalytic oxidation of multi-component organochlorine mixtures in water, in: *Photocatalytic Purification and Treatment of Water and Air,* Ollis, D. F. and Al-Ekabi, H. (Eds.), Elsevier Science Publishers, Amsterdam, 1993, pp. 613–618.

37. Murabayashi, M., Itoh, K., Furushima, H., and Chen, D., Photocatalytic degradation of chloroform with rutile-anatase mixed TiO$_2$ powder, *Denki Kagaku,* 59, 524–525, 1991.

38. Murabayashi, M., Itoh, K., Kuroda, S., Huda, R., Masuda, R., Takahashi, W., and Kawashima, K., Photocatalytic degradation of chloroform with TiO₂ coated glass fiber cloth, *Denki Kagaku,* 60, 741–742, 1992.

39. Murabayashi, M., Itoh, K., Suzuki, S., and Kawashima, K., Photocatalytic degradation, *Interface,* Vol. 2, 183rd Electrochem. Soc. Meeting (May 1993), Honolulu, Hawaii, Abstr. 977, p. 1417.

40. Murabayashi, M., Itoh, K., Kawashima, K., Masuda, R., and Suzuki, S. Photocatalytic degradation of chloroform with TiO₂ coated glassfiber cloth, in: *Photocatalytic Purification and Treatment of Water and Air,* Ollis, D. F. and Al-Ekabi, H. (Eds.), Elsevier Science Publishers, Amsterdam, 1993, pp. 783–788.

41. Bahnemann, D. W., Bockelmann, D., and Goslich, R., Mechanistic studies of water detoxification in illuminated TiO₂ suspensions, *Solar Energy Mater., 24,* 564–583, 1991.

42. Kormann, C., Bahnemann, D. W., and Hoffmann, M. R., Photolysis of chloroform and other organic molecules in aqueous TiO₂ suspensions, *Environ. Sci. Technol.,* 25, 494–500, 1991.

43. Kondo, M. M. and Jardim, W. F., Photodegradation of chloroform and urea using Ag-loaded titanium dioxide as catalyst, *Water Res.,* 25, 823–827, 1991.

44. Alfassi, Z. B., Mosseri, S., and Neta, P., Reactivities of chlorine atoms and peroxyl radicals formed in the radiolysis of dichloromethane, *J. Phys. Chem.,* 93, 1380–1385, 1989.

45. Halmann, M., Hunt, A. J., and Spath, D., Photodegradation of dichloromethane, tetrachloroethylene and 1,2-dibromo-3-chloropropane in aqueous suspensions of TiO₂ with natural, concentrated and simulated sunlight, *Solar Energy Mater. Solar Cells,* 26, 1–16, 1992.

46. Mao, Y., Schöneich, C. and Asmus, K.-D., Influence of TiO₂ surface on 1,2-chlorine shift in β-chlorine substituted radicals as studied by radiation chemistry and photocatalysis, *J. Phys. Chem.,* 96, 8522–8529, 1992.

47. Kenekke, J. F., Ferry, J. L., and Glaze, W. H., The TiO₂-mediated Photocatalytic degradation of chloroalkenes in water, in: *Photocatalytic Purification and Treatment of Water and Air,* Ollis, D. F. and Al-Ekabi, H. (Eds.), Elsevier Science Publishers, Amsterdam, 1993, pp. 179–191.

48. Peyton, G. R., Huang, F. Y., Burleson, J. L., and Glaze, W. H., Destruction of pollutants in water with ozone in combination with ultraviolet radiation. 1. General principles and oxidation of tetrachloroethylene, *Environ. Sci. Technol.,* 16, 448–453, 1982.

49. Neumann-Spallart, M. and Getoff, N., Photolysis and radiolysis of monochloroacetic acid in aqueous solution, *Radiat. Phys. Chem.,* 13, 101, 1979.

50. Lee, T. K., Kim, D. H., and Auh, P. C., The preparation of platinized TiO₂ photocatalyst for the application of solar detoxification or organics in water phase, *Proc. 7th Int. Symp. Solar Thermal Concentrating Technologies,* September 26–30, 1994, Moscow, Russia.

51. Milano, J. C., Bernat-Escallon, C., and Vernet, J. L., Hydrolytic and photolytic degradation of chloroethyl ether in water, *Environ. Technol. Lett.,* 10, 291–300, 1989.

52. Watts, R. J., Kong, S., Orr, M. P., and Miller, G. C., Titanium dioxide-mediated photocatalysis of a biorefractory chloroether in secondary wastewater effluent, *Environ. Technol.,* 15, 469–475, 1994.

53. Hilgendorff, M., Hilgendorff, M., and Bahnemann, D. W., Reductive elimination of tetrachloromethane on platinized titanium dioxide in aqueous suspensions, *Environmental Aspects of Electrochemistry and Photoelectrochemistry,* Proc. 183rd Meet. Electrochem. Soc., Yoneyama, H., Hori, Y., Haynes, R., and Tomkiewicz, M. (Eds.), 1993.

54. Dillert, R. and Bahnemann, D., Photocatalytic degradation of organic pollutants: Mechanisms and solar applications, *EPA Newslett.,* 52, 33–52, 1994.

55. Kuhler, R. J., Santo, G. A., Caudill, T. R., Betterton, E. A., and Arnold, R. G., Photoreductive dehalogenation of bromoform with titanium dioxide–cobalt macrocycle hybrid catalysts, *Environ. Sci. Technol.,* 27, 2104–2111, 1993.

56. Hong, A. P., Bahnemann, D. W., and Hoffmann, M. R., Cobalt(II) tetrasulfo phthalocyanine on titanium dioxide: a new efficient electron relay for the photocatalytic formation and depletion of hydrogen peroxide in aqueous suspensions, *J. Phys. Chem.,* 91, 2109–2117, 1987.

57. Lal, M., Schöneich, C., Mönig, J., and Asmus, K.-D., Rate constants for the reactions of halogenated organic radicals, *Int. J. Radiat. Biol.,* 54, 773–785, 1988.

58. Mönig, J., Bahnemann, D., and Asmus, K.-D., One electron reduction of CCl_4 in oxygenated solutions: a $CCl_3 O_2^-$ free radical mediated formation of Cl^- and CO_2, *Chem. Biol. Interact.,* 47, 15–27, 1983.

59. Brault, D., Neta, P., and Patterson, L. K., The lipid peroxidation model for halogenated hydrocarbon toxicity. Kinetics of peroxyl radical processes involving fatty acids and Fe(III) porphyrins, *Chem. Biol. Interact.,* 54, 289–297, 1985.

60. Bahnemann, D. W., Mönig, J., and Chapman, R., Efficient photocatalysis of the irreversible one-electron and two-electron reduction of halothane on platinized colloidal titanium dioxide in aqueous suspension, *J. Phys. Chem.,* 91, 3782–3788, 1987.

61. Alfassi, Z. B., Mosseri, S., and Neta, P., Halogenated alkylperoxyl radicals as oxidants: effects of solvents and of substituents on rates of electron transfer, *J. Phys. Chem.,* 91, 3383–3385, 1987.

62. Milano, J. C. and Vernet, J. L., Photolytic degradation of traces of 1,2-dibromopropane in water. Influence of hydrogen peroxide, *Chemosphere,* 17, 943–962, 1988.

63. Milano, J. C., Bernat-Escallon, C., and Vernet, J. L., Photolysis of 1,2-dibromo-3-chloropropane in water, *Water Res.,* 24, 557–564, 1990.

64. Reed, N. R., Olsen H. E., Marty, M., Beltran, L. M., McKone, T. E., Bogen, E. T., Tablante, N. L., and Hsieh, D. P. H., Health risk assessment of 1,2-dibromo-3-chloropropane (DBCP) in California drinking water, *University of California Davis, Dept. of Environ. Toxicology, Technical Report UCD/ET-87/1 (PB88 169 693)* 1987.

65. Castro, C. E., Mayorga, S., and Belser, N. O., Photohydrolysis of 1,2-dibromo-3-chloropropane, *J. Agric. Food Chem.,* 35, 865–870, 1987.

66. Calderwood, T. S. and Sawyer, D. T., Oxygenation by superoxide ion of 1,2-dibromo 1,2-diphenylethane, 2,3-dibromobutane, ethylene dibromide (EDB), and 1,2-dibromo 3-chloropropane (DBCP), *J. Am. Chem. Soc.,* 106, 7185–7186, 1984.

67. Dureja, P., Casida, J. E., and Ruzo, L. O., Superoxide-mediated dehydrogenation reactions of the pyrethroid permethrin and other chlorinated pesticides, *Tetrahedr. Lett.,* 23, 5003–5004, 1982.

68. Muszkat, L., Halmann, M., Raucher, D., and Bir, L., Solar photodegradation of xenobiotic contaminants in polluted well water, *J. Photochem. Photobiol. A: Chem.*, 65, 409–417, 1992.

69. Ogura, K., Kobayashi, W., and Migita, C. T., Complete photodecomposition of CFC-113, $CHCl_3$ and CCl_4, and scavenging of generated reactive species, *Environ. Technol.*, 13, 81–88, 1992.

70. Sabin, F., Türk, T., and Vogler, A., Photooxidation of organic compounds in the presence of titanium dioxide: determination of the efficiency, *J. Photochem. Photobiol. A: Chem.*, 63, 99–106, 1992.

71. Ohtani, B., Ueda, Y., Nishimoto, S.-I., Kagiya, T., and Hachisuka, H., Photocatalytic oxidative decomposition of fluoroalkenes by titanium dioxide, *J. Chem. Soc. Perkin Trans. 2*, 1955–1960, 1990.

72. Eliasson, B., Kogelschatz, U., and Stein, H. J., New trends in high intensity UV generation, *EPA Newslett.*, 32, 29–40, 1988.

73. Loraine, G. A. and Glaze, W. H., Destruction of vapor phase halogenated methanes by means of ultraviolet photolysis, in: *47th Purdue Industrial Waste Conference Proceedings*, Lewis Publishers, Boca Raton, FL, 1992, pp. 309–316.

74. Loraine, G. A., Short wavelength ultraviolet photolysis of aqueous carbon tetrachloride, *Haz. Waste Haz. Mater.*, 10, 185–194, 1993.

75. Krijgsheld, K. R. and Van der Gen, A., Assessment of the impact of the emission of certain organochlorine compounds on the environment. Part I: Monochlorophenols and 2,4-dichlorophenol, *Chemosphere*, 15, 825–860, 1986.

76. Kochany, J. and Bolton, J. R., Mechanism of photodegradation of aqueous organic pollutants. 2. Measurement of the primary rate constants for reaction of OH· radicals with benzene and some halobenzenes using an EPR spin-trapping method following the photolysis of H_2O_2, *Environ. Sci. Technol.*, 26, 262–265, 1992.

77. Stegeman, M. H. L., Peijnenburg, W. J. G. M., and Verboom, H., A quantitative structure-activity relationship for the direct photohydrolysis of meta-substituted halobenzene derivatives, *Chemosphere*, 26, 837–849, 1993.

78. Peijnenburg, W. J. G. M., de Beer, K. G. M., de Haan, M. W. A., den Hollander, H. A., Stegeman, M. H. L., and Verboom, H. H., The development of a structure-reactivity relationship for the photohydrolysis of substituted aromatic halides, *Environ. Sci. Technol.*, 26, 2116–2121, 1992.

79. Matthews, R. W., Photocatalytic oxidation of chlorobenzene in aqueous suspensions of titanium dioxide, *J. Catal.*, 97, 565–568, 1986.

80. Pelizzetti, E., Barbeni, M., Pramauro, E., Serpone, N., Borgarello, E., Jamieson, M. A., and Hidaka, H., Sunlight photodegradation of haloaromatic pollutants catalyzed by semiconductor particulate materials, *Chim. Ind. (Milano)*, 67, 623–625, 1986.

81. Al-Ekabi, H. and Serpone, N., Kinetic studies in heterogeneous photocatalysis. 1. Photocatalytic degradation of chlorinated phenols in aerated aqueous solutions over TiO_2 supported on a glass matrix, *J. Phys. Chem.*, 92, 5726–5731, 1988.

82. Al-Ekabi, H., Serpone, N., Pelizzetti, E., Minero, C., Fox, M. A., and Draper, R. B., Kinetic studies in heterogeneous photocatalysis. 2. TiO_2-mediated degradation of 4-chlorophenol alone and in a three-component mixture of 4-chlorophenol, 2,4-dichlorophenol, and 2,4,5-trichlorophenol in air-equilibrated aqueous media, *Langmuir,* 5, 250–255, 1989.

83. Sehili, T., Bonhomme, G., and Lemaire, J., Transformation of chlorinated aromatic compounds photocatalyzed by zinc oxide. I. Behavior of 3-chlorophenol, *Chemosphere,* 17, 2207–2218, 1988.

84. Sehili, T., Boule, P., and Lemaire, J., Photocatalyzed transformation of chloroaromatic derivatives on zinc oxide. II. Dichlorobenzenes, *J. Photochem. Photobiol. A: Chem.,* 50, 103–116, 1989.

85. Sehili, T., Boule, P., and Lemaire, J., Photocatalyzed transformation of chloroaromatic derivatives on zinc oxide. III. Chlorophenols, *J. Photochem. Photobiol. A: Chem.,* 50, 117–127, 1989.

86. Rao, N. N., Dube, S., and Natarajan, P., Photocatalytic degradation of some chlorohydrocarbons in aqueous suspensions of MO_3/TiO_2 (M = Mo or W), in: *Photocatalytic Purification and Treatment of Water and Air,* Ollis, D. F. and Al-Ekabi, H. (Eds.), Elsevier Science Publishers, Amsterdam, 1993, pp. 695–700.

87. Sabate, J., Anderson, M. A., Kikhawa, H., Edwards, M., and Hill, C. G., A kinetic study of the photocatalytic degradation of 3-chlorosalicylic acid over TiO_2 membranes supported on glass, *J. Catal.,* 127, 167–177, 1991.

88. Sabate, J., Anderson, M. A., Kikhawa, H., Xu, Q., Cervera-March, S., and Hill, C. G., Nature and properties of pure and niobium-doped titania ceramic membranes affecting the photocatalytic degradation of 3-chlorosalicylic acid as a model of halogenated organic compounds, *J. Catal.,* 134, 36–46, 1992.

89. Chu, W. and Jafvert, C. T., Photodechlorination of polychlorobenzene congeners in surfactant micelle solutions, *Environ. Sci. Technol.,* 28, 2415–2422, 1994.

90. Soumillion, J. P., Vandereecken, P., and De Schryver, F. C., Photodechlorination of chloroaromatics by electron transfer from an anionic sensitizer, *Tetrahedron Lett.,* 30, 697–700, 1989.

91. Soumillion, J.-P., Photoinduced electron transfer employing organic anions, *Topics Current Chem.,* 168, 93–141, 1993.

92. Herrmann, J. M., Kerzhentsev, M., Tahiri, H., and Ait Ichou, Y., Photocatalytic degradation of some pesticides in aqueous medium, *Abstr. 2nd Int. Symp. New Trends Photoelectrochemistry,* University of Tokyo, Japan, March 1994, p. 10.

93. Milano, J. C., Yassin-Hussan, S., and Vernet, J. L., Photooxidations of aromatic halogenized ethers: 2-bromo-ethyl-phenyl, 2-bromo-ethyl-4-bromo-phenyl, and para-bromo-phenyl-ether. Influence of hydrogen peroxide, *Environ. Technol.,* 13, 507–519, 1992.

94. Milano, J. C., Yassin-Hussan, S., and Vernet, J. L., Photochemical degradation of 4-bromodiphenylether: influence of hydrogen peroxide, *Chemosphere,* 25, 353–360, 1992.

95. Grabowski, Z. R., Photochemical reactions of several aromatic halogen compounds, *Z. Physik. Chem. (Neue Folge),* 27, 239–252, 1961.

96. Omura, K. and Matsuura, T., Photoinduced reactions — L-photolysis of halogenophenols in aqueous alkali and in aqueous cyanide, *Tetrahedron,* 27, 3101–3109, 1971.

97. Boule, P., Guyon, C., and Lemaire, J., Photochemistry and environment. VI. Direct phototransformation of chlorophenols and interactions with phenol on UV exposure in aqueous solution, *Toxicol. Environ. Chem.,* 7, 97–110, 1984.

98. Lipczynska-Kochany, E. and Bolton, J. R., Flash photolysis–HPLC method applied to the study of photodegradation reactions, *J. Chem. Soc., Chem. Commun.,* 1596–1597, 1990.

99. Lipczynska-Kochany, E. and Bolton, J. R., Flash photolysis/HPLC method for studying the sequence of photochemical reactions: applications to 4-chlorophenol in aerated aqueous solution, *J. Photochem. Photobiol. A: Chem.,* 58, 315–322, 1991.

100. Durand, A. P. Y., Brattan, D., and Brown, R. G., Mechanism of the primary photoreaction in the aqueous photochemistry of 4-chlorophenol, *Chemosphere,* 25, 783–792, 1992.

101. Oudjehani, K. and Boule, P., Photoreactivity of 4-chlorophenol in aqueous solution, *J. Photochem. Photobiol. A: Chem.,* 68, 363–373, 1992.

102. Lipczynska-Kochany, E. and Bolton, J. R., Flash photolysis/HPLC applications. 2. Direct photolysis vs. hydrogen peroxide mediated photodegradation of 4-chlorophenol as studied by a flash photolysis/HPLC technique, *Environ. Sci. Technol.,* 26, 259–262, 1992.

103. Lipczynska-Kochany, E. and Kochany, J., Electron paramagnetic spin trapping detection of free radicals generated in direct photolysis of 4-bromophenol in aqueous solution, *J. Photochem. Photobiol. A: Chem.,* 73, 23–33, 1993.

104. Lipczynska-Kochany, E., Direct photolysis of 4-bromophenol and 3-bromophenol as studied by a flash photolysis HPLC technique, *Chemosphere,* 24, 911–918, 1992.

105. Lipczynska-Kochany, E., Kochany, J., and Bolton, J. R., Electron paramagnetic resonance spin trapping detection of short-lived radical intermediates in the direct photolysis of 4-chlorophenol in aerated aqueous solution, *J. Photochem. Photobiol. A: Chem.,* 62, 229–240, 1992.

106. Esplugas, S., Yue, P. L., and Pervez, M. I., Degradation of 4-chlorophenol by photocatalytic oxidation, *Water Res.,* 28, 1323–1328, 1994.

107. Kochany, J. and Bolton, J. R., Mechanism of photodegradation of aqueous organic pollutants. 1. EPR spin-trapping technique for the determination of ·OH rate constants in the photooxidation of chlorophenols following the photolysis of H_2O_2, *J. Phys. Chem.,* 95, 5116–5120, 1991.

108. Kawaguchi, H., Oxidation efficiency of hydroxyl radical in the photooxidation of 2-chlorophenol using ultraviolet radiation and hydrogen peroxide, *Environ. Technol.,* 14, 289–293, 1993.

109. Oudjehani, K. and Boule, P., Phototransformation of 2-halogenophenols induced by excitation of hydroquinone in aqueous solution, *N. J. Chem.,* 17, 567–571, 1993.

110. Boule, P., Rossi, A., Pilichowski, J. F., and Grabner, G., Photoreactivity of hydroquinone in aqueous solution, *N. J. Chem.,* 16, 1053, 1992.

111. Miskoski, S. and García, N. A., Effect of chlorophenolic pesticides on the photochemistry of riboflavin, *Toxicol. Environ. Chem.,* 25, 33–43, 1989.

112. Palumbo, M. C., García, N. A., Gutiérrez, M. I., and Luiz, M., Singlet molecular oxygen-mediated photooxidation of monochloro and mononitrophenols. A kinetic study, *Toxicol. Environ. Chem.,* 29, 85–94, 1990.

113. Gsponer, H. E., Previtali, C. M., and García, N. A., Kinetics of the photosensitized oxidation of polychlorophenols in alkaline aqueous solutions, *Toxicol. Environ. Chem.,* 16, 23–37, 1987.

114. Palumbo, M. C. and García, N. A., On the mechanism of quenching of singlet oxygen by chlorinated phenolic pesticides, *Toxicol. Environ. Chem.,* 17, 103–116, 1988.

115. Bertolotti, S. G., Gsponer, H. E., and García, N. A., Surfactant effect on the sensitized photooxidation of polychlorophenolic pesticides, *Toxicol. Environ. Chem.,* 22, 229–237, 1989.

116. Li, X., Fitzgerald, P., and Bowen, L., Sensitized photo-degradation of chlorophenols in a continuous flow reactor system, *Water Sci. Technol.,* 26, 367–376, 1992.

117. Jakob, L., Hashem, T. M., Bürki, S., Guindy, N. M., and Braun, A. M., Vacuum-ultraviolet (VUV) photolysis of water: Oxidative degradation of 4-chlorophenol, *J. Photochem. Photobiol.,* 75, 97–103, 1993.

118. Fenton, H. J. H., Oxidation of tartaric acid in presence of iron, *J. Chem. Soc.,* 65, 899–910, 1894.

119. Haber, F. and Weiss, J., The catalytic decomposition of hydrogen peroxide by iron salts, *Proc. R. Soc.,* A147, 332–351, 1934.

120. Potter, F. J. and Roth, J. A., Oxidation of chlorinated phenols using Fenton reagent, *Haz. Waste Haz. Mater.,* 10, 151–170, 1993.

121. Ruppert, G., Bauer, R., and Heisler, G., The photo-Fenton reaction — an effective photochemical wastewater treatment process, *J. Photochem. Photobiol. A: Chem.,* 73, 75–78, 1993.

122. Ruppert, G., Bauer, R., Heisler, G., and Novalic, S., Mineralization of cyclic organic water contaminants by the photo-Fenton reaction — influence of structure and substituents, *Chemosphere,* 27, 1339–1347, 1993.

123. Ruppert, G., Hofstadler, K., Bauer, R., and Heisler, G., Heterogeneous and homogeneous photoassisted wastewater treatment, *Proc. Indian Acad. Sci. Chem. Sci.,* 105, 393–397, 1993.

124. Ruppert, G., Bauer, R., and Heisler, G., UV-O_3, UV-H_2O_2, UV-TiO_2, and the photo-Fenton reaction. Comparison of advanced oxidation processes for water treatment, *Chemosphere,* 28, 1447–1454, 1994.

125. Bauer, R., Application of solar irradiation for photochemical wastewater treatment, *Chemosphere,* 29, 1225–1233, 1994.

126. Kawaguchi, H. and Inagaki, A., Photochemical generation rates of hydroxyl radical in aqueous solutions containing Fe(III)-hydroxy complex, *Chemosphere,* 27, 2381–2387, 1993.

127. Kawaguchi, H. and Inagaki, A., Kinetics of ferric ion promoted photodecomposition of 2-chlorophenol, *Chemosphere,* 28, 57–62, 1994.

128. Pichat, P., D'Oliveira, J.-C., Maffre, J.-F., and Mas, D., Destruction of 2,4-dichlorophenoxyethanoic acid (2,4-D) in water by TiO_2-UV, H_2O_2-UV or direct photolysis, in: *Photocatalytic Purification and Treatment of Water and Air,* Ollis, D. F. and Al-Ekabi, H. (Eds.), Elsevier Science Publishers, Amsterdam, 1993, pp. 683–688.

129. Pignatello, J. J., Dark and photoassisted Fe^{3+}-catalyzed degradation of chlorophenoxy herbicides by hydrogen peroxide, *Environ. Sci. Technol.*, 26, 944–951, 1992.

130. Sun, Y. F. and Pignatello, J. J., Photochemical reactions involved in the total mineralization of 2,4-D by $Fe^{3+}/H_2O_2/UV$, *Environ. Sci. Techn.*, 27, 304–310, 1993.

131. Sun, Y. F. and Pignatello, J. J., Activation of hydrogen peroxide by iron(III) chelates for abiotic degradation of herbicides and insecticides in water, *J. Agric. Food Chem.*, 41, 308–312, 1993.

132. Balzani, V. and Carassiti, V., *Photochemistry of Coordination Compounds*, Academic Press, New York, 1970, pp. 167–192.

133. Pelizzetti, E., Minero, C., Sega, M., and Vincenti, M., Formation and disappearance of biphenyl derivatives in the photocatalytic transformation of 1,2,4-trichlorobenzene on titanium dioxide, in: *Photocatalytic Purification and Treatment of Water and Air,* Ollis, D. F. and Al-Ekabi, H. (Eds.), Elsevier Science Publishers, Amsterdam, 1993, pp. 291–300.

134. D'Oliveira, J.-C., Al-Sayyed, G., and Pichat, P., Photodegradation of 2- and 3-chlorophenol in TiO_2 aqueous suspensions, *Environ. Sci. Technol.*, 24, 990–996, 1990.

135. Cunningham, J. and Sedlák, P., Interrelationship between pollutant concentration, extent of adsorption, TiO_2-sensitized removal, photon flux and levels of electron or hole trapping additives. I. Aqueous monochlorophenol–TiO_2 (P25) suspensions, *J. Photochem. Photophys. A: Chem.*, 77, 255–263, 1994.

136. Al-Sayyed, G., D'Oliveira, J.-C., and Pichat, P. Semiconductor-sensitized photodegradation of 4-chlorophenol in water, *J. Photochem. Photobiol. A: Chem.*, 58, 99–114, 1991.

137. Cunningham, J. and Al-Sayyed, G., Factors influencing efficiencies of TiO_2-sensitized photodegradation. Part 1. Substituted benzoic acids — discrepancies with dark-adsorption parameters, *J. Chem. Soc. Faraday Trans. 1,* 86, 3935–3941, 1990.

138. Mills, A., Morris, S., and Davies, R., Photomineralisation of 4-chlorophenol sensitized by titanium dioxide: a study of the intermediates, *J. Photochem. Photobiol. A: Chem.*, 70, 183–191, 1993.

139. Mills, A. and Morris, S., Photomineralization of 4-chlorophenol sensitized by titanium dioxide: a study of the initial kinetics of carbon dioxide photogeneration, *J. Photochem. Photobiol. A: Chem.*, 71, 75–83, 1993.

140. Mills, A. and Morris, S., Photomineralization of 4-chlorophenol by titanium dioxide. A study of annealing the photocatalyst at different temperatures, *J. Photochem. Photobiol. A: Chem.*, 71, 285–289, 1993.

141. Stafford, U., Gray, K. A., and Kamat, P. V., Radiolytic and TiO_2-assisted photocatalytic degradation of 4-chlorophenol: a comparative study, *J. Phys. Chem.*, 98, 6343–6351, 1994.

142. Gray, K. A. and Stafford, U., Probing photocatalytic reaction in semiconductor systems. Study of the chemical intermediates in 4-chlorophenol degradation by a variety of methods, *Res. Chem. Intermed.*, 20, 835–853, 1994.

143. Murabayashi, M., Itoh, K., Breuer, H. D., and Müller, T. S., Photocatalytic degradation of chloroorganic compounds by using TiO_2 thin-film catalyst, *Abstr. 2nd Int. Symp. New Trends Photoelectrochemistry,* University of Tokyo, Japan, March 1994, p. 8.

144. Alberici, R. M. and Jardim, W. F., Photocatalytic degradation of phenol and chlorinated phenols using Ag–TiO$_2$ in a slurry reactor, *Water Res.,* 28, 1845–1849, 1994.

145. Hofstadler, K., Bauer, R., Novalic, S., and Heisler, G., New reactor design for photocatalytic wastewater treatment with TiO$_2$ immobilized on fused-silica glass fibers: photomineralization of 4-chlorophenol, *Environ. Sci. Technol.,* 28, 670–674, 1994.

146. Gray, K. A., Stafford, U., Dieckmann, M. S., and Kamat, P. V., Mechanistic studies in TiO$_2$ systems: photocatalytic degradation of chloro- and nitrophenols, in: *Photocatalytic Purification and Treatment of Water and Air,* Ollis, D. F. and Al-Ekabi, H. (Eds.), Elsevier Publishers, Amsterdam, 1993, pp. 455–472.

147. Minero, C., Aliberti, C., Pelizzetti, E., Terzian, R., and Serpone, N., Kinetic studies in heterogeneous photocatalysis. 6. AM1 simulated sunlight photodegradation over titania in aqueous media. A 1st case of fluorinated aromatics and identification of intermediates, *Langmuir,* 7, 928–936, 1991.

148. O'Shea, K. E. and Cardona, C., Hammett study of the TiO$_2$-catalyzed photooxidation of para-substituted phenols. A kinetic and mechanistic analysis, *J. Org. Chem.,* 59, 5005–5009, 1994.

149. Guillard, C., Amalric, L., D'Oliveira, J.-C., Delprat, H., Hoang-Van, C., and Pichat, P., Heterogeneous photocatalysis: use in water treatment and involvement in atmospheric chemistry, in: *Aquatic and Surface Photochemistry,* Helz, G. R., Zepp, R. G., and Crosby, D. G. (Eds.), Lewis Publishing, Boca Raton, FL, 1994, pp. 369–386.

150. D'Oliveira, J.-C., Guillard, C., Maillard, C., and Pichat, P., Photocatalytic destruction of hazardous chlorine- or nitrogen- containing aromatics in water, *J. Environ. Health,* A28, 941–962, 1993.

151. Pelizzetti, E., Carlin, V., Minero, C., and Grätzel, M., Enhancement of the rate of photocatalytic degradation on TiO$_2$ of 2-chlorophenol, 2,7-dichloro-dibenzodioxin and atrazine by inorganic oxidizing species, *N. J. Chem.,* 15, 351–359, 1991.

152. Draper, R. B. and Fox, M. A., Titanium dioxide photosensitized reactions studied by diffuse reflectance flash photolysis in aqueous suspensions of TiO$_2$ powder, *Langmuir,* 6, 1396–1402, 1990.

153. Ku, Y. and Hsieh, C. B., Photocatalytic decomposition of 2,4-dichlorophenol in aqueous TiO$_2$ suspensions, *Water Res.,* 26, 1451–1456, 1992.

154. D'Oliveira, J.-C., Minero, C., Pelizzetti, E., and Pichat, P., Photodegradation of dichlorophenols and trichlorophenols in TiO$_2$ aqueous suspensions: kinetic effects of the positions of the Cl atoms and identification of the intermediates, *J. Photochem. Photobiol. A: Chem.,* 72, 261–267, 1993.

155. Pichat, P., Guillard, C., Maillard, C., Amalric, L., and D'Oliveira, J.-C., TiO$_2$ photocatalytic destruction of water aromatic pollutants: intermediates; properties–degradability correlation; effects of inorganic ions and TiO$_2$ surface area; comparison with H$_2$O$_2$ processes, in: *Photocatalytic Purification and Treatment of Water and Air,* Ollis, D. F. and Al-Ekabi, H. (Eds.), Elsevier Science Publishing, Amsterdam, 1993, pp. 207–223.

156. Davis, A. P. and Huang, C. P., Removal of phenols from water by a photocatalytic oxidation process, *Water Sci. Technol.,* 21, 455–464, 1989.

157. Davis, A. P. and Huang, C. P., The photocatalytic oxidation of toxic organic compounds, in: *Physicochemical and Biological Detoxification of Hazardous Wastes,* Wu, Y. C. (Ed.), Vol. 1, Technomic, Lancaster, PA, 1989, pp. 352–357.

158. Davis, A. P. and Huang, C. P., The removal of substituted phenols by a photocatalytic oxidation process with cadmium sulfide, *Water Res.,* 24, 543–550, 1990.

159. Li Puma, G. and Yue, P. L., Photodegradation of pentachlorophenol, in: *Photocatalytic Purification and Treatment of Water and Air,* Ollis, D. F. and Al-Ekabi, H. (Eds.), Elsevier Science Publishers, Amsterdam, 1993, pp. 689–694.

160. Vollmuth, S., Zajc, A., and Niessner, R., Formation of polychlorinated dibenzo-*p*-dioxins and polychlorinated dibenzofurans during the photolysis of pentachlorophenol-containing water, *Environ. Sci. Technol.,* 28, 1145–1149, 1994.

161. Barbeni, M., Pramaura, E., Pelizzetti, E., Borgarello, E., and Serpone, N., Photodegradation of pentachlorophenol catalyzed by semiconductor particles, *Chemosphere,* 14, 195–208, 1985.

162. Mills, G. and Hoffmann, M. R., Photocatalytic degradation of pentachlorophenol on TiO_2 particles. Identification of intermediates and mechanism of reaction, *Environ. Sci. Technol.,* 27, 1681–1689, 1993.

163. Tennakone, K., Kiridena, W. C., and Punchihewa, S., Photodegradation of visible-light-absorbing organic compounds in the presence of semiconductor catalysts, *J. Photochem. Photobiol. A: Chem.,* 68, 389–393, 1992.

164. Vinodgopal, K., Hotchandani, S., and Kamat, P. V., Electrochemically assisted photocatalysis. TiO_2 particulate film electrodes for photocatalytic degradation of 4-chlorophenol, *J. Phys. Chem.,* 97, 9040–9044, 1993.

165. Haque, I. U. and Rusling, J. F., Photodegradation of 4-chlorophenol to carbon dioxide and HCl using high surface area titanium dioxide anodes, *Chemosphere,* 26, 1301–1309, 1993.

166. Vinodgopal, K., Stafford, U., Gray, K. A., and Kamat, P. V., Electrochemically assisted photocatalysis. 2. The role of oxygen and reaction intermediates in the degradation of 4-chlorophenol on immobilized TiO_2 particulate films, *J. Phys. Chem.,* 98, 6797–6803, 1994.

167. Kesselman, J. M., Kumar, A., and Lewis, N. S., Fundamental photoelectrochemistry of TiO_2 and $SrTiO_3$ applied to environmental problems, in: *Photocatalytic Purification and Treatment of Water and Air,* Ollis, D. F. and Al-Ekabi, H. (Eds.), Elsevier Science Publishers, Amsterdam, 1993, pp. 19–37.

168. Koester, C. J. and Hites, R. A., Photodegradation of polychlorinated dioxins and dibenzofurans adsorbed to fly-ash, *Environ. Sci. Technol.,* 26, 502–507, 1992.

169. Tysklind, M. and Rappe, C., Photolytic transformation of polychlorinated dioxins and dibenzofurans in fly ash, *Chemosphere,* 23, 1365–1375, 1991.

170. Dickson, L. C., Lenoir, D., and Hutzinger, O., Quantitative comparison of *de novo* and precursor formation of polychlorinated dibenzo-*p*-dioxins under simulated solid waste incinerator postcombustion conditions, *Environ. Sci. Technol.,* 26, 1822–1828, 1992.

171. Lamparski, L. L., Stehl, R. H., and Johnson, R. L., Photolysis of pentachlorophenol-treated wood. Chlorinated dibenzo-*p*-dioxin formation, *Environ. Sci. Technol.,* 14, 196–200, 1980.

172. Hawari, J., Demeter, A., and Samson, R., Sensitized photolysis of polychlorobiphenyls in alkaline 2-propanol: dechlorination of Aroclor 1254 in soil samples by solar radiation, *Environ. Sci. Technol.*, 26, 2022–2027, 1992.

173. Brown, M. B., McCready, T. L., and Bunce, N. J., Factors affecting the toxicity of dioxin-like toxicants: a molecular approach to risk assessment of dioxins, *Toxicol. Lett.*, 61, 141–147, 1992.

174. Kieatiwong, S., Nguyen, L. V., Hebert, V. R., Hackett, M., Miller, G. C., Miille, M. J., and Mitzel, R., Photolysis of dioxins in organic solvents and on soils, *Environ. Sci. Technol.*, 24, 1575–1580, 1990.

175. Nestrick, T. J., Lamparski, L. L., and Townsend, D. I., Identification of tetrachlorodibenzo-*p*-dioxin isomers at the 1-ng level by photolytic degradation and pattern recognition techniques, *Anal. Chem.*, 52, 1865–1874, 1980.

176. Miller, G. C., Hebert, V. R., Miille, M. J., Mitzel, R., and Zepp, R. G., Photolysis of octachlorodibenzo-*p*-dioxin on soils: Production of 2,3,7,8-TCDD, *Chemosphere*, 18, 1265–1274, 1989.

177. McPeters, A. L. and Overcash, M. R., Demonstration of photodegradation by sunlight of 2,3,7,8-tetrachlorodibenzo-*p*-dioxin in 6 cm soil columns, *Chemosphere*, 27, 1221–1234, 1993.

178. Choudhry, G. G. and Webster, G. R. B., Photochemical quantum yields and sunlight half-lives of polychlorinated dibenzo-*p*-dioxins in aquatic systems, *Chemosphere*, 15, 1935–1940, 1986.

179. Choudhry, G. G. and Webster, G. R. B., Environmental photochemistry of polychlorinated dibenzofurans (PCDFs) and dibenzo-*p*-dioxins (PCDDs): a review, *Toxicol. Environ. Chem.*, 14, 43–61, 1987.

180. Choudhry, G. G. and Webster, G. R. B., Environmental photochemistry of PCDDs. 2. Quantum yields of the direct phototransformation of 1,2,3,7-tetra-, 1,3,6,8-tetra-, 1,2,3,4,6,7,8-hepta-, and 1,2,3,4,6,7,8,9-octachlorodibenzo-*p*-dioxin in aqueous acetonitrile and their sunlight half-lives, *J. Agric. Food Chem.*, 37, 254–261, 1989.

181. Nohr, R. S., MacDonald, J. G., Kogelschatz, U., Mark, G., Schuchmann, H.-P., and von Sonntag, C., Application of excimer incoherent-UV sources as a new tool in photochemistry: photodegradation of chlorinated dibenzodioxins in solution and adsorbed on aqueous pulp sludge, *J. Photochem. Photobiol. A: Chem.*, 79, 141–149, 1994.

182. Pelizzetti, E., Borgarello, M., Minero, C., Pramaura, E., Borgarello, E., and Serpone, N., Photocatalytic degradation of polychlorinated dioxins and polychlorinated biphenyls in aqueous suspensions of semiconductors irradiated with simulated solar light, *Chemosphere*, 17, 499–510, 1988.

183. Bunnett, J. F. and Wamser, C. C., Radical-induced deiodination of aryl iodides in alkaline methanol, *J. Am. Chem. Soc.*, 89, 6712–6718, 1967.

184. Dung, M. H. and O'Keefe, P. W., Comparative rates of photolysis of polychlorinated dibenzofurans in organic solvents and in aqueous solutions, *Environ. Sci. Technol.*, 28, 549–554, 1994.

185. Friesen, K. J. and Foga, M. M., Kinetics of sunlight photodegradation of 2,3,4,7,8-pentachlorodibenzofuran in natural water, *Proc. Indian Acad. Sci. — Chem. Sci.*, 105, 399–403, 1993.

186. Bunce, N. J., Di Diodato, G., and Safe, S. H., Photolysis of nitropolychloro-dibenzo-*p*-dioxins with base as potential methodology for the destruction of polychlorinated dibenzo-*p*-dioxins, *Chemosphere,* 24, 433–438, 1992.

187. Miller, G. C. and Zepp, R. G., 2,3,7,8-TCDD (2,3,7,8-tetrachlorodibenzo-*p*-dioxin). Environmental chemistry, Report, EPA/600/D87/086, 1987. *Chem. Abstr.,* 107, 222343p.

188. Pelizzetti, E., Minero, C., Carlin, V., and Borgarello, E., Photocatalytic soil decontamination, *Chemosphere,* 25, 343–351, 1992.

189. Zhang, P.-C., Scrudata, R. J., Pagano, J. J., and Roberts, R. N., Photodecomposition of PCBs in aqueous systems using TiO_2 as catalyst, *Chemosphere,* 26, 1213–1223, 1993.

190. Zhang, P., Scrudata, R. J., Pagano, J. J., and Roberts, R. N., Photocatalytic decomposition of PCBs in aqueous systems with solar light, in: *Photocatalytic Purification and Treatment of Water and Air,* Ollis, D. F. and Al-Ekabi, H. (Eds.), Elsevier Science Publishers, Amsterdam, 1993, pp. 619–624.

191. Zabik, M. J., Leavitt, A., and Su, G. C. C., Photochemistry of bioactive compounds. A review of pesticide photochemistry, *Annu. Rev. Entomol.,* 21, 61–79, 1976.

192. Hartley, D. and Kidd, H. (Eds.), *The Agrochemicals Handbook,* Royal Society of Chemistry, Nottingham, England, 1987.

193. Hidaka, H., Jou, H., Nohara, K., and Zhao, J., Photocatalytic degradation of the hydrophobic pesticide Permethrin in fluoro surfactant/TiO_2 aqueous dispersions, *Chemosphere,* 25, 1589–1597, 1992.

194. Hidaka, H., Nohara, K., Zhao, J., Serpone, N., and Pelizzetti, E., Photooxidative degradation of the pesticide permethrin catalyzed by irradiated titania semiconductor slurries in aqueous media, *J. Photochem. Photobiol. A: Chem.,* 64, 247–254, 1992.

195. Hustert, K., Kotzias, D., and Korte, F., Photocatalytic decomposition of organic compounds on titanium dioxide, *Chemosphere,* 12, 55–58, 1983.

196. McBee, E. T., Roberts, C. W., Idol, J. D., Jr., and Earle, R. H., Jr., An investigation of the chlorocarbon, $C_{10}Cl_{12}$, m.p. 485° and the ketone, $C_{10}Cl_{10}O$, m.p. 349°, *J. Am. Chem. Soc.,* 78, 1511–1512, 1956.

197. Alley, E. G., Use of mirex in control of the imported fire ant, *J. Environ. Qual.,* 2, 52–61, 1973.

198. Williams, D. F. and Porter, S. D., Fire ant control, *Science,* 264, 1653, 1994.

199. Aley, E. G., Layton, B. R., and Minyard, J. Jr., Identification of the photoproducts of the insecticides mirex and kepone, *J. Agric. Food Chem.,* 22, 442–445, 1974.

200. Mudambi, A. R. and Hassett, J. P., Photochemical activity of mirex associated with dissolved organic matter, *Chemosphere,* 17, 1133–1146, 1988.

Chapter V

Organic Nitrogen Compounds

Organic nitrogen compounds include several groups of widely used pesticides and the intermediates in their manufacture and in their partial degradation. Among these, important groups are the triazine herbicides, carbamate pesticides, and nitroaromatic compounds. Other major groups of organo-N compounds include synthetic dyestuffs and the intermediates in their production, and industrial and military explosives. Also, the degradation of all organic matter in natural waters and in municipal effluents involves nitrogen-containing organic compounds. Some of these compounds are carcinogenic, mutagenic, or otherwise toxic, or impart bad taste or odor to water. European Community directives (80/778/EEC) mandate a maximal permitted concentration in drinking water of 0.1 µg/L for each pesticide separately and 0.5 µg/L for all pesticides together.[1,2] Interest in the photochemistry of organic nitrogen compounds has been stimulated by the observed degradation of nitrogen pesticides such as the halogenated triazines under sunlight. Earlier work on the direct photolysis of pesticides has been reviewed.[3]

In a broad screening of the photocatalytic degradation of a large variety of organic nitrogen compounds in aqueous suspensions of TiO_2, the following compounds were found to release either NH_4^+ or NO_3^- ions: n-pentylamine, piperidine, pyridine, phenylalanine, thioridazine, penicillamine, isosorbide dinitrate, 4-nitrocatechol, 2,4-dinitrophenol, cyclophosphamide, 5-fluorouracil, atrazine, ethylenediamine tetraacetic acid (EDTA), and tetrabutylammonium phosphate. Among the ring compounds n-pentylamine, piperidine, and pyridine, the rate of NH_4^+ production was in the order n-pentylamine $\gg$ pyridine $>$ piperidine, while the rate of NO_3^- formation was in the order pyridine $=$ piperidine $\gg$ n-pentylamine.[4] In the photocatalytic oxidation with TiO_2 of theophylline, proline, pyridine, and pyridine, the rate of formation of NH_4^+ and NO_3^- ions was faster than the release of CO_2, even when the nitrogen atom was part of an aromatic ring. The opening of rings, both aromatic and saturated, was considered not to be the rate-determining step. Since the rate of mineralization was slower than the disappearance of the cyclic nitrogen compounds, the formation of reaction intermediates had to be assumed.[5]

Organic nitrogen compounds in acidic pH underwent TiO_2-promoted photodegradation, releasing a large part of their nitrogen as ammonium ions, which are quite resistant to oxidation to nitrate. Ammonium ions are very toxic for fish, and if large concentrations should be produced, it may be necessary to remove the ammonium ions by microbial denitrification or by ion exchange.[6]

Pyridine in aqueous dispersions of TiO_2 (Degussa P25; 2.5 g/L) under illumination with a high-pressure Hg lamp ($\lambda > 340$ nm) was readily decomposed, with the initial rate observing apparent first-order kinetics. In a comparison with several other nitrogen compounds, the relative order of reactivity was pyridine > benzamide > *N*-phenylethanamide > nitrobenzene. Many intermediates of the degradation of pyridine were detected. Among these, identified intermediate products included 2-hydroxypyridine, 4-cyanopyridine, 2-cyanopyridine, 2,2′-dipyridyl, 2,4′-dipyridyl, formamide, formate, and acetate. The nitrogen atoms were primarily released as NH_4^+ ions, which were slowly oxidized to nitrate. After an ultraviolet (UV) irradiation period about 2.5 times longer than required to eliminate pyridine, all the nitrogen-containing intermediates were mineralized. The remaining organic products were acetate and formate.[7]

The reactions of nitrogen-containing organic phosphorus compounds are discussed in Chapter VI.

I. *s*-TRIAZINES

A. Homogeneous Photolysis

With the important *s*-triazine pesticides atrazine, simazine, and promazine,

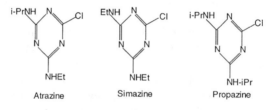

| Atrazine | Simazine | Propazine |

SCHEME V.1

the predominant photochemical reaction in aqueous solutions was replacement of halogen atoms by hydroxyl groups, leading to 2-hydroxy derivatives.[8] Yao and Haag[9] observed that the direct (dark) action of ozone on *s*-triazines is very slow, with second-order rate constants in the range of 5 to 6 M^{-1} s^{-1}. Much-enhanced rates of degradation were achieved with the combination O_3–H_2O_2 in the dark. In this reaction system, as was shown by Hoigné and Bader,[10] the hydrogen peroxide-initiated decomposition of ozone generated hydroxyl radicals, which were the primary oxidants.[10] Rate constants for the reaction of ·OH radicals with triazines were determined by Haag and Yao[11] using the "photo Fenton" reaction to produce ·OH radicals,[11] and by Chramosta et al.,[2] generating these radicals with the O_3–H_2O_2 decomposition. Competition kinetics were used in mixtures of each triazine in aqueous solution with chlorobenzene, for which the rate constant with ·OH was known, that is, 4.3×10^9 M^{-1} s^{-1}.[12] The second-order rate constants for the reaction of hydroxyl radicals with atrazine, simazine, and propazine were determined to be 1.7×10^9, 2.1×10^9, and 1.2×10^9 M^{-1} s^{-1}, respectively. The slower rate with propazine relative to that with atrazine and simazine was explained by steric hindrance of the two exocyclic -NHCH(CH$_3$)$_2$ groups in propazine, which impede approach of the hydroxyl radicals on the triazine molecule.[2]

See also Section I.B on the heterogeneous photocatalytic degradation of the triazine herbicides.

1. Propazine

Propazine is a preemergence herbicide, of low toxicity to mammals.[13] Its direct photolysis in distilled water was carried out under illumination with simulated sunlight, using a xenon lamp. The main degradation product was hydroxypropazine, due to replacement of a chlorine atom by a hydroxyl group. Other reactions included dealkylation and dehalogenation. The reactions could be accounted for by the tentative pathway[14] shown in Scheme V.2. These transformations in direct photolysis were similar to those of the other chlorotriazines, such as atrazine.

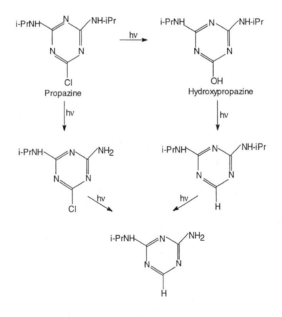

SCHEME V.2

2.   Atrazine

The rate of direct photolysis of atrazine in dilute aqueous solution (3 to 33 μM) under UV irradiation with a low-pressure Hg lamp (mainly 254 nm) was independent of pH in the range of pH 3 to 11, with apparent first-order rate constants k_1 of about 2×10^{-4} s^{-1}. The photodegradation was considerably accelerated at low concentrations of hydrogen peroxide (≤ 2 mM), but only in acidic and neutral solutions (pH 3 and 7), resulting in values of $k_1 = (20–50)$ $\times 10^{-4}$ s^{-1}. At higher initial concentrations of H_2O_2 (20 mM), a rapid "dark" degradation of atrazine was observed. The products of the direct photolysis in the absence of added hydrogen peroxide included 2-OH-atrazine. In the presence of H_2O_2, the dealkylation products desethyl atrazine and desisopropyl atrazine were identified.[15]

The rate of the homogeneous-phase photodecomposition of the triazine herbicides atrazine, ametryn, prometon, and prometryn was enhanced by the presence of 20 to 300 μM of Fe^{3+}. The mechanism probably involved the intermediate formation of ·OH radicals. In the absence of oxygen, the reaction was much slower.[16]

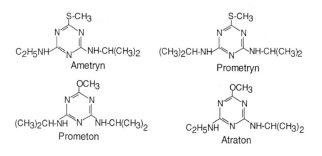

SCHEME V.3

The direct photolysis of atrazine in aqueous solution under illumination at 254 nm was very strongly accelerated by hydrogen peroxide. The optimal concentration of H_2O_2 was about 10 mM; at higher concentrations the rate of degradation of atrazine decreased. The presence of humic substances inhibited the rate of photodegradation in the presence of H_2O_2.[17]

The riboflavin-photosensitized degradation of the triazine herbicides atrazine, ametryn, and atraton (see Scheme V.3) by illumination with sunlight resulted in the formation of several reaction intermediates.[18] The primary products identified were N^6-dealkylated s-triazines and oxidized products, the 6-acetamido-s-triazines. More prolonged irradiation led to the formation of 4,6-diamino-s-triazine derivatives. Thus, for the photosensitized decomposition of atrazine, the products shown in Scheme V.4 were observed. These reactions were very different from those in the direct photolysis of the s-triazines, which led, as noted earlier, to 2-hydroxyl derivatives. The proposed

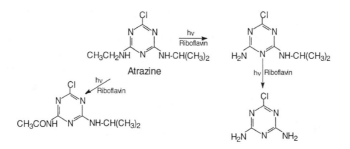

SCHEME V.4

mechanism of sensitized photodegradation of the *s*-triazines may involve excited-state complexes (exciplexes), formed from the excited sensitizer, and resulting in oxidation of the alkylamino groups. Since oxygen was required for these reactions, hydroperoxide intermediates may also be involved.[18] The photolysis of atrazine in aqueous solutions under illumination with simulated sunlight (>340 nm) was markedly accelerated in the presence of humic substances.[19] As discussed in Chapter VIII, Section I, humic acids play an important role as photosensitizers in the natural aquatic environment.

In the presence of H_2O_2 and Fe(III) chelates, such as with Fe(III)–gallate, –picolinate, and –rhodizonate, the disappearance of atrazine was rapid even in the dark, but involved only dechlorination, dealkylation and deamination, while leaving the *s*-triazine ring intact, presumable by transformation to cyanuric acid. Among these chelators, gallic acid was most effective.[20]

3. Metazachlor

The direct photolysis of the herbicide metazachlor at 254 nm in water was found to be unaffected by the solution pH. With low initial concentration of the herbicide (about 40 μ*M*), the apparent first-order rate constant for its disappearance was 2.2×10^{-4} s^{-1} at pH 3–11. In the presence of H_2O_2 ($\leq$2 m*M*), the apparent first-order rate constant increased up to about 150×10^{-4} s^{-1}.[15]

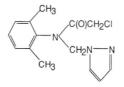

SCHEME V.5

B. Heterogeneous Photodegradation

Hustert et al.[21] observed the TiO$_2$-photocatalyzed disappearance of atrazine under UV light or sunlight.

In illuminated aqueous suspensions of TiO$_2$, the photodegradation of the *s*-triazine herbicides involved substitution of the chlorine atoms with hydroxyl groups and the dealkylation of the alkylamino side chains, followed by deamination. The final product in the photodegradation of atrazine, simazine, trietazine, prometon, and prometryn was cyanuric acid, which was resistant to further phototransformation. Atrazine was decomposed to a concentration of less than 0.1 ppb. The intermediates (a) to (d) identified during the TiO$_2$-photocatalyzed degradation of atrazine are presented in Scheme V.6.[22–30]

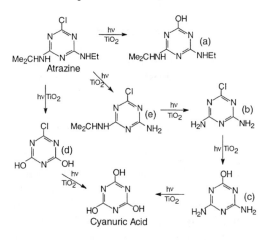

SCHEME V.6

In a kinetic study carried out in an annular batch photoreactor of quartz illuminated with a 250-W medium-pressure Hg lamp, in aerated aqueous suspensions of TiO$_2$, atrazine (initially 10 mg/L) was 98% degraded within 10 to 20 min. Much longer irradiation times were required for the degradation of the dealkylated intermediates (desethyl, desisopropyl, and desethyl-desisopropyl).[31] The photooxidation of atrazine was also performed using titania colloids and microbeads prepared by the sol-gel technique. These photocatalysts included colloidal sols with 0.05 μm diameter and microspheres with 1 mm diameter. The observed first-order rate constants for the disappearance of atrazine with the colloids, the microspheres, and Degussa P25 were 0.0076, 0.0068, and 0.18 min^{-1}, respectively. The final stable product was in all cases cyanuric acid. Although the photoactivity of the microspheres was smaller than that of the commercial product (Degussa P25), such microbeads should be very useful in practical flow-through photoreactors.[27]

An annular flow-through photoreactor was used for the photodecomposition of atrazine. The photoactive surface was the inner Pyrex® glass wall of the photoreactor, which was coated with a TiO$_2$ membrane (0.1 μm thick; prepared by dipping the wall in a 4 wt% solution of TiO$_2$ colloid and annealing at

400°C). Since the degradation rates were small, the reaction was carried out by recycling. In order to prevent mass transfer limitation of the process, high recycle flow rates had to be used. The apparent quantum yield was 0.49 mmol einstein^{-1}.[32]

With ZnO as photocatalyst, the degradation of atrazine was slower than with TiO$_2$, and the distribution of the intermediates was different. The predominant route with ZnO was dealkylation, with initial formation of the intermediates (b) and (c), which after 30 h were converted to a large extent to compound (d) (see Scheme V.6).[30]

The combined effect of photocatalysis with TiO$_2$ and dye sensitization with rose bengal was tested on the *s*-triazine herbicide terbutylazine (terbuthylazine, 2- *tert*-butylamino-4-chloro-6-ethylamino-1,3,5-triazine). Under illumination with visible light ($\lambda > 420$ nm), in the absence of the dye sensitizer, TiO$_2$ was ineffective as photocatalyst since it was photoexcited only below about 380 nm. In the presence of both TiO$_2$ (1 g/L) and rose bengal (10 μM), there occurred a slow photodegradation of the herbicide. Simultaneously, the rose bengal was bleached. Since the herbicide was not degraded in the absence of the TiO$_2$, a singlet oxygen mechanism had to be excluded. In the proposed mechanism, light was absorbed by a surface-adsorbed dye molecule. This sensitized dye molecule then injected an electron into the conduction band of the n-type TiO$_2$. Such conduction band electron may be captured by O$_2$ molecules, forming O$_2^-$ radical-anions, which may be the reactive species attacking the herbicide molecules.[33] This approach of combined semiconductor photocatalysis with dye sensitization could be useful for the solar photodegradation of pesticides.

1. Vacuum-UV Photolysis of Atrazine

While near-UV illumination of atrazine in aqueous dispersions of TiO$_2$ led to almost quantitative formation of cyanuric acid, vacuum-UV photolysis with an Xe excimer lamp ($\lambda = 172$ nm) led to partial mineralization (about 65%) of the herbicide. Surprisingly, this mineralization was larger in oxygen-free solutions than in the presence of oxygen. In argon-, air-, and oxygen-saturated solutions, 10, 30, and 50% of the initial atrazine was converted to cyanuric acid. Since vacuum-UV photolysis of water produces mainly ·OH and ·H radicals, it may be possible that a reductive process could be responsible for the partial mineralization of atrazine. Cyanuric acid itself in aqueous solutions was not affected by 5 h of illumination with the Xe excimer lamp.[34]

2. Cyanuric Acid

Since cyanuric acid is toxic, its accumulation in photocatalytically treated wastewater should be unacceptable. Photocatalysis may, however, be a useful pretreatment prior to biodegradation. Cyanuric acid and other *s*-triazine ring compounds were completely mineralized by stable bacterial mixed cultures.[35]

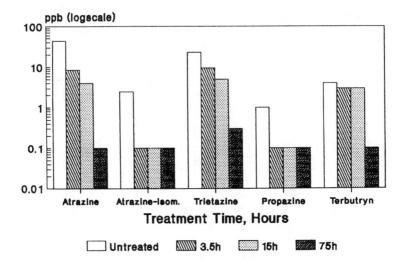

FIGURE V.1.   Solar photodegradation of triazine herbicides in polluted well water in the presence of Fe^{3+} (0.3 mM) + TiO_2 (1 g/L) + H_2O_2 (0.1 M). Effect of treatment time.

3.   Solar Photodegradation of Triazine Herbicides in Ground Water

In experiments on the sunlight photodegradation of pesticides from a heavily polluted well, the treatment with several mixtures of Fe(III), TiO_2, and H_2O_2 on the disappearance of triazine herbicides was tested. The effect of illumination time on treatment with Fe^{3+} + TiO_2 + H_2O_2 is presented in Figure V.1. Atrazine and terbutryn had completely disappeared (to less than 0.1 ppb) within 75 h, while for propazine even 3.5 h was sufficient.[36]

4.   Bentazone

The toxic contact herbicide bentazone (3-isopropyl-2,1,3-benzothiadiazin-4-one-2,2-dioxide)

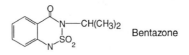

SCHEME V.7

is used for the postemergence control of many annual weeds.[13] Bentazone in aqueous dispersions of TiO_2 underwent rapid degradation under illumination with simulated sunlight, leading to complete mineralization, as indicated by the quantitative formation of CO_2 and SO_4^{2-}. The half-life of its disappearance at pH 2 and pH 7 was about 30 and 4 min, respectively.[37]

II. AMINES, AMIDES, AND CARBAMATES

Amides and carbamate derivatives include several very important pesticides and plant growth regulators. Their toxic action is mainly due to the inhibition of acetylcholine esterase. Many of these compounds undergo slow hydrolysis (dark reaction). Their degradation is considerably enhanced by UV irradiation, but often the reaction intermediates formed are more stable, and sometimes more toxic. With carbamate pesticides, the main homogeneous-phase photochemical reaction is photohydrolysis, leading to phenols.[3] Aminopolycarboxylates are used on a very large scale in domestic and industrial detergent compositions, to sequester di- and trivalent ions and for water softening. Their release into wastewater effluents causes a heavy load on water treatment plants.

The semiconductor promoted photodegradation has been found to be very effective in the decomposition of various carbamate pesticides in dilute aqueous solutions.

A. Anilines

The rate of the direct photolysis of aniline in aqueous solutions is strongly dependent on pH. At pH 5, at which it exists in the protonated form, aniline was resistant to illumination by a 200-W Hg lamp ($\lambda > 300$ nm). At pH 7, the half-lives of aniline, *p*-chloroaniline, *p*-nitroaniline, and 4-bromo-3-chloroaniline by direct photolysis were 180, 4.6, 880, and 61 min, respectively. The photodegradation of the anilines was very strongly accelerated in the presence of riboflavin (5 μM). *p*-Toluidine, which was resistant to direct photolysis, was rapidly decomposed in the presence of the sensitizer ($t_{1/2}$ ~0.7 min). The photosensitized degradation of the anilines was retarded by the presence of oxygen. This indicated that singlet oxygen was not an intermediate, but that a "Type I" sensitized reaction occurred, as in the riboflavin sensitized photodegradation of phenols (see Chapter III, Section II.A.5). In this mechanism, energy may be transferred directly from the photoexcited sensitizer to the anilines.[38]

1. *p*-Phenylenediamines

The carcinogenic and toxic *p*-phenylenediamines undergo slow oxidation in the dark by peroxydisulfate. The oxidation of *N*-substituted *p*-phenylenediamines by peroxydisulfate results in the formation of the intensely colored *p*-semiquinonediimines, a process that is important in photography. This reaction was found to be strongly accelerated by irradiation with near-UV light ($\lambda > 300$ nm) in the presence of tri-2–2′ -bipyridylylruthenium(II), $Ru(bpy)_3^{2+}$, which acted as both photosensitizer and photocatalyst. The catalytic cycle involved the excitation of $Ru(bpy)_3^{2+}$ to $Ru^*(bpy)_3^{2+}$, which reacted with $S_2O_8^{2+}$, with formation of $Ru(bpy)_3^{3+}$, the sulfate radical anion $\cdot SO_4^-$, and the stable SO_4^{2-} anion. The *p*-phenylenediamine was then oxidized by $Ru(bpy)_3^{3+}$, producing the colored *p*-semiquinonediimine and recycling $Ru(bpy)_3^{2+}$. A further oxidation step led to the *p*-quinonediimine, which absorbs only in the UV region. However, if the initial *p*-phenylenediamine was in large excess, it reacted with the *p*-quinonediimine to form additional *p*-semiquinonediimine.[39]

B. Aminopolycarboxylates

The direct photolysis of Fe(III)–EDTA under illumination with a Xe lamp was studied at pH 4.5, 6.9, and 8.5. The maximal rate of photolysis was at pH 4.5. The photodegradation products included CO_2, HCHO, *N*-carboxymethyl-*N,N'*-ethylenediglycine, *N,N'*-ethylenediglycine, iminodiacetic acid, and glycine.[40]

Fe(III) is one of the more abundant cations in natural waters, and since it forms a very stable complex with nitrilotriacetic acid ($N[CH_2COOH]_3$, NTA), with $\log K_c = 15.9$, NTA will occur in such waters essentially as the Fe(III)–NTA complex. The Fe(III)–NTA complex has an absorption spectrum that extends to the visible region, and the complex undergoes photodegradation under sunlight illumination. The main products of this direct photolysis were Fe(II), CO_2, iminodiacetic acid (IDA), and glycine.[41–43] Iminodiacetate is relatively resistant to further photolysis and may thus accumulate in natural waters.[44]

The sunlight photolysis in natural surface waters (at 60° N, Stockholm latitude) of the Fe(III) complexes of nitrilotriacetic acid, ethylenediamine tetraacetate (EDTA), and diethylenetriamine pentaacetate (DTPA) in aqueous solutions resulted in estimated half-lives of 43, 11, and 8 min, respectively. The quantum yields for photoconversion of these compounds in sunlight (rate of conversion/rate of light absorption) were estimated to be 0.013, 0.024, and 0.008, respectively.[42] Photolysis of Fe(III) nitrilotriacetate in acidic aqueous solutions resulted in Fe(II), formaldehyde, and CO_2. Under illumination at 365 nm, equimolar amounts of Fe(II) and HCHO were formed, suggesting a redox reaction between of Fe(III) and a COOH group. The quantum yield of Fe(III) reduction in such solutions was 0.18. Prolonged illumination of acidic solutions yielded iminodiacetic acid and glycine:

$$Fe(III)-OC(O)CH_2N(CH_2COO^-)_2 + h\nu \rightarrow$$
$$Fe(II) + \cdot OOCCH_2N(CH_2COO^-)_2 \tag{1}$$

$$\cdot OOCCH_2N(CH_2COO^-)_2 \rightarrow \cdot CH_2N(CH_2COO^-)_2 + CO_2 \tag{2}$$

$$\cdot CH_2N(CH_2COO^-)_2 + H_2O \rightarrow HCHO + HN(CH_2COO^-)_2 \qquad (3)$$

$$HN(CH_2COO^-)_2 \rightarrow H_2NCH_2COOH \qquad (4)$$

Such reactions can be presumed to occur also in natural waters at pH 5 to 6 under sunlight.[43]

Iminodiacetic acid, $HN(CH_2COOH)_2$, the major intermediate product in the photodegradation of NTA, may in natural waters be converted into a carcinogenic nitroso derivative. Iminodiacetic acid formed a very stable Fe(III) complex at pH ≤ 3.5. At neutral and alkaline pH the complex was not maintained, and hydrous ferric oxide was precipitated. The direct photolysis of the Fe(III)–iminodiacetate complex with a high-pressure Hg lamp resulted in the reduction of Fe(III) to Fe(II) and the formation of glycine and glyoxylic acid:

$$\left[FeHN(CH_2COO)_2\right]^+ + h\nu \rightarrow$$
$$Fe^{2+} + NH_3^+ - CH_2 - COO^- + OHC - COOH \qquad (5)$$

The quantum yield was remarkably high, 0.2 at 405 nm, rising to 0.53 at 254 nm. The presence of oxygen had little influence on the quantum yields.[45]

The direct phototransformation of copper(II) nitrilotriacetate [Cu(II)–NTA] in aqueous solutions was generally similar to that of Fe(III)–NTA complexes. Cu(II)–NTA absorbs in the UV-visible region, with $\lambda_{max} = 238$ nm, and with an absorption tail reaching to 380 nm. On illumination with both low- and high-pressure Hg lamps (at $\lambda = 254$, 296, or 313 nm), the photolysis reaction was similar in yielding as products Cu(I), HCHO, CO_2, and iminodiacetic acid. Cu(I) and HCHO were formed in a 1:1 ratio, suggesting a mechanism involving charge transfer between Cu(II) and one carboxylate group of NTA. The quantum yield of the transformation of Cu(II)–NTA at 254 and 313 nm in deaerated solutions (at pH 7) was 0.011 and 0.0014, respectively. The presence of oxygen had only minor effects on the initial rate of the phototransformation and on the quantum yields, but it caused rapid reoxidation of the insoluble Cu(I) into the water-soluble Cu(II). Thus, Cu ions act as catalysts in the photooxidation of NTA.[46]

While EDTA itself is nontoxic, the heavy metal chelates of EDTA are toxic. These chelates are very stable, and difficult to degrade by conventional water treatment methods. Under illumination with a 500-W high-pressure Hg arc through a Pyrex® glass window, in aqueous dispersions of TiO_2 (2.8 g/L), the metal chelates were efficiently photooxidized. The order of reactivity was Fe > Cu > Pb > Na > Zn > Co > Ni > Cd > Mn. This order of reactivity was the same as that of the association constants of the chelates. Thus, it was proposed that the reactivity of the chelates was determined by the electronegativity of the metals, or by their redox potentials. The photodegradation of the chelates resulted in deposition of the metals on the Ti.O$_2$.[47]

The photodegradation of the Fe(III) complex of ethylenediamine tetra(methylenephosphonic acid) is described in Chapter VI, Section I.

C. Benzamide

The direct photolysis of benzamide in aerated aqueous solution under irradiation with a 125-W high-pressure Hg lamp ($\lambda > 220$ nm) was very slow, and 25 h was required for complete disappearance of benzamide. At $\lambda > 340$ nm there was no direct photolysis at all. In the presence of TiO$_2$ (2.5 g/L), at $\lambda > 340$ nm, the photocatalytic reaction was very rapid, causing the complete disappearance of benzamide within 110 min. However, while the disappearance of benzamide was fast, the complete mineralization was much slower, and about 150 h of illumination was required for total conversion of all the nitrogen atoms of benzamide into NO$_3^-$. Primary reaction intermediates were the three hydroxybenzamides, *p*-benzoquinone, and one unidentified intermediate. These five intermediates underwent further TiO$_2$-photocatalyzed degradation at about the same rate as benzamide. Since the complete mineralization of the carbon atoms of benzamide was slow, it was necessary to assume the formation of some intermediate containing one carbon and one nitrogen atom. A tentative mechanism was proposed for the photocatalytic degradation of benzamide,[1] as in Scheme V.8.

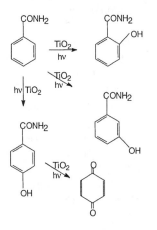

SCHEME V.8

The initial rate of disappearance of benzamide under illumination at $\lambda \geq 220$ nm in the presence of TiO$_2$ (2.5 g/L) or H$_2$O$_2$ (1.37 m*M*) at pH 6.5 was approximately similar, with $t_{1/2} = 5$ to 6 min, and 99% disappearance after 35 to 40 min. The combination of TiO$_2$ and H$_2$O$_2$ was not more effective than either oxidant alone. On the other hand, by the "dark" Fenton reaction, in the presence of 0.15 m*M* FeSO$_4$ and using the molar ratio H$_2$O$_2$/benzamide = 8, at pH 3.4, the oxidation of benzamide was very much slower, requiring 24 h for 99% disappearance.[48]

D. Isoxaben

Isoxaben, *N*-[3-(1-ethyl-1-methylpropyl)-isoxazol-5-yl]-2,6-dimethoxyl–benzamide, is a selective preemergence herbicide used for the control of broad-leaved weeds in winter cereals.[13] The photolysis of isoxaben was studied in dilute aqueous solutions, as well as in an adsorbed layer on silica gel plates and on glass plates. The major products of simulated-sunlight irradiation of isoxaben were three isomers of isoxaben, as well as 2,6-dimethoxybenzamide and 2,6-dimethoxybenzonitrile. The half-life for the disappearance of isoxaben in water under natural sunlight illumination was estimated to be about 14 d. The main degradation reactions both in water and in solid phase were proposed to be ring opening and rearrangement at the isoxazole ring, fission of the amide bond, and fission of the aliphatic chain. A key intermediate in the photodegradation may be an azirine derivative (see Scheme V.9).[49]

E. Carbetamide

Carbetamide, 2-phenylcarbamoyloxy-*N*-ethyl-propionamide, is a selective herbicide adsorbed through the roots and leaves, and acts by the inhibition of cell division in young tissues of roots and shoots. It is of moderate toxicity to mammals.[13] A study of the kinetics of photolysis of carbetamide in aqueous solutions containing 1% acetone indicated a radical mechanism.[50] In the presence of acetophenone, carbetamide was photodegraded to form *N*-ethylacetamide 4-aminobenzoate, due to a Fries rearrangement, presumably by triplet transfer between acetophenone and carbetamide.[51] In direct homogeneous-phase photolysis of carbetamide in water by UV light, the main degradation products were aniline, the dealkylation product C$_6$H$_5$–NH–CO–O–CH(CH$_3$)–CO–NH$_2$, the hydrolysis product HO–CH(CH$_3$)–CO–NH–CH$_2$–CH$_3$, and a product of hydroxylation on the aromatic ring.[52] The products of photodegradation of aqueous solutions of carbetamide in the presence of TiO$_2$ included cyclic structures, possibly obtained by radical-cation formation on the surface of TiO$_2$.[53]

The photodegradation of carbetamide adsorbed on kaolin or silica may provide a model of the fate of this herbicide after spraying on soil. There was no marked difference in the spectroscopic properties of the compound when adsorbed on kaolin, bentonite, silica, or soil (λ_{max} = 280 nm, triplet state = 278 kJ) or when in aqueous solution (λ_{max} = 271 nm, triplet state = 313 kJ). However, the kinetics of photodegradation were quite different. In solution, the mineralization of the herbicide followed first-order kinetics, while in the adsorbed state second-order kinetics were observed — even for small quantities of the compound adsorbed on a solid support. Presumably, in the adsorbed

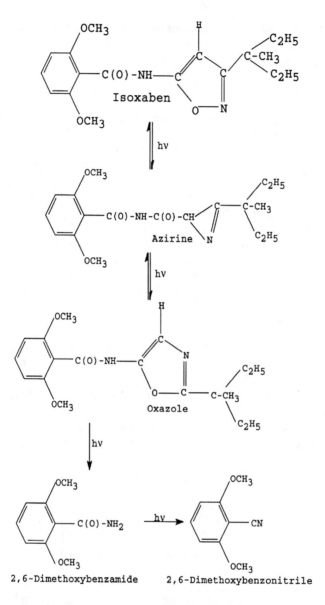

SCHEME V.9

phase, photoexcited carbetamide interacts with another molecule of the herbicide in the rate-determining step.[54]

F. Aldicarb, Carbaryl, Carbofuran, and Baygon

Aldicarb, 2-methyl-2-(methylthio)propanal *O*-[(methylamino)-carbonyl] oxime, is used as an insecticide, acaricide, and nematicide. It is a contact poison with

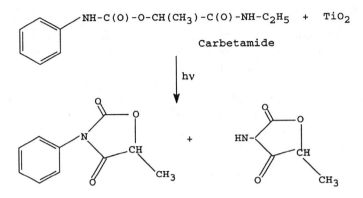

SCHEME V.10

systemic action, functioning as an acetylcholinesterase inhibitor. It is extremely toxic to mammals. Carbaryl, 1-naphthyl methyl carbamate, is used as an insecticide and as a plant growth regulator. It is an acetylcholine esterase inhibitor, but of low mammalian toxicity. Carbofuran, 2,3-dihydro-2,2-dimethylbenzofuran-7-yl methyl carbamate, serves as insecticide, acaricide, and nematicide. It acts as stomach and contact poison by acetylcholinesterase inhibition and is very toxic to mammals.[13]

The homogeneous-phase photodegradation of the carbamate pesticides aldicarb, carbaryl, and carbofuran was tested in distilled water and in seawater, without and with added humic acids. The main photodegradation product from aldicarb was aldicarb sulfoxide. From carbaryl, which is quite labile to "dark" hydrolysis, the main reaction product was 1-naphthol. From carbofuran, one main unidentified hydrolysis product was detected.[55]

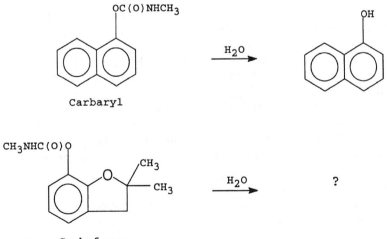

SCHEME V.11

The rates of photodegradation of these carbamate pesticides in distilled water under illumination with a xenon arc lamp were in the order aldicarb > carbaryl > carbofuran. With carbaryl and carbofuran, the degradation was photosensitized by addition of humic acids. In a comparison of different water sources, the order of degradation for carbofuran was artificial seawater with humic acids > artificial seawater without humic acids > pond water > distilled water.[55]

In the presence of hydrogen peroxide and Fe(III) chelates, such as Fe(III)–gallate, the carbamate insecticides carbaryl and baygon were completely transformed even in the dark within a few minutes.[20]

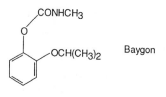

SCHEME V.12

G. Urea and Uracil Derivatives

1. Urea

Urea (initially 0.1 g/L) in aqueous dispersions of Ag (1% w/w)/TiO$_2$ (anatase form) after 12 min of illumination with a 125-W medium-pressure Hg lamp was mineralized to about 83%. With the "bare" TiO$_2$ catalyst, only 16% of the urea was oxidized under the same conditions.[56]

2. Monuron

The herbicide monuron [3-(4-chlorophenyl)-1,1-dimethylurea] is widely used for weed control, acting as an inhibitor of photosynthesis. Its "dark" hydrolysis as well as its direct photolysis is very slow. Rapid complete mineralization of monuron was achieved by photocatalytic degradation in the presence of suspended TiO$_2$ (anatase form). Other semiconductors were less effective, and the order of activity was TiO$_2$ $\gg$ ZnO $\sim$ WO$_3$ > SnO$_2$. Thus, by illumination with simulated AM1 sunlight (λ > 310 nm) and starting with an aerated aqueous solution of 20 mg/L of monuron and 0.1 g/L of TiO$_2$ at pH 5.5, the herbicide was completely oxidized within 30 to 40 min, with the chlorine atoms released as chloride ions. The nitrogen atoms were initially released as ammonium ions, and about 6 h was required for their conversion to nitrate ions. The photodegradation followed pseudo first-order kinetics, and the initial rates of degradation could be related to the initial concentrations of monuron by a Langmuir–Hinshelwood type equation. In the pH range 3 to 9, complete mineralization was achieved, while at pH 1.0 and 11.0, mineralization was partial even after prolonged irradiation. The mechanism of the degradation was clarified by identifying several reaction intermediates. The major intermediates included 4-chlorophenyl isocyanate, 4-chloro-2-benzoxazolone, and 4-chlorophenol, but these were also degraded within about 30 min.[57,58]

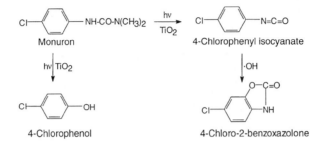

SCHEME V.13

3. Bromacil and Terbacil

Dye-sensitized phototransformation was used for the degradation of two uracil herbicides. Bromacil (5-bromo-3-*sec*-butyl-6-methyluracil) is a nonselective herbicide for total weed and brush control in noncropped areas and is very effective against perennial grasses. Terbacil (3-*tert*-butyl-5-chloro-6-methyluracil) is a selective herbicide for control of annual weeds and perenial grasses. Both herbicides are taken up mainly by the roots and are inhibitors of photosynthesis.[13]

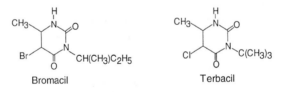

SCHEME V.14

Direct solar photolysis of bromacil (without sensitizer) was shown to be extremely slow, forming in very low yield one detectable product, 5-bromo-6-methyluracil. Most of the bromacil was recovered unchanged.[59] In the methylene blue (MB) photosensitized oxygenation of bromacil with sunlight, bromacil was dehalogenated and rearranged to one major product, 3-*sec*-butyl-5-acetyl-5-hydroxyhydantoin, isolated in 83% yield.[60]

The most effective dye sensitizer for the solar photooxidation of bromacil was riboflavin, followed by methylene blue, rose bengal, humic acids, and chlorophyll. The reaction was faster at alkaline pH, and was very slow in acidic solutions.[61]

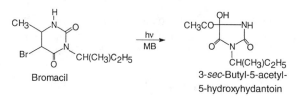

Bromacil   3-*sec*-Butyl-5-acetyl-
                                                5-hydroxyhydantoin

SCHEME V.15

The uracil herbicide terbacil in aqueous solutions (pH 3 to 9) underwent photosensitized decomposition under sunlight in the presence of either methylene blue (3 ppm) or riboflavin (10 ppm), yielding mainly *tert*-butyl-5-acetyl-5-hydroxyhydantoin, as well as some other products.[62]

See also Chapter VIII, Section V, on the methylene blue-sensitized photooxidation of bromacil under highly concentrated sunlight. Chapter VII discusses the photodegradation and photoisomerization of metanilic acid.

4. Isoproturon

The urea derivative isoproturon is used as a selective pre- and postemergence herbicide.[13]

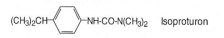

SCHEME V.16

Direct photolysis of aqueous solutions of isoproturon resulted in photoproducts that were due to *N*-demethylation, *N*-oxidation, ring hydroxylation, and dimerization. The half-life under sunlight illumination was about 1 d.[63]

H. Propachlor, Alachlor, and Pendimethalin

1. Propachlor

The preemergence herbicide propachlor (2-chloro-*N*-isopropyl acetanilide) is used for the control of annual weeds and grasses.[13] Its direct photolysis under UV illumination led within 5 h to about 80% decomposition. The products identified included *N*-isopropyloxindole, *N*-isopropyl-1,3-hydroxyoxindole, and a spirocompound. On the other hand, photosensitized degradation of propachlor by visible light in the presence of riboflavin resulted in essentially complete destruction of the herbicide within 12 h, with only traces of a product identified as *m*-hydroxy-propachlor. Both the UV photolysis and the visible light sensitized photodegradation required the presence of oxygen (see Scheme V.17).[64]

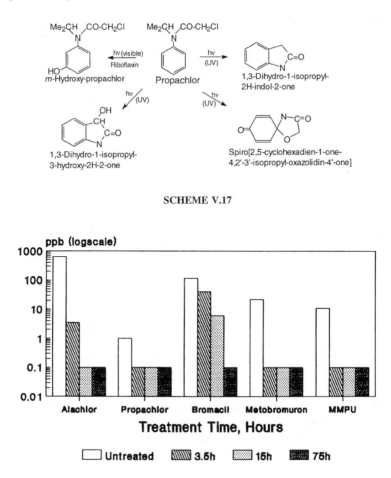

SCHEME V.17

FIGURE V.2. Solar photodegradation of acetamide and bromouracil herbicides and urea derivatives in polluted well water in the presence of Fe^{3+} (0.3 mM) + TiO_2 (1 g/L) + H_2O_2 (0.1 M). Effect of treatment time.

The photodegradation under sunlight of several acetamide and bromouracil herbicides in heavily polluted well water is presented in Figure V.2, showing the pollutant disappearance with time in a mixture of Fe^{3+} (0.3 mM) + TiO_2 (1 g/L) + H_2O_2 (0.1 M). Propachlor, metobromuron, and 1-methoxy-1-methyl-3-phenylurea (MMPU) were brought to the limit of detection (0.1 ppm) within 3.5 h. Alachlor had disappeared within 15 h, and bromacil within 75 h.[36]

2. Alachlor and Pendimethalin

Alachlor (Lasso) is used as a pre- and postemergence herbicide, acting by protein synthesis inhibition. Pendimethalin is a selective herbicide used for preemergence treatment. It inhibits cell division and cell elongation. Both herbicides are toxic to fish.[13]

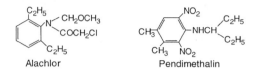

Alachlor   Pendimethalin

SCHEME V.18

The photocatalytic degradation of these important herbicides in aqueous slurries of TiO$_2$ by UV illumination (>290 nm) followed first-order kinetics. The degradation of alachlor occurred by oxidation of the side chain and of the aromatic ring, while pendimethalin decomposed by N-dealkylation and elimination of the nitro group.[65]

I.   Dequalinium Chloride

The antibacterial and antifungal drug dequalinium chloride is used for the treatment of throat and mouth infections. Dequalinium chloride (4,4'-diamino-2,2'-dimethyl-N,N-decamethylene diquinolinium chloride) in aqueous solution absorbs in the UV region (λ_{max} = 220, 240, and 320 nm). The compound was rapidly degraded under UV irradiation, at rates observing first-order kinetics. First-order rate constants in pure water (pH 5.9) and in water containing hydrogen peroxide (0.05%, v/v) were 3.55 and 20.6 min^{-1}, respectively. The rate enhancement in the presence of hydrogen peroxide, a source of hydroxyl radicals, suggested that the photolysis of the compound was hydroxyl radical mediated. The only nitrogen-containing product of the photolysis of dequalinium chloride in aqueous solution was 4-aminoquinaldine.[66]

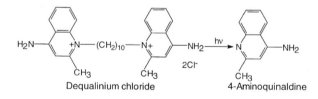

Dequalinium chloride   4-Aminoquinaldine

SCHEME V.19

J.   Propyzamide

The photosynthesis inhibitor propyzamide, 3,5-dichloro-N-(1,1-dimethyl-2-proynyl) is used for herbicidal control of many annual weeds.[13] Its photocatalytic degradation under illumination with a 400-W xenon lamp (λ > 300 nm) was

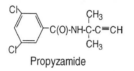

Propyzamide

SCHEME V.20

studied in aqueous suspensions of microcrystalline colloidal TiO_2 (1 or 2 g/L, prepared via a sol-gel process by hydrolyzing titanium tetraisopropoxide). With this "naked" photocatalyst, the release of chloride ions from propyzamide was complete within 2 h, but the mineralization was much slower, and after 6 h less than 30% of the CO_2 had been released. Very considerable enhancement of the photodegradation was achieved by using a photocatalyst obtained by precipitating colloidal TiO_2 on activated carbon particles. With 80 wt%-loaded TiO_2/active carbon, the mineralization of propyzamide to CO_2 was complete after only 1 h of illumination. The higher activity of this TiO_2/active carbon photocatalyst was explained by the higher adsorption capacity of the carbon-containing catalyst.[67]

K. Metalaxyl

The fungicide metalaxyl serves for control of foliar diseases, functioning by systemic action.[13]

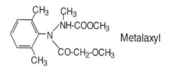

Metalaxyl

SCHEME V.21

Under UV illumination at 290 nm for 2.5 h, metalaxyl in aqueous solution, in the presence of either H_2O_2 or suspended TiO_2, was degraded to 29 and 84%, respectively. The degradation followed apparent first-order kinetics. In the TiO_2-promoted reaction, deacylation products were identified.[68]

L. *p*-Aminophenol

p-Aminophenol is used as a color photographic developer. In order to elucidate the primary mechanism of the photocatalytic degradation of this toxic compound in aqueous dispersions of TiO_2, the detection of free radicals by electron spin resonance (ESR) was carried out, using DMPO as a spin trap. The ESR spectrum of the dispersion, under illumination with a 200-W high-pressure Hg lamp, was consistent with the formation of the spin traps of only two intermediates, the aminophenoxyl radical, p-H_2N-C_6H_4-$O\cdot$, and the $\cdot H$ atom.[69]

M. Aliphatic Diamines

Illumination of aliphatic diamines in aqueous suspensions of platinized TiO_2 resulted in their conversion to cyclic secondary diamines.[70] With the α,ω-diaminocarboxylic acids, a similar ring closure occurred under illumination (λ > 300 nm) for about 40 h in the presence of platinized TiO_2 or CdS in an oxygen-free aqueous suspension (initial pH about 9.7). From L-ornithine and L-lysine, optically active proline and pipecolinic acid were obtained, as well as H_2 and NH_3. More prolonged irradiation caused degradation of the reaction products.[71]

N. Rhodamine 6G

The photostable and toxic rhodamine 6G (basic red 1) is used as a laser dye.

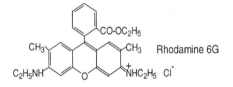

SCHEME V.22

The photosensitized bleaching of rhodamine 6G in oxygenated aqueous dispersions of TiO_2 (0.1 g/L) at pH 3 under illumination with blacklight UV lamps (λ = 350 to 360 nm) occurred with pseudo first-order kinetics, and with a formal quantum yield of about 0.3%. No reaction occurred in the absence of either TiO_2 or oxygen. The only identified intermediate was the deesterified form of rhodamine 6G, which is rhodamine 19.[72]

III. NITROBENZENE AND NITROPHENOLS

Nitroaromatic compounds are components of dyestuffs, explosives, and other industrial products. Nitrophenols are formed in the atmosphere by the gas-phase reaction of phenol and cresols with hydroxyl and NO_3 radicals.[73] Also, nitrophenols are intermediates in the manufacture of various pesticides such as parathion, and are released during their partial degradation. Nitrophenols are priority pollutants in many contaminated waters. Nitroaromatic compounds are fairly resistant to biodegradation.[74]

A. Homogeneous Photolysis

1. Nitrobenzene and Nitrotoluenes

In the homogeneous photolysis of nitrobenzene in aqueous solutions at pH 5.5 illuminated at 253.7 nm, the quantum yield for the disappearance of the compound was 3×10^{-4}, both in aerated and in desoxygenated solutions. The products were phenol, nitrite, and nitrate ions in the aqueous phase, and NO in the vapor phase. The primary reaction was thus a photohydrolysis.[75]

$$C_6H_5-NO_2 + h\nu \rightarrow C_6H_5-OH + H^+ + NO_2^- \tag{6}$$

In homogeneous solutions, the photooxidation of 2,4 -dinitrotoluene by UV light was achieved in the presence of hydrogen peroxide. The degradative reaction involved the initial oxidation of the methyl group, forming 1,3-dinitrobenzene. Subsequent oxidation caused hydroxylation of the benzene ring to hydroxynitrobenzenes, which underwent further degradation by ring opening to carboxylic acids and aldehydes, and finally to carbon dioxide and nitric acid.[76]

The photooxidation of various aromatic compounds is promoted by both natural or artificial dye sensitizers, involving singlet oxygen as the oxidant.[77-79]

2. Nitrobenzyl Derivatives

In the photooxidation of nitrobenzyl derivatives, there were marked differences between the three isomers. With *m*-nitrobenzyl derivatives, including alcohols, alkyl ethers, and amines, the major isolated product was *m*-nitrobenzaldehyde. This reaction in aqueous solution was H^+ catalyzed, and reached maximal quantum efficiencies ($\phi = 0.3$ to 0.4) in the 20 to 50% sulfuric acid range. On the other hand, the photooxidation of *p*-nitrobenzyl alcohol was hydroxide ion catalyzed, and yielded *p*-nitrosobenzaldehyde as the major product. With *o*-nitrobenzaldehyde in organic solvents, the major product was *o*-nitrosobenzaldehyde. The minimal structural requirement for this photooxidation of *m*-nitrobenzyl derivatives was the presence of at least one α-hydrogen and one α-heteroatom in the substrate.[80]

3. Nitrophenols

The homogeneous photolysis of 4-nitrophenol is very slow and is not appreciably affected by the presence of oxygen. With illumination at $\lambda = 365$ nm, the quantum yields for its degradation in both aerated and degassed aqueous solutions were 4.5×10^{-5}, 3.0×10^{-5}, and 1.8×10^{-5} at pH 2, 5.5, and 8.3, respectively. The half-time for disappearance of 4-nitrophenol under natural sunlight illumination was estimated to be 60 and 16 h for acidic and alkaline solutions, respectively. The main products of photodegradation were hydroquinone, nitrate and nitrate ions, nitric oxide, nitrohydroquinone, 4-nitrocatechol, 4-nitrosophenol, and in aerated solutions also benzoquinone. The mechanism proposed involved a strong polarization of the C–N bond of the excited singlet state of 4-nitrophenol, thus leading to a heterolytic fission of the C–N bond, followed by a secondary dismutation reaction of nitrite ions to NO and nitrate ions.[75]

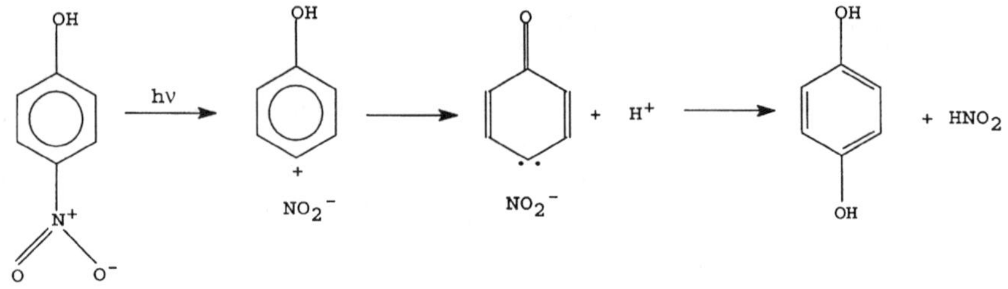

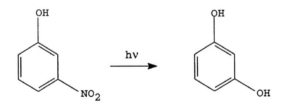

SCHEME V.23

The phototransformation of 3-nitrophenol in homogeneous aqueous solutions is also very slow. In deoxygenated solutions illuminated at $\lambda = 365$ nm, the quantum yields for the decomposition were only 4.5×10^{-6} and 1.0×10^{-6} at pH 2.2 and pH 8.2, respectively. The identified reaction products were resorcinol, 4-nitrosoresorcinol, 2,4-dinitrosoresorcinol, 3-nitrocatechol, 4-nitrocatechol, nitrohydroquinone, and 2,3′-dihydroxy-4-nitrobiphenyl. These photoeffects were due to the direct absorption of light in the absorption bands of the molecular ($\lambda_{max} = 273$ and 331 nm) and the anionic forms ($\lambda_{max} = 278$ and 390 nm), respectively. The production of resorcinol from photoexcited 3-nitrophenol was explained by a heterolytic mechanism of C–N bond fission, because the presence of added ethanol or *n*-butanol did not the affect the rate of formation of resorcinol + 4-nitrosoresorcinol (4-nitrosoresorcinol being a secondary product from resorcinol). Ethanol and *n*-butanol are efficient scavengers of hydroxyl radicals.[81]

SCHEME V.24

Ethanol and *n*-butanol did inhibit the formation of nitrocatechols and nitrohydroquinone during the photolysis of 3-nitrophenol, and these products were thus proposed to be formed via secondary reactions with hydroxyl radicals. The hydroxyl radicals were presumably produced by the UV excitation of nitrite or nitrate ions. Hydroxyl radicals were probably also involved in the formation of the nitrodiphenols. This was confirmed by a heterogeneous photocatalysis experiment, with added ZnO. When 3-nitrophenol was illuminated in the 300 to 400 nm range in the presence of suspended ZnO, the three nitrodiphenols were produced. The formation of the dimerization product 2,3′-dihydroxy-4-nitrobiphenyl (DHNBP) was proposed not to occur via hydroxyl radicals.[81]

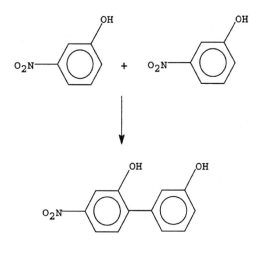

SCHEME V.25

While the direct photolysis of nitrobenzene and of the nitrophenols in aqueous solutions was extremely inefficient and slow, it became quite rapid in the presence of a source of hydroxyl radicals. A comparison was made of the rate of transformation of nitrobenzene and of several nitrophenols (initially 0.1 mM), either under illumination with a 150-W Hg–Xe lamp in the presence of hydrogen peroxide (initially 8 mM), or by the "dark" Fenton reaction, with H_2O_2 (initially 8 mM) and $FeCl_2$ (initially 0.035 mM). Half-times for disappearance by the two methods were for nitrobenzene 260 and 250 min, for 2-nitrophenol 480 and 60 min, for 2,4-dinitrophenol 200 and 13 min, and for 4-nitrophenol 440 and 23 min, respectively. Thus, the rates for photolysis in the presence of hydrogen peroxide were similar to the Fenton reaction only for nitrobenzene, but considerably slower for the nitrophenols. Both methods yielded 2-, 3-, and 4-nitrophenol as intermediates in the degradation of nitrobenzene. The distribution of nitrophenol isomers produced from nitrobenzene by the H_2O_2-enhanced photolysis was *ortho,* 43%, *meta,* 27%, and *para,* 30%. This seemed to indicate that the hydroxylation of nitrobenzene had the character of a nucleophilic substitution. The degradation of 4-nitrophenol resulted in hydroquinone, *p*-benzoquinone, and 4-nitrocatechol.[82]

4. Dye-Sensitized Photodegradation

The dye-sensitized photooxygenation of mono- and dinitrophenols in homogeneous aqueous solution was shown to involve their interaction with singlet molecular oxygen, $O_2(^1\Delta_g)$, produced by a sensitizer such as rose bengal (RB), methylene blue, or eosin:

$$^3RB^* + {}^3O_2 \rightarrow O_2\left(^1\Delta_g\right) + RB \qquad (7)$$

Using competition kinetics experiments with known singlet oxygen quenchers, quantum yields (Φ) were derived for the photosensitized oxidation of 2-nitrophenol (0.0068), 3-nitrophenol (0.0033), 4-nitrophenol (0.0026), 2-nitroanisol ($<10^{-4}$), and 4-methyl-2-nitrophenol (0.0046).[83]

Since surfactants are often used in pesticide formulations, the dye-sensitized photooxygenation of several mononitrophenols was studied in the presence of the cationic micellar surfactant cetyl trimethylammonium bromide (CTAB). The quantum yields for the rose bengal-sensitized photooxidation of 4-nitrophenol, 2-nitrophenol, and 4-methyl-2-nitrophenol in aerated alkaline aqueous solutions in the absence of the surfactant were 26, 68, and 6%, and in the presence of 0.06 M CTAB were 43, 52, and 100%, respectively. Thus, the micellar surfactant significantly enhanced the rates of photodegradation of 4-nitrophenol and of 4-methyl-2-nitrophenol.[84]

B. Heterogeneous Photodegradation

1. Nitrobenzene

The photocatalytic degradation of nitrobenzene under illumination with a 1500-W xenon lamp ($\lambda > 340$ nm) in aqueous dispersions of TiO_2 (0.2 g/L) or ZnO (0.2 g/L) resulted in the disappearance of the nitrobenzene, with half-lives of about 3 min and 80 min, and >90% mineralization within 1 h and 100 h, respectively. The initial inorganic nitrogen species released with both catalysts was the nitrite anion, NO_2^-, which after more prolonged irradiation was converted into nitrate and ammonium ions. The intermediate organic products identified were 2-, 3-, and 4-nitrophenol and dihydroxybenzenes. These intermediates by further hydroxylation and ring cleavage reactions were completely mineralized during more extended illumination. The nitrophenols were formed in approximately equal yields, which is surprising since the nitro group was expected to cause predominant *meta* orientation.[30,85,86]

In a further study, UV irradiation of aqueous solutions of nitrobenzene in the presence of suspended TiO_2 resulted in the identification of several additional aromatic intermediates: nitrohydroquinone, *p*-benzoquinone, phenol, the three dinitrobenzene isomers, nitrosobenzene, and an unidentified product. Acetic and formic acids were also formed. All these intermediates underwent complete mineralization. The production of the dinitrobenzenes was presumably due to nitration of nitrobenzene by the NO_2 released from nitrate or nitrate ions. The photocatalytic degradation of nitrobenzene observed apparent first-order kinetics. The presence of $Pb(NO_3)_2$ (0.1 mM) in solutions of nitrobenzene at pH 6 caused a sixfold decrease in the rate of photodegradation of nitrobenzene. However, such a treatment simultaneously eliminated the highly toxic Pb^{2+} ions from solution, possibly by precipitation on the TiO_2 as PbO_2. At pH 1.3, Pb^{2+} ions had less effect on the rate of photodegradation of nitrobenzene. At this strongly acidic pH, the Pb^{2+} ions were not eliminated from solution by irradiation.[87]

2. Trinitrotoluene

2,4,6-Trinitrotoluene (TNT) is an important explosive and is also very toxic.[88] Its manufacture results in wastewater streams containing both TNT and various other nitroaromatic compounds. Microbial biodegradation results in only partial degradation, leading to aminodinitrotoluenes, which are resistant to further biodegradation. More complete degradation was achieved by UV irradiation combined with either ozone or hydrogen peroxide.[89] Very effective oxidation was obtained by heterogeneous photocatalysis. TNT (50 mg/L) and TiO_2 (0.25 g/L) in a stirred oxygen-saturated aqueous dispersion was illuminated with a 450-W Hg lamp ($\lambda > 340$ nm). The disappearance of TNT followed first-order kinetics, with a pseudo first-order rate constant of 0.07 min^{-1}. The TNT had essentially disappeared (>99%) within 60 min, while the total organic carbon (TOC) content had decreased to <10% within 2 h. In a similar experiment by direct photolysis, without TiO_2, the pseudo first-order rate constant was 0.016 min^{-1}. Two hours was needed to reach disappearance of TNT (>99%), but then there still remained 80% of the TOC. The photocatalytic degradation of TNT with TiO_2 occurred also in nitrogen-purged dispersions, but at a slower rate than in oxygenated mixtures. This photocatalytic method of disposal of TNT and of other nitroaromatic compounds with TiO_2 may be particularly attractive as a wastewater pretreatment prior to biodegradation. Subsequent nitrification and denitrification would then remove the inorganic nitrate and ammonia formed.[90]

3. Nitrophenols

The kinetics of the photocatalytic degradation of 2-, 3-, and 4-nitrophenol in oxygenated aqueous dispersions containing titanium dioxide has been studied as a function of the solution pH and of the initial concentration of the nitrophenols. Under illumination with a 500-W medium-pressure Hg lamp, complete mineralization to CO_2, NO_2^-, and NO_3^- was achieved within about 6 h. This photooxidation was accompanied by a decrease in the pH of the solution. For 2-nitrophenol and 4-nitrophenol, the photocatalyzed reaction had rate maxima at pH 3 and 4.5, respectively. The initial rates of photodegradation decreased with increasing initial concentration of the nitrophenols. This was explained as due to the adsorption of intermediate degradation products on the active sites of the photocatalyst, which compete with adsorption by the parent nitrophenol. Oxygen molecules were assumed to be adsorbed on separate sites on the TiO_2 surface, where they acted as traps for the photogenerated electrons. The photodegradation observed pseudo first-order kinetics with respect to the substrate concentration, and could be described by a modification of the Langmuir–Hinshelwood model, with distinctive active sites for the substrates and for oxygen. The order of reactivity of the three isomers of nitrophenol was 4-nitrophenol > 2-nitrophenol > 3-nitrophenol.[91–93] Iron doping was found to decrease the activity of TiO_2 for the photodegradation of 4-nitrophenol.[94]

Nitrophenols adsorbed on TiO_2 surfaces underwent rapid photodegradation under illumination. This was ascribed to the semiconducting properties of this photocatalyst. The non-semiconducting support materials Al_2O_3 and SiO_2 were inert as photocatalyst. The half-lives for degradation on TiO_2 were only 10 min for 4-nitrophenol, 70 min for 3-nitrophenol, about 2 h for dinitrophenols, and more than 8 h for trinitrophenol (picric acid).[95]

In oxygen-saturated aqueous slurries of TiO_2 at pH 8.5 illuminated with visible light (from a halogen lamp), 4-nitrophenol adsorbed on the photocatalyst sensitized the photodegradation of 4-chlorophenol, which was completely degraded within 90 min. In such a mixture, a synergistic effect was observed. The initial rates of photodegradation of 4-nitrophenol and of 4-chlorophenol separately were 7.7×10^{-3} and 1.7×10^{-2} min^{-1}, and in a mixture of both components were 1.9×10^{-2} and 2.8×10^{-2} min^{-1}, respectively. The mechanism proposed was the formation of an excited state of the colored compound, by absorption of visible light. Interaction between the excited state and the semi-conductor surface enabled charge injection into the conduction band, forming a cationic species, which underwent hydrolytic degradation.[96]

4. 2,6-Dichloroindophenol

The photocatalytic oxidation of 2,6-dichloroindophenol (DCIP) was achieved in aqueous dispersions of TiO_2 (P25). Illumination with a medium-pressure Hg lamp ($\lambda > 280$ nm) caused rapid change in the color of the reaction solution from the initial blue to pink and then to white, with disappearance of the initial absorption peak ($\lambda_{max} = 600$ nm) of the 2,6-dichloroindophenol. The photodecompositon of 2,6-dichloroindophenol could be described by formal first-order kinetics. The complete reaction involved stepwise photochemical dechlorination, via monochloroindophenol (CIP) to indophenol (IP), followed by oxidation of the indophenol to carboxylic acids, and their decarboxylation by a Kolbe photoreaction (see Scheme V.26).[97,98]

IV. BROMOXYNIL AND CHLOROXYNIL

The herbicide bromoxynil (3,5-dibromo-4-hydroxybenzonitrile) is used for postemergent control of annual broad-leaf weeds, primarily as a contact herbi-cide. Its mode of action is due to inhibition of the second light reaction of photosynthesis, and also to uncoupling oxidative phosphorylation of respira-tion.[13] The direct photolysis at 313 nm of bromoxynil in aqueous solutions was found to be quenched by the presence of chloride, carbonate, bicarbonate, and nitrite anions. In aqueous solutions, bromoxynil photolysis yielded two major products, 3-bromo-4-hydroxybenzonitrile and 4-hydroxybenzonitrile. With added NaCl (0.01 M), additional products due to photochlorination were identified: 3-bromo-5-chloro-benzonitrile and 3-chloro-4-hydroxybenzonitrile.[99,100]

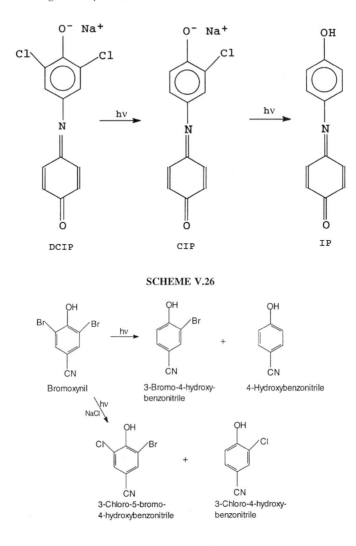

SCHEME V.26

SCHEME V.27

In acidic aqueous solutions (pH 4.5), the photolysis of bromoxynil was sensitized by the presence of Fe(III) and Mn(II) ions. A synergistic effect was observed in the presence of both Fe(III) and Mn(II) ions: the sensitization by a mixture of both cations was larger than that by separate solutions of the same concentration.[101]

The solar photodegradation of the herbicides chloroxynil and bromoxynil in polluted well water in the presence of Fe^{3+} (0.3 mM) + TiO_2 (1 g/L) + H_2O_2 (0.1 M) is presented in Figure V.3. Bromoxynil was brought to the limit of detection (0.1 ppm) within 3.5 h, and chloroxynil within 15 h.[36]

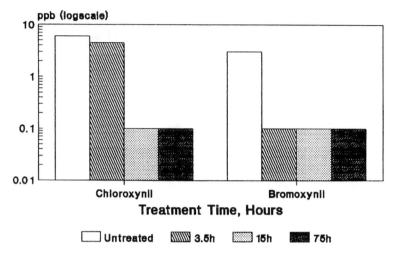

FIGURE V.3.   Solar photodegradation of the herbicides chloroxynil and bromoxynil in polluted well water in the presence of Fe^{3+} (0.3 mM) + TiO_2 (1 g/L) + H_2O_2 (0.1
M). Effect of treatment time.

V. THYMINE

The photoinduced oxidation of thymine (5-methyluracil) has often been studied as a model for the light-promoted damage to DNA and related compounds.
Its photochemical transformations are also of interest in connection with the
phototherapy of tumor cells. Under far-UV irradiation (mainly 185 nm) with
a low-pressure Hg arc, thymine in aqueous solutions underwent direct photolysis, which observed pseudo first-order kinetics, with apparent rate constants in
aerated and deaerated media of 9.4×10^{-4} and 4.4×10^{-4} s^{-1}, and with quantum
efficiencies of 0.42 and 0.30, respectively. Under near-UV illumination (mainly
254 nm), the apparent rates were about 1/10 smaller. Thymine has absorption
bands both in the region of the 185-nm and the 254-nm lines of the mercury
arc. It was estimated that at 185 nm, 66% of the photons were absorbed by
thymine (0.5 mM) and 34% by water, while at 254 nm, the light absorption was
only by thymine, which undergoes a π–π^* transition of the *cis*-olefinic bond
(λ_{max} = 260 nm). Under N_2-saturated conditions at 185 nm, the main photoproducts were 5,6-dihydrothymine (DHT), 5-hydroxymethyluracil (HMU), and 6-
hydroxy-5,6-dihydrothymine (HOT), which were not formed under irradiation
at 254 nm, but which were the main species formed by γ-radiolysis of thymine.
In oxygenated solutions irradiated at 185 nm (and also in γ-radiolysis), the
main products were *cis*- and *trans*-5,6-dihydroxy-5,6-dihydrothymine (TG), 6-
hydroperoxy-5-hydroxy-5,6-dihydrothymine (HTP), and N^1-formyl-N^2-
pyruvylurea (FPU), as well as hydrogen peroxide. These products could be
accounted for by assuming photolysis at 185 nm of water to ·H atoms and ·OH
radicals, and their reactions with thymine forming hydrothyminyl and thyminyl
radicals, respectively.[102]

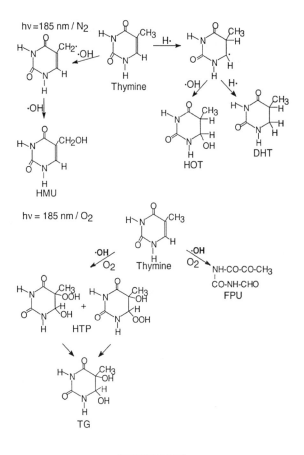

SCHEME V.28

In the presence of oxygen in aqueous suspensions of TiO_2, under illumination with a 400-W high-pressure Hg arc ($\lambda > 300$ nm), thymine underwent oxygenation to *trans-* and *cis-*5,6-dihydroxy-5,6-dihydrothymines (TG), as well as N^1-formyl-N^2-pyruvylurea (FPU). These photoproducts were thus similar to those of the direct photolysis in oxygenated solutions at 185 nm. With a mixture of TiO_2 and platinum black, the reaction rate was three times larger than only with TiO_2. The mechanism proposed is formation of radical intermediates such as 5-hydroxy-6-thyminyl, as well as its O_2 adduct.[103]

VI. TRICLOPYR

The herbicide triclopyr (3,5,6-trichloro-2-pyridinyloxyacetic acid) is used for selective postemergence control of annual and perennial broad-leaf weeds and woody plants. It is of moderate toxicity.[13] Its direct homogeneous-phase photolysis was studied under natural and simulated sunlight — both in sterile pH 7 (phosphate buffer) solution from distilled water and in natural river water.

Pseudo first-order decay was observed in both waters, but the half-lives in the two waters were 0.5 d and 1.3 d, respectively. Also there was a difference in the distribution of photoproducts. Photolysis in the sterile pH 7 water resulted in replacement of two chlorine atoms by hydroxyl groups, yielding 5-chloro-3,6-dihydroxy-2-pyridinyloxyacetic acid (MDPA) as the major product, with minor amounts of oxamic acid and some low-molecular-weight acids. In river water, photolysis yielded oxamic acid as the main product, and also small amounts of some low-molecular-weight acids.[104]

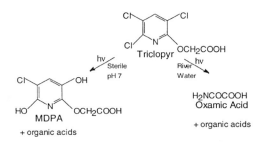

SCHEME V.29

The lower rate of photolysis of triclopyr in the natural water relative to sterile water was ascribed to quenching by dissolved organic matter in the river water. The different product distribution was explained as being possibly due to effects of dissolved humic and fulvic acids in the natural water.[104]

VII. FENARIMOL

The fungicide fenarimol is widely used for the control of powdery mildew. It acts by inhibition of ergosterol biosynthesis.[13] With sunlight illumination under environmental conditions, fenarimol in aqueous solution decomposed slowly over a period of days. Because of the low solubility in water, the photolysis degradation products were studied in organic solvents. The products tentatively identified were due to cleavage of the bond between the quaternary carbon and either the chloroaromatic or the pyrimidine ring.[105]

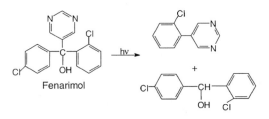

SCHEME V.30

VIII. FLAVINS

Flavin derivatives provide crucial functions in biological redox systems. Flavin coenzymes serve as redox switches both in two- and in three-electron cellular metabolism. This is in contrast to nicotinamide coenzymes (NAD, NADH, NADP, and NADPH), which are uniquely active in one-electron-transfer reactions. While nicotinamide coenzymes are inert to oxygen, flavin coenzymes react with oxygen, reducing it to hydrogen peroxide. The oxidation products of flavins, such as the semiquinones, are relatively stable to further oxidation.[106] The photooxidation of the aromatic amino acid tryptophan was performed using several flavin derivatives as sensitizers, including riboflavin, biopterin, 7-carboxypterin, pterin, lumazin, xanthopterin, and isoxanthopterin. The oxidation involved $^1O_2$, and is thus a Type II photodynamic reaction.[107]

IX. CATECHOLAMINES

With the biologically important catecholamines noradrenaline (norepinephrine), dopamine (3-hydroxytyramine), and the highly toxic adrenaline (epinephrine),

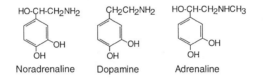

Noradrenaline   Dopamine   Adrenaline

SCHEME V.31

the involvement of $^1O_2$ as a reactive intermediate in the dye-sensitized photooxidation was proven by trapping it with a spin trap, 2,2,6,6-tetramethyl 4-piperidone (TMP). TMP was thus converted to the stable 2,2,6,6-tetramethyl 4-piperidone-*N*-oxyl (TMPNO). The stimulatory effect on the photooxidation rate of replacing H_2O by D_2O and the inhibitory effects of adding either KN_3 or β-limonene confirmed the involvement of singlet oxygen. The catechols were oxidized to semiquinones and quinones.[108]

X. DYES

A. Azo Dyes

Azo dyes are produced on a very large scale and may be released to the environment in the waste stream of dyeing operations. Hustert and Zepp[109] reported the photocatalytic degradation of aqueous solutions of several commercial azo dyes in the presence of suspended TiO_2 by illumination with a high-pressure Hg lamp ($\lambda > 290$ nm). The compounds studied were Solvent

Red 1, Acid Orange 7 (Orange II), and Orange G, as well as a model azo derivative, 4-hydroxyazobenzene.

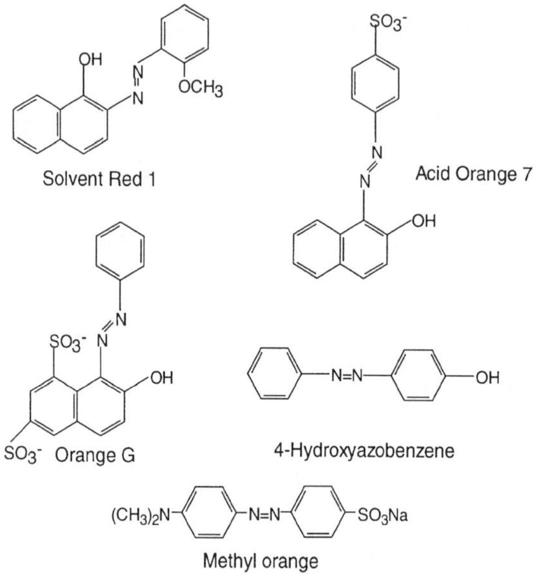

SCHEME V.32

In the dilute aqueous dye solutions used in these experiments (about 40 mg/L), the photocatalytic degradation of Acid Orange 7, of Orange G, and of 4-hydroxyazobenzene observed first-order kinetics for several half-lives. Solvent Red 1 was very insoluble in water. Its photodegradation in methanol or aqueous acetonitrile did not follow first-order kinetics. With Orange G, the half-lives at pH 7 and 12 were 40 and 10 min, respectively. In the photodegradation of 4-hydroxyazobenzene, the main breakdown product was hydroquinone, as well as small amounts of 4-aminophenol. From Acid Orange 7, the primary photoproduct was identified as 4-hydroxybenzenesulfonic acid. With this dye, photodegradation was rapid, but mineralization was much slower, and 20% of the original TOC still remained after 3 h. With Solvent Red 1, the only product identified was *o*-anisidine, presumably formed by azo cleavage.[109] Solvent Red 1 and 4-hydroxyazobenzene also underwent photodegradation by UV light (λ > 350 nm) in a solid system, adsorbed onto TiO$_2$ and exposed to an oxygen atmosphere. This degradation was followed by changes in the diffuse reflectance spectra of the dyes, which indicated that the C–N bond was the site for oxidative attack. A slower photodegradation was observed with solid SiO$_2$, and none with Al$_2$O$_3$ as support for the dyes.[110]

A reductive process was shown to cause the photocatalytic decolorization of the commercial textile azo dyes Acid Orange 7 and Direct Blue 1 with colloidal TiO$_2$ and WO$_3$.[111]

SCHEME V.33

The WO_3 and TiO_2 colloids have steep absorption onsets below 380 nm (their band gap is <3.0 eV). Under band-gap irradiation, both semiconductor colloids turned from colorless to blue, with formation of broad absorption bands in the red to near-infrared region. These bands were assigned to trapped electron states. In the presence of the semiconductors colloids, as well as of ethanol as sacrificial hole scavenger, both Acid Orange 7 and Direct Blue 1 under steady-state UV illumination were rapidly bleached. This was due to scavenging of the trapped electrons of the semiconductors by the dyes, forming the colorless semireduced dye products. The quantum efficiencies of the reduction of Acid Orange 7 and of Direct Blue 1 in WO_3 colloids were 4.7 and 4.4%, respectively. The kinetics of the interfacial electron transfer between the semiconductors and the azo dyes was determined by nanosecond laser flash photolysis. The rate constants observed for the heterogeneous electron transfer between the photoexcited WO_3 colloid and these dyes were of the order of 10^8 M^{-1} s^{-1}.[111]

The hydroxyl radical promoted degradation of azo dyes was also achieved in "dark" reactions of Fe^{3+} ions with H_2O_2. The proposed mechanism was addition of $\cdot OH$ to the azo-bond-bearing carbon atom of a hydroxy- or amino-substituted aromatic ring.[112]

In an air–solid interface reaction, the photodegradation of Acid Orange 7 (Orange II) was performed with the dye adsorbed on TiO_2 particles (Degussa P25). Under illumination with visible light from a 250-W halogen lamp, the orange color and the diffuse reflectance absorption spectrum disappeared practically completely within an hour.[113]

Methyl orange in both alkaline and acidic aqueous solutions containing suspended TiO_2 underwent bleaching under illumination with a 1000-W super-high-pressure mercury arc lamp. In acidic solutions, the rate of bleaching was very fast in oxygenated solutions, and much slower in the absence of oxygen. On the other hand, in alkaline solutions, the rate of bleaching was unaffected by the presence of oxygen. At an initial pH of 1.8 and with oxygen bubbling, methyl orange was completely bleached to colorless compounds within 40 min. ZnO was also very effective as photocatalyst for the bleaching of methyl orange, but ZnO was quite rapidly corroded and dissolved. Other powdered semiconductors tested, $BaTiO_3$, Fe_2O_3, MoO_3, and WO_3, were practically inactive. The nature of the bleached products was not determined.[114]

Reactive Black 5 in oxygenated aqueous dispersions of TiO_2 (0.12 g/L) under illumination with a 250-W Xe lamp was bleached and oxidized in a reaction following pseudo first-order kinetics and fitting a Langmuir–Hinshelwood type model. Photobleaching of the dye ($t_{1/2}$ = 90 min) was much faster than CO_2 release ($t_{1/2}$ = 6 h), indicating the formation of relatively stable intermediates. The complete mineralization occurred according to the equation

$$C_{26}H_{21}N_5Na_4O_{19}S_6 + 38O_2 \rightarrow$$
$$26CO_2 + 4H_2O + 4NaNO_3 + HNO_3 + 6H_2SO_4$$

(8)

The rate of reaction was found to be proportional to the square root of the light intensity.[115]

The photocatalytic degradation of azo dyes was also performed with semiconducting iron compounds.[116]

See Chapter VIII, Section I, on the colloidal TiO_2-mediated photoreduction of oxazine dyes by naturally occurring humic acids.

B. Tannery Dyes

The photodegradation with dispersed TiO_2 of concentrated wastewater polluted by a silk tannery dying operation was tested in a batch reactor. With "acid dye" (pH 4), practically total disappearance of color from these deeply colored wastes occurred within 8 h. With the "reactive dye" (pH 10), discoloration required only 1 h. This removal of color was accompanied by only a partial decrease in the total organic carbon (TOC) content of the solutions — indicating that the dyes had been partially converted to colorless organic intermediates. More rapid total mineralization of these organic compounds was possible after dilution of the concentrated dye slurries.[117]

C. Dyes in Municipal Wastewater

The photocatalytic decolorization of intensely colored discharges of a municipal wastewater treatment facility receiving effluents from textile dying processes was studied in a batch illumination reactor. Conventional microbial biodegradation is ineffective against the very stable modern textile dyes. Using dispersed TiO_2 (Degussa P25) under continuous aeration, the first-order rate constants of decolorization increased linearly with the concentration of TiO_2 up to 0.1 wt% of the catalyst, but leveled off at a higher catalyst concentration, reaching a plateau at about 0.4 wt%. Dilution with water of the very highly colored initial effluents resulted in considerably increased rates of decolorization. The lower rates with the more concentrated effluents were explained as due to the higher concentration of heavy metals such as copper and various organic compounds in the raw dye-waste, which may adsorb to the photocatalyst and thus interfere in the degradation of the dyes. The rates of photodecolorization were little affected by temperature in the range of 25 to 70°C, resulting in activation energies of only 3 to 6 kJ mol⁻¹. The decolorization of the wastewater was accompanied by a substantial decrease in the chemical oxygen demand (COD), indicating mineralization of the dissolved organic compounds.[118]

A comparison of several advanced oxidation processes (AOPs) for the mineralization of wastewater from a dye house (initial TOC = 275 mg/L; pH 10) resulted in complete removal of the intense red color by ozonation in the dark within 1 h. UV irradiation (UV/O$_3$) did not enhance the rate of decolorization. Mineralization required several hours of ozonation, and was accelerated by UV irradiation (with a 150-W high-pressure Hg lamp). About 75% mineralization (TOC removal) was achieved after 90 min by the photo Fenton reaction (UV/H$_2$O$_2$/Fe^{2+}) and after 150 min with UV/ozone. Slower bleaching of the dye-waste and its mineralization were obtained by UV/TiO$_2$ or UV/H$_2$O$_2$ treatments. Decolorization presumably involved as initial steps the cleavage of aromatic rings and of azo groups.[119]

D. Methyl Violet

The photodegradation of methyl violet (initially 0.1 mM) in oxygen-purged aqueous suspensions of TiO$_2$ was followed by the decrease in the intense color of the dye (λ = 580 nm), which observed first-order kinetics.[120]

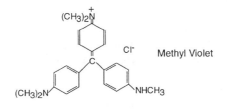

SCHEME V.34

E. Methyl Viologen

In a laser flash photolysis study of the photooxidation of methyl viologen dichloride in aqueous oxygenated suspensions of TiO$_2$ at pH 5, transient species were detected by diffuse reflectance spectroscopy. Following excitation at 355 nm, transient absorption reflectance peaks were observed at 390, 480, and 600 nm. The maxima at 390 and 600 nm were assigned to the methyl viologen cation radical ·MV$^+$, previously known from pulse radiolysis. The new transient absorption band at 480 nm was proposed to be due to an ·OH radical adduct of methyl viologen.[121]

XI. POLYCYCLIC AROMATIC NITROGEN HETEROCYCLES

Polyaromatic aromatic nitrogen heterocycles are components of fossil fuels and are released during combustion. Many of these compounds are toxic, carcinogenic, or irritant. They are generally more water-soluble than the polycyclic aromatic hydrocarbons, and are quite widely distributed in the aquatic environment. Due to the unpaired electron on the nitrogen atom, the electron density on the aromatic system of these compounds is higher than that on the polycyclic aromatic hydrocarbons. Protonation of the nitrogen atom at lowered pH in solution affects both their spectroscopic and photochemical properties. These compounds strongly absorb light in the UV region of the solar spectrum (>290 nm), and they are thus susceptible to both direct and sensitized photodegradation.[122]

The direct photolysis of quinoline, benzo[f]quinoline, and 9*H*-carbazole in pure water under illumination at 313 nm occurred with half-lives of 25, 14, and 2.9 h, and with quantum yields of 3.3×10^{-4}, 1.4×10^{-2}, and 2.9×10^{-3}, respectively. For 7*H*-dibenzo[c,g]carbazole illuminated at 366 nm, the half-life and quantum yield were 2.2 min and 2.8×10^{-3}, respectively.[123] For quinoline, the rates of photolysis at 313 nm in pure water, in lake water, and in pH 7.0 phosphate buffer were similar, with $t_{1/2} = 22$ to 25 h. However, at pH 4.5, in phosphate buffer, the rate increased, with $t_{1/2} = 15.0$ h. Also, the addition of $NaNO_3$ (0.5 m*M*) or dissolved organic matter (DOM, 1 mg/L) led to decreased half-lives. These additives presumably initiated the production of hydroxyl radicals, thus enhancing the rate of photodegradation. The photoproducts identified from the photolysis of quinoline at 313 nm were 2-hydroxyquinoline (2-quinolinol; carbostyril) and 8-hydroxyquinoline (8-quinolinol).[124]

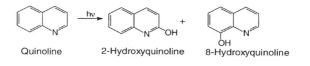

Quinoline   2-Hydroxyquinoline   8-Hydroxyquinoline

SCHEME V.35

REFERENCES

1. Maillard, C., Guillard, C., Pichat, P., and Fox, M. A., Photodegradation of benzamide in TiO_2 aqueous suspensions, *N. J. Chem.*, 16, 821–825, 1992.
2. Chramosta, N., De Laat, J., Dore, M., Suty, H., and Pouillot, M., Rate constants for reaction of hydroxyl radicals with triazines, *Environ. Technol.*, 14, 215–226, 1993.
3. Zabik, M. J., Leavitt, A., and Su, G. C. C., Photochemistry of bioactive compounds. A review of pesticide photochemistry, *Annu. Rev. Entomol.*, 21, 61–79, 1976.

4. Low, G. K. C., McEvoy, S. R., and Matthews, R. W., Formation of nitrate and ammonium ions in titanium dioxide mediated photocatalytic degradation of organic compounds containing nitrogen atoms, *Environ. Sci. Technol.*, 25, 460–467, 1991.

5. Low, G. K. C., McEvoy, S. R., and Matthews, R. W., Formation of ammonium and nitrate ions from photocatalytic oxidation of ring nitrogenous compounds over titanium dioxide, *Chemosphere*, 19, 1611–1621, 1989.

6. Pichat, P., Guillard, C., Maillard, C., Amalric, L., and D'Oliveira, J.-C., TiO$_2$ photocatalytic destruction of water aromatic pollutants: intermediates; properties–degradability correlation; effects of inorganic ions and TiO$_2$ surface area; comparison with H$_2$O$_2$ processes, in: *Photocatalytic Purification and Treatment of Water and Air,* Ollis, D. F. and Al-Ekabi, H. (Eds.), Elsevier Science Publishers, Amsterdam, 1993, pp. 207–223.

7. Maillard-Dupuy, C., Guillard, C., Courbon, H., and Pichat, P., Kinetics and products of the TiO$_2$ photocatalytic degradation of pyridine in water, *Environ. Sci. Technol.*, 28, 2176–2183, 1994.

8. Pape, B. E. and Zabik, M. J., Photochemistry of bioactive compounds. Solution phase photochemistry of symmetrical triazines, *J. Agric. Food Chem.*, 20, 315–320, 1972.

9. Yao, C. C. D. and Haag, W. R., Rate constants for direct reaction of ozone with several drinking water contaminants, *Water Res.*, 25, 761–773, 1991.

10. Hoigné, J. and Bader, H., Rate constants for reaction of ozone with organic and inorganic compounds in water. I. Non-dissociating organic compounds, *Water Res.*, 17, 173–183, 1983.

11. Haag, W. R. and Yao, C. C. D., Rate constants for reaction of hydroxyl radicals with several drinking water contaminants, *Environ. Sci. Technol.*, 26, 1005–1013, 1992.

12. Kochany, J. and Bolton, J. R., Mechanism of photodegradation of aqueous organic pollutants. 2. Measurement of the primary rate constants for reaction of OH· radicals with benzene and some halobenzenes using an EPR spin-trapping method following the photolysis of H$_2$O$_2$, *Environ. Sci. Technol.*, 26, 262–265, 1992.

13. Hartley, D. and Kidd, H. (Eds.), *The Agrochemicals Handbook,* Royal Society of Chemists, Nottingham, England, 1987.

14. Barceló, D., Durand, G., De Bertrand, N., and Albaigés, J., Determination of aquatic photodegradation products of selected pesticides by gas-chromatography mass-spectrometry and liquid-chromatography mass-spectrometry, *Sci. Total Environ.*, 132, 283–296, 1993.

15. Hessler, D. P., Gorenflo, V., and Frimmel, F. H., Degradation of aqueous atrazine and metazachlor solutions by UV and UV/H$_2$O$_2$ — influence of pH and herbicide concentration, *Acta Hydrochim. Hydrobiol.*, 21, 209–214, 1993.

16. Larson, R. A., Schlauch, M. B., and Marley, K. A., Ferric ion promoted photodecomposition of triazines, *J. Agric. Food Chem.*, 39, 2057–2062, 1991.

17. Beltran, F. J., Ovejero, G., and Acedo, B., Oxidation of atrazine in water by ultraviolet radiation combined with hydrogen peroxide, *Water Res.*, 27, 1013–1021, 1993.

18. Rejto, M., Saltzman, S., Acher, A. J., and Muszkat, L., Identification of sensitized photooxidation products of *s*-triazine herbicides in water, *J. Agric. Food Chem.*, 31, 138–142, 1983.

19. Minero, C., Pramaura, E., Pelizzetti, E., Dolci, M., and Marchesini, A., Photosensitized transformations of atrazine under simulated sunlight in aqueous humic acid solution, *Chemosphere,* 24, 1597–1606, 1992.

20. Sun, Y. F. and Pignatello, J. J., Activation of hydrogen peroxide by iron(III) chelates for abiotic degradation of herbicides and insecticides in water, *J. Agric. Food Chem.,* 41, 308–312, 1993.

21. Hustert, K., Kotzias, D., and Korte, F., Photocatalytic decomposition of organic compounds on titanium dioxide, *Chemosphere,* 12, 55–58, 1983.

22. Pelizzetti, E., Minero, C., Pramauro, E., Barbeni, M., Maurino, V., and Tosato, M. L., Photocatalytic degradation of atrazine at ppb levels under solar light and in the presence of titania particles, *Chim. Ind. (Milano),* 69, 88–89, 1987.

23. Pelizzetti, E., Maurino, V., Minero, C., Carlin, V., Tosata, M. L., and Pramaura, E., Photocatalytic degradation of atrazine and other *s*-triazine herbicides, *Environ. Sci. Technol.,* 24, 1559–1565, 1990.

24. Pelizzetti, E., Carlin, V., Minero, C., and Grätzel, M., Enhancement of the rate of photocatalytic degradation on TiO_2 of 2-chlorophenol, 2,7-dichlorodibenzodioxin and atrazine by inorganic oxidizing species, *N. J. Chem.,* 15, 351–359, 1991.

25. Pelizzetti, E., Minero, C., Carlin, V., Vincenti, M., Pramaura, E., and Dolci, M., Identification of photocatalytic degradation pathways of 2-Cl-*s*-triazine herbicides and detection of their decomposition intermediates, *Chemosphere, 24,* 891–910, 1992.

26. Pelizzetti, E., Carlin, V., Minero, C., Pramaura, E., and Vincenti, M., Degradation pathways of atrazine under solar light and in the presence of TiO_2 colloidal particles, *Sci. Total Environ.,* 123/124, 161–169, 1992.

27. Pelizzetti, E., Minero, C., Borgarello, E., Tinucci, L., and Serpone, N., Photocatalytic activity and selectivity of titania colloids and particles prepared by the sol-gel technique: photooxidation of phenol and atrazine, *Langmuir, 9,* 2995–3001, 1993.

28. Pelizzetti, E. and Minero, C., Mechanism of the photo-oxidative degradation of organic pollutants over TiO_2 particles, *Electrochim. Acta,* 38, 47, 1993.

29. Pelizzetti, E., Minero, C., and Carlin, V., Photoinduced degradation of atrazine over different metal oxides, *N. J. Chem.,* 17, 315, 1993.

30. Pelizzetti, E., Minero, C., Piccinini, P., and Vincenti, M., Phototransformations of nitrogen containing organic compounds over irradiated semiconductor metal oxides. Nitrobenzene and atrazine over TiO_2 and ZnO, *Coord. Chem. Rev.,* 125, 183–194, 1993.

31. Yue, P. L. and Allen, D., Photocatalytic degradation of atrazine, in: *Photocatalytic Purification and Treatment of Water and Air,* Ollis, D. F. and Al-Ekabi, H. (Eds.), Elsevier Publishing, Amsterdam, 1993, pp. 601–611.

32. Chester, G., Anderson, M., Read, H., and Esplugas, S., A jacketed annular membrane photocatalytic reactor for wastewater treatment. Degradation of formic acid and atrazine, *J. Photochem. Photobiol. A: Chem.,* 71, 291–297, 1993.

33. Ross, H., Bendig, J., and Hecht, S., Sensitized photocatalytic oxidation of terbutylazine, *Solar Energy Mater. Solar Cells,* 33, 475–481, 1994.

34. Gonzalez, M. C., Braun, A. M., Prevot, A. B., and Pelizzetti, E., Vacuum-ultraviolet (VUV) photolysis of water: mineralization of atrazine, *Chemosphere,* 28, 2121–2127, 1994.

35. Mandelbaum, R. T., Wackett, L. P., and Allan, D. L., Mineralization of the *s*-triazine ring of atrazine by stable bacterial mixed cultures, *Appl. Environ. Microbiol.,* 59, 1695–1701, 1993.

36. Muszkat, L., Halmann, M., Raucher, D., and Bir, L., Solar photodegradation of xenobiotic contaminants in polluted well water, *J. Photochem. Photobiol. A: Chem.,* 65, 409–417, 1992.

37. Pelizzetti, E., Maurino, V., Minero, C., Zerbinati, O., and Borgarello, E., Photocatalytic degradation of bentazon by titanium dioxide particles, *Chemosphere,* 18, 1437–1445, 1989.

38. Larson, R. A., Ellis, D. D., Ju, H.-L., and Marley, K. A., Flavin-sensitized photodecomposition of anilines and phenols, *Environ. Toxicol. Chem.,* 8, 1165–1170, 1989.

39. Nickel, U., Chen, Y.-H., Schneider, S., Silva, M. I., Burrows, H. D., and Formosinho, S. J., Mechanism and kinetics of the photocatalyzed oxidation of *p*-phenylenediamines by peroxydisulfate in the presence of tri-2,2'-bipyridylruthenium(II), *J. Phys. Chem.,* 98, 2883–2888, 1994.

40. Lockhart, H. B., Jr. and Blakeley, R. V., Aerobic photodegradation of Fe(III)-(ethylenedinitrilo)tetraacetate, *Environ. Sci. Technol.,* 9, 1035–1038, 1975.

41. Trott, T., Henwood, W., and Langford, C. H., Sunlight photochemistry of ferric nitrilotriacetate complexes, *Environ. Sci. Technol.,* 6, 367–368, 1972.

42. Svenson, A., Kaj, L., and Björndal, H., Aqueous photolysis of the iron(III) complexes of NTA, EDTA, and DTPA, *Chemosphere,* 18, 1805–1808, 1989.

43. Andrianirinaharivelo, S. L., Pilchowski, J. F., and Bolte, M., Nitrilotriacetic acid transformation photoinduced by iron(III) in aqueous solution, *Transition Met. Chem. (London),* 18, 37–41, 1993.

44. Stolzberg, R. J. and Hume, D. N., Rapid formation of iminodiacetate from photochemical degradation of Fe(III)nitriloacetate solutions, *Environ. Sci. Technol.,* 9, 654–656, 1975.

45. Andrianirinaharivelo, S. L. and Bolte, M., Photochemical transformation of iminodiacetic complexes with iron(III) in aqueous solution, *Chemosphere,* 24, 953–958, 1992.

46. Andrianirinaharivelo, S. L. and Bolte, M., Photochemical behaviour of copper(II) nitrilotriacetate in aqueous solution, *Chemosphere, J. Photochem. Photobiol. A: Chem.,* 73, 213–216, 1993.

47. Tanaka, K., Chen, Y. S., and Hisanaga, T., Photocatalytic degradation of metal chelate of EDTA in aqueous TiO$_2$, *Abstr. 2nd Int. Symp. New Trends in Photoelectrochemistry,* University of Tokyo, Japan, March 1994, p. 11.

48. Maillard, C., Guillard, C., and Pichat, P., Comparative effects of the TiO$_2$–UV, H$_2$O$_2$–UV, H$_2$O$_2$–Fe^{2+} systems on the disappearance rate of benzamide and 4-hydroxybenzamide in water, *Chemosphere,* 24, 1085–1094, 1992.

49. Schmitt, Ph., Mamouni, A., Mansour, M., and Schiavon, M., Photodecomposition of isoxaben in aqueous systems and solid phase, *Sci. Total Environ.,* 123, 171–182, 1992.

50. Méallier, P., Mamouni, A., and Mansour, M., Photodegradation of phytosanitary molecules. VII. Photodegradation of carbetamide alone or in the presence of formulation adjuvants, *Chemosphere,* 20, 267–273, 1990.

51. Méallier, P., Mamouni, A., and Mansour, M., Effect of acetophenone on the photostability of formulations: case of carbetamide, *Chemosphere,* 21, 913–917, 1990.

52. Méallier, P., Mamouni, A., and Mansour, M., Photodegradation of pesticides. VII. Photodegradation of carbetamide — photoproducts, *Chemosphere,* 26, 1917–1923, 1993.

53. Percherancier, J. P., Perichet, G., and Pouyet, B., Photocatalysed oxidation of aqueous solutions of pesticides, *XIII IUPAC Symp. Photochemistry,* University of Warwick, England, Abstract P340, 1990.

54. Emmelin, C., Guittonneau, S., Lamartine, R., and Méallier, P., Photodegradation of pesticides on adsorbed phases. Photodegradation of carbetamid, *Chemosphere,* 27, 757–763, 1993.

55. De Bertrand, N. and Barceló, D., Photodegradation of the carbamate pesticides aldicarb, carbaryl and carbofuran in water, *Anal. Chim. Acta,* 254, 235–244, 1991.

56. Kondo, M. M. and Jardim, W. F., Photodegradation of chloroform and urea using Ag-loaded titanium dioxide as catalyst, *Water Res.,* 25, 823–827, 1991.

57. Pramaura, E., Vincenti, M., Augugliaro, V., and Palmisano, L., Photocatalytic degradation of monuron in aqueous TiO_2, *Environ. Sci. Technol.,* 27, 1790–1795, 1993.

58. Augugliaro, V., Marci, G., Palmisano, L., Pramauro, E., and Biancoprevot, A., Kinetics of heterogeneous photocatalytic decomposition of monuron over anatase titanium dioxide powder, *Res. Chem. Intermed.,* 19, 839–853, 1993.

59. Moilanen, K. W. and Crosby, D. G., Photodecomposition of bromacil, *Arch. Environ. Contam. Toxicol.,* 2, 3–8, 1974.

60. Acher, A. J. and Dunkelblum, E., Identification of sensitized photooxidation products of bromacil in water, *J. Agric. Food Chem.,* 27, 1164–1167, 1979.

61. Acher, A. J. and Saltzman, S., Dye-sensitized photooxidation of bromacil in water, *J. Environ. Qual.,* 9, 190–194, 1980.

62. Acher, A. J., Saltzman, S., Brates, N., and Dunkelblum, E., Photosensitized decomposition of terbacil in aqueous solutions, *J. Agric. Food Chem.,* 29, 707–711, 1981.

63. Dureja, P., Walia, S., and Sharma, K. K., Photolysis of isoproturon in aqueous solution, *Toxicol. Chem. Environ. Chem.,* 34, 65–71, 1991.

64. Rejto, M., Saltzman, S., and Acher, A. J., Photodecomposition of propachlor, *J. Agric. Food Chem.,* 32, 226–230, 1984.

65. Moza, P. N., Hustert, K., Pal, S., and Sukul, P., Photocatalytic decomposition of Pendimethalin and Alachlor, *Chemosphere,* 25, 1675–1682, 1992.

66. Patel, R. and Sugden, J. K., Photodegradation of aqueous solutions of dequalinium chloride, *Pharmazie,* 47, 113–115, 1992.

67. Uchida, H., Itoh, S., and Yoneyama, H., Photocatalytic decomposition of propyzamide using TiO_2 supported on activated carbon, *Chem. Lett.,* 1995–1998, 1993.

68. Moza, P. N., Sukul, P., Hustert, K., and Kettrup, A., Photooxidation of metalaxyl in aqueous solution in the presence of hydrogen peroxide and titanium dioxide, *Chemosphere,* 28, 341–347, 1994.

69. Chen, C. P., Ren, X. M., Lu, D. H., and Xu, G. Z., Study on free radicals produced during the photolysis of *p*-aminophenol in aqueous TiO_2 suspension, *Chin. Sci. Bull.,* 38, 41–46, 1994.

70. Nishimoto, S., Ohtani, B., Yoshikawa, T., and Kagiya, T., Photocatalytic conversion of primary amines to secondary amines and cyclization of polymethylene-α,ω-diamines by an aqueous suspension of TiO$_2$/Pt, *J. Am. Chem. Soc.,* 105, 7180, 1983.

71. Ohtani, B., Tsuru, S., Nishimoto, S., and Kagiya, T., Photocatalytic one-step synthesis of cyclic iminoacids by aqueous semiconductor suspensions, *J. Org. Chem.,* 55, 5551–5553, 1990.

72. Mills, A., Belghazi, A., Davies, R. H., Worsley, D., and Morris, S., A kinetic study of the bleaching of rhodamine 6G photosensitized by titanium dioxide, *J. Photochem. Photobiol. A: Chem.,* 79, 131–139, 1994.

73. Atkinson, R., Aschmann, S. M., and Arey, J., Reactions of OH and NO$_3$ radicals with phenol, cresols, and 2-nitrophenol at 296K ± 2K, *Environ. Sci. Technol.,* 26, 1397–1403, 1992.

74. O'Connor, O. A. and Young, L. L., Toxicity and anaerobic biodegradability of substituted phenols under methanogenic conditions, *Environ. Toxicol. Chem.,* 8, 853–862, 1989.

75. Alif, A., Boule, P., and Lemaire, J., Photochemical behavior of 4-nitrophenol in aqueous solution, *Chemosphere,* 16, 2213–2223, 1987.

76. Ho, P. C., Photooxidation of 2,4-dinitrotoluene in aqueous solution in the presence of hydrogen peroxide, *Environ. Sci. Technol.,* 20, 260–267, 1986.

77. Haag, W. R. and Hoigné, J., Singlet oxygen in surface waters. 3. Photochemical formation and steady-state concentrations in various types of waters, *Environ. Sci. Technol.,* 20, 681, 1986.

78. Scully, F. E., Jr. and Hoigné, J., Rate constants for reactions of singlet oxygen with phenols and other compounds in water, *Chemosphere,* 16, 681–694, 1987.

79. Tratnyek, P. G. and Hoigné, J., Environmental photooxidation of phenolic compounds by singlet oxygen, *XIII IUPAC Symp. on Photochemistry,* University of Warwick, England, Abstract p. 264, 1990.

80. Rafizadeh, K. and Yates, K., Acid-catalyzed photooxidation of *m*-nitrobenzyl derivatives in aqueous solution, *J. Org. Chem.,* 51, 2777–2781, 1986.

81. Alif, A., Boule, P., and Lemaire, J., Photochemistry and environment. XII: Phototransformation of 3-nitrophenol in aqueous solution, *J. Photochem. Photobiol. A: Chem.,* 50, 331–342, 1990.

82. Lipczynska-Kochany, E., Degradation of nitrobenzene and nitrophenols by means of advanced oxidation processes in a homogeneous phase. Photolysis in the presence of hydrogen peroxide versus the Fenton reaction, *Chemosphere,* 24, 1369–1380, 1992.

83. Palumbo, M. C., García, N. A., Gutiérrez, M. I., and Luiz, M., Singlet molecular oxygen-mediated photooxidation of monochloro and mononitrophenols, *Toxicol. Environ. Chem.,* 29, 85–94, 1990.

84. Luiz, M., Gutiérrez, M. I., Bocco, G., Bertolotti, S. G., and García, N. A., Sensitized photooxygenation of mononitrophenols in micellar media, *Toxicol. Environ. Chem.,* 35, 115–123, 1992.

85. D'Oliveira, J.-C., Guillard, C., Maillard, C., and Pichat, P., Photocatalytic destruction of hazardous chlorine- or nitrogen-containing aromatics in water, *J. Environ. Health,* A28, 941–962, 1993.

86. Minero, C., Pelizzetti, E., Piccinini, P., and Vincenti, M., Photocatalyzed transformation of nitrobenzene on TiO$_2$ and ZnO, *Chemosphere,* 28, 1229–1244, 1994.

87. Maillard-Dupuy, C., Guillard, C., and Pichat, P., The degradation of nitroben-zene in water by photocatalysis over TiO_2: kinetics and products; simultaneous elimination of benzamide or phenol or Pb^{2+} cations, *N. J. Chem.*, 18, 941–948, 1994.

88. Yinon, J., *Toxicity and Metabolism of Explosives*, CRC Press, Boca Raton, FL, 1990.

89. Kearney, P. C., Qiang, Z., and Ruth, J. M., Oxidative pretreatment accelerates TNT metabolism in soils, *Chemosphere*, 12, 1583–1597, 1983.

90. Schmelling, D. C. and Gray, K. A., Feasibility of photocatalytic degradation of TNT as a single or integrated treatment process, in: *Photocatalytic Purification and Treatment of Water and Air*, Ollis, D. F. and Al-Ekabi, H. (Eds.), Elsevier Science Publishing, Amsterdam, 1993, pp. 625–631.

91. Palmisano, L., Augugliaro, V., Schiavello, M., and Sclafani, A., Influence of acid-base properties on photocatalytic and photochemical processes, *J. Mol. Catal.*, 56, 284–295, 1989.

92. Augugliaro, V., López-Muñoz, M. J., Palmisano, L., and Soria, J., Photocatalytic degradation of nitrophenols in aqueous titanium dioxide dispersions, *Appl. Catal.*, 69, 323–340, 1991.

93. Augugliaro, V., López-Muñoz, M. J., Palmisano, L., and Soria, J., Influence of pH on the degradation kinetics of nitrophenol isomers in a heterogeneous photocatalytic system, *Appl. Catal.*, 101, 7–13, 1993.

94. Palmisano, L., Schiavello, M., Sclafani, A., Martin, I., and Rives, V., Surface properties of iron-titania photocatalysts employed for 4-nitrophenol photodegradation in aqueous TiO_2 dispersion, *Catal. Lett.*, 24, 303–315, 1994.

95. Dieckmann, M. S., Gray, K. A., and Kamat, P. V., Photocatalyzed degradation of adsorbed nitrophenolic compounds on semiconductor surfaces, *Water Sci. Technol.*, 25, 277–279, 1992.

96. Gray, K. A., Stafford, U., Dieckmann, M. S., and Kamat, P. V., Mechanistic studies in TiO_2 systems: photocatalytic degradation of chloro- and nitrophenols, in: *Photocatalytic Purification and Treatment of Water and Air*, Ollis, D. F. and Al-Ekabi, H. (Eds.), Elsevier Publishing, Amsterdam, 1993, pp. 455–472.

97. Brezová, V., Ceppan, M., Vesely, M., and Lapcik, L., Photocatalytic oxidation of 2,6-dichloroindophenol in the titanium dioxide aqueous suspension, *Chemicke Zvesti (Chemical Papers)*, 45, 233–246, 1991.

98. Brezová, V., Stasko, A., Ceppan, M., Mikula, M., Blecha, J., Vesely, M., Blazkova, A., Panák, J., and Lapcík, L., Photocatalytic activity and the forma-tion of radical intermediates, in: *Photocatalytic Purification of Water and Air*, Ollis, D. F. and Al-Ekabi, H. (Eds.), Elsevier, Amsterdam, 1993, pp. 659–664.

99. Kochany, J., Choudhry, G. G., and Webster, G. R. B., Environmental phototransformation of the herbicide bromoxynil (3,5-dibromo-4-hydroxybenzotrile) in aquatic systems containing sodium chloride, *Environ. Sci. Res.*, 42, 259–276, 1991.

100. Kochany, J., Effects of carbonates on the aquatic photodegradation rate of bromoxynil (3,5-dibromo-4-hydroxybenzotrile) herbicide, *Chemosphere*, 24, 1119–1126, 1992.

101. Kochany, J., Effects of iron(III) and manganese(II) ions on the aquatic photodegradation rate of bromoxynil (3,5-dibromo-4-hydroxybenzotrile) her-bicide, *Chemosphere*, 25, 261–270, 1992a.

102. Ohtani, B., Nagasaki, H., Nishimoto, S., Sakano, K., and Kagiya, T., Far ultraviolet induced decomposition of thymine in deaerated and aerated aqueous solutions, *Can. J. Chem.,* 64, 2297–2300, 1986.

103. Ohtani, B., Nagasaki, H., Sakano, K., Nishimoto, S., and Kagiya, T., Photoinduced oxygenation of thymine in an aqueous suspension of titanium dioxide, *J. Photochem. Photobiol. A: Chem.,* 41, 141–143, 1987.

104. Woodburn, K. B., Batzer, F. R., White, F. H., and Schultz, M. R., The aqueous photolysis of Triclopyr, *Environ. Toxicol. Chem.,* 12, 43–55, 1993.

105. Mateus, M. C. D. A., Silva, A. M., and Burrows, H. D., Environmental and laboratory studies of the photodegradation of the pesticide fenarimol, *J. Photochem. Photobiol. A: Chem.,* 80, 409–416, 1994.

106. Walsh, C., Flavin coenzymes: at the crossroads of biological redox chemistry, *Acc. Chem. Res.,* 13, 148, 1980.

107. Momzikoff, A., Santus, R., and Giraud, M., A study of the photosensitizing properties of seawater, *Mar. Chem.,* 12, 1–14, 1983.

108. Kruk, I., The identification by electron spin resonance spectroscopy of singlet oxygen formed in the photooxidation of catecholamines, *Z. Phys. Chem. (Leipzig),* 266, 1239–1242, 1985.

109. Hustert, K. and Zepp, R. G., Photocatalytic degradation of selected azo dyes, *Chemosphere,* 24, 335–342, 1992.

110. Dieckmann, M. S., Gray, K. A., and Zepp, R. G., The sensitized photocatalysis of azo dyes in a solid system: a feasibility study, *Chemosphere,* 28, 1021–1034, 1994.

111. Vinodgopal, K., Bedja, I., Hotchandani, S., and Kamat, P. V., A photocatalytic approach for the reductive decolorization of textile azo dyes in colloidal semiconductor suspensions, *Langmuir,* 10, 1767–1771, 1994.

112. Spadaro, J. T., Isabelle, L., and Renganathan, V., Hydroxyl radical mediated degradation of azo dyes. Evidence for benzene generation, *Environ. Sci. Technol.,* 28, 1389–1393, 1994.

113. Kamat, P. V. and Vinodgopal, K., TiO_2 mediated photocatalysis using visible light: photosensitization approach, in: *Photocatalytic Purification and Treatment of Water and Air,* Ollis, D. F. and Al-Ekabi, H. (Eds.), Elsevier Publishing, Amsterdam, 1993, pp. 83–94.

114. Chen, L.-C. and Chou, T.-C., Photobleaching of methyl orange in titanium dioxide suspended in aqueous solution, *J. Mol. Catal.,* 85, 201–214, 1993.

115. Mills, A. and Davies, R., The photomineralization of Reactive Black 5 sensitized by titanium dioxide: a study of the initial kinetics of dye photobleaching, in: *Photocatalytic Purification and Treatment of Water and Air,* Ollis, D. F. and Al-Ekabi, H. (Eds.), Elsevier Science Publishing, Amsterdam, 1993, pp. 595–600.

116. Hustert, K. and Moza, P. N., Photocatalytic degradation of azo dyes by semiconducting iron compounds, *Fresenius Environ. Bull.,* 3, 762, 1994.

117. Tinucci, L., Borgarello, E., Minero, C., and Pelizzetti, E., Treatment of industrial wastewaters by photocatalytic oxidation on TiO_2, in: *Photocatalytic Purification and Treatment of Water and Air,* Ollis, D. F. and Al-Ekabi, H. (Eds.), Elsevier Science Publishing, Amsterdam, 1993, pp. 585–594.

118. Davis, R. J., Gainer, J. L., O'Neal, G., and Wu, I. W., Photocatalytic decolorization of wastewater dyes, *Water Environ. Res.,* 66, 50–53, 1994.

119. Ruppert, G., Bauer, R., and Heisler, G., UV–O_3, UV–H_2O_2, UV–TiO_2, and the photo-Fenton reaction. Comparison of advanced oxidation processes for water treatment, *Chemosphere* , 28, 1447–1454, 1994.

120. Tennakone, K., Punchiheva, S., Wickremanayaka, S., and Tantrigoda, R. U., Titanium dioxide catalysed photo-oxidation of methyl violet, *J. Photochem. Photobiol. A: Chem.*, 46, 247–252, 1989.

121. Draper, R. B. and Fox, M. A., Titanium dioxide photosensitized reactions studied by diffuse reflectance flash photolysis in aqueous suspensions of TiO_2 powder, *Langmuir,* 6, 1396–1402, 1990.

122. Kochany, J. and Maguire, R. J., Abiotic transformations of polynuclear aromatic hydrocarbons and polynuclear aromatic nitrogen heterocycles in aquatic environments, *Sci. Total Environ.,* 144, 17–31, 1994.

123. Mill, T., Mabey, W. R., Lan, B. Y., and Baraze, A., Photolysis of polycyclic aromatic hydrocarbons in water, *Chemosphere,* 10, 1281–1290, 1981.

124. Kochany, J. and Maguire, R. J., Photodegradation of quinoline in water, *Chemosphere,* 28, 1097–1110, 1994.

Chapter VI

Organic Phosphorus Compounds

Organic phosphorus compounds are essential components and intermediates in all living organisms. Natural aquatic systems contain such compounds from the breakdown of biomolecules, and also from residues of widely used herbicides, pesticides, and other agrochemicals. Organophosphorus esters have many industrial applications as plasticizers, lubricant additives, and hydraulic fluids. The photochemical degradation of organophosphorus compounds is also important in the context of the environmental stability of nerve gases, and in the detoxification of such compounds, which act by inhibition of cholinesterase.[1] The literature on the photochemical reactions of organophosphorus compounds prior to 1967 has been reviewed.[2]

I. HOMOGENEOUS PHOTOLYSIS

With alkyl phosphates, the observed photolytic pathways differ from the thermal hydrolytic reactions by the nature of the reaction products, indicating the intermediate formation of highly reactive species. These direct photodegradations, without added sensitizers, occur only with UV light shorter than about 300 nm. For photocatalyzed reactions in the visible region, photosensitization with either dyes or semiconductors is required.

A. Alkyl Phosphates

1. Ethyl Dihydrogen Phosphate

Ultraviolet illumination of ethyl dihydrogen phosphate in the absence of oxygen resulted in the production of equimolar amounts of orthophosphate, hydrogen, and acetaldehyde:

$$CH_3CH_2O-PO_3H_2 + H_2O + h\nu \rightarrow CH_3CHO + H_2 + H_3PO_4 \qquad (1)$$

The proposed mechanism to account for this stochiometry was clarified using tracer and scavenger techniques:[3]

$$CH_3CH_2O-PO_3H_2 + h\nu \rightarrow \left(CH_3CHO-PO_3H_2\right) \cdot + H \cdot \qquad (2)$$

$$\left(CH_3CHO-PO_3H_2\right) \cdot + H_2O \rightarrow CH_3CH(OH)O-PO_3H_2 + H \cdot \qquad (3)$$

$$H\cdot + H\cdot \rightarrow H_2 \tag{4}$$

$$CH_3CH(OH)O-PO_3H_2 \rightarrow CH_3CHO + H_3PO_4 \tag{5}$$

2. Trimethyl Phosphate

The photodegradation of trimethyl phosphate may be considered a model reaction for the light-induced decomposition of organophosphorus pesticides and nerve gases. Products of the UV photolysis of trimethyl phosphate were those of stepwise degradation, leading to dimethyl phosphate, methyl phosphate, and orthophosphate, as well as to formaldehyde, formic acid, CO, and CO_2. The initial step probably involved hydrogen abstraction,[4]

$$(CH_3O)_3 PO + h\nu \rightarrow (CH_3O)_2 POCH_2\cdot + H\cdot \tag{6}$$

as indicated by the detection by mass spectrometry of a small amount of the dimer,

$$(CH_3O)_2 POCH_2 - CH_2 OP(OCH_3)_2 \tag{7}$$

In the presence of oxygen, photolysis yields were much enhanced, possibly because of involvement of $\cdot OH$ radicals:[4]

$$(CH_3O)_3 PO + \cdot OH \rightarrow (CH_3O)_2 POCH_2\cdot + H_2O \tag{57}$$

B. Phophonofluoridates

Photooxidation provided a route to the degradation of the extremely toxic nerve agent isopropyl methylphosphonofluoridate (code named GB), $(Me_2CHO)MeP(O)F$. Under oxygen, and in the presence of hydrogen peroxide, the compound was degraded at $\lambda < 300$ nm first to $MeP(O)(ONa)OCHMe_2$ and $MeP(O)(ONa)_2$, and eventually to the harmless orthophosphate and acetone.[5]

C. Dimethyl Vinyl Phosphate and Dichlorvos

The insecticide and acaricide dimethyl 2,2-dichlorovinyl phosphate [DDVP; dichlorvos; $(MeO)_2P(O)-O-CH=CCl_2$] is widely used for the control of aphids and caterpillars, as well as against household pests. It acts as a contact and stomach poison, with fumigant and penetrant effects.[6] The photolytic reactions of dimethyl vinyl phosphate were studied as a model of the sunlight-induced degradation of enol phosphate insecticides such as dichlorvos. The direct photolysis of $CH_2=CH-O-P(O)(OMe)_2$ at 254 nm in the absence of oxygen resulted in $(MeO)_2P(O)OH$ as the major product, as well as in formation of a dimer. The postulated intermediate was $(MeO)_2P(O)CH(OH)Me$, which decomposed to dimethyl phosphate and acetaldehyde.[7]

See Section II on the TiO_2-photocatalyzed degradation of dichlorvos.

D. Glycerophosphates

Both $CH_2OH-CHOH-CH_2OPO_3H_2$ (Glycero-1-phosphate) and $CH_2OH-CHOPO_3H_2-CH_2OH$ (Glycero-2-phosphate) by UV illumination at 254 nm gave the same products of phosphate elimination,

$$CH_2-CH-CH_2OH \text{ (glycidol)}$$
$$\diagdown \diagup$$
$$O$$

as well as $CH_2OH-CHOH-CHO$ (glyceraldehyde), $CH_2OH-CO-CH_2OH$ (dihydroxyacetone), and $CH_2OH-COOH$ (glycolic acid).[8]

E. Sugar Phosphates

α-D-Glucose 6-phosphate in aqueous solutions under illumination at 254 nm in the presence of oxygen yielded only an acid-labile ester, possibly a triose phosphate. In the absence of oxygen, the main photolysis products were 6-phophogluconate, arabinose 5-phosphate, phosphoglycerate, phosphoglycolate, carbon dioxide, and orthophosphate. Glucose was not formed.[9] The primary process in this sugar phosphate is presumably due to light absorption in its shallow band in the 240 to 280 nm region, which has been attributed to an internal electronic transition.[10]

With α-D-glucose 1-phosphate in aqueous solutions at 254 nm, the quantum yields of orthophosphate release in anoxic and oxygenated media were 0.010 and 0.036, respectively. In both media, orthophosphate release was the predominant reaction, but glucose was not formed. Under argon the important organic products were deoxy sugars and malonaldehyde, while under oxygen the major products were CO_2, CO, gluconate, arabinose, glyoxal, and glycolate.[11]

Irradiation of fructose 6-phosphate in water at 254 nm in the absence and presence of oxygen resulted in quantum yields of orthophosphate release of $\Phi = 0.50$ and 0.60, respectively.[12] These values were much higher than those observed for glucose 6-phosphate[9] and for glucose 1-phosphate.[11] The major products were CO and orthophosphate. Organic intermediates that were identified included 2-deoxyerythrose 4-phosphate, 2,4-dihydroxy cyclobutyl phosphate, and glycolaldehyde phosphate. In the presence of oxygen, glyceraldehyde was also formed. The primary process was proposed to be excitation of the free carbonyl group of the sugar phosphate, leading to CO elimination by a Norrish type II reaction. Orthophosphate release was considered to be due to secondary degradation reactions, leading to hydrolytically labile organic phosphates.[12]

In contrast to the "dark" acid-catalyzed hydrolysis of adenosine 5-phosphate (ADP), adenosine triphosphate (ATP), and adenosine tetraphosphate (which mainly led to cleavage of the side chain, producing AMP and orthophosphate), photolytic degradation produced condensed phosphates, by fission of the $5'$–C–O–P bond. The mechanism proposed depends upon direct excitation of the ribose phosphate group.[13]

F.   Vamidothion

The cholinesterase inhibitor vamidothion is a systemic insecticide and acaricide that is used particularly for control of aphids. It is quite toxic to mammals and very toxic to bees.[6] In aqueous solution (containing 2 to 4% methanol) vamidothion was found to be very resistant to direct photolysis under simulated sunlight, and only 2% was degraded during 6 h of illumination. Adding acetone (5%) as a photosensitizer, the photooxidation rate was considerably enhanced, and after 5 h only 5% of the starting compound remained. The main transformation product was vamidothion sulfoxide, which is even more toxic than vamidothion.[14]

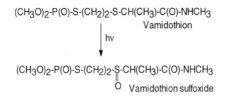

$(CH_3O)_2$-P(O)-S-$(CH_2)_2$-S-CH(CH_3)-C(O)-NHCH$_3$
Vamidothion

hv

$(CH_3O)_2$-P(O)-S-$(CH_2)_2$-S-CH(CH_3)-C(O)-NHCH$_3$
O   Vamidothion sulfoxide

SCHEME VI.1

G.   Ethylenediamine Tetra (methylenephosphonic Acid)

Ethylenediamine tetra (methylenephosphonic acid) (EDTMP) is used as an industrial sequestering agent, in textile processing, in detergent formulations, and for softening boiler water. It is quite resistant to biodegradation. EDTMP forms stable Fe(III) complexes, which absorb light in the visible region. The ligand/Fe^{3+} ratio in these complexes was determined to be between $1/1.29$ and $1/1.86$, and these complexes also contained bound water. Their direct photolysis was studied in dilute aqueous solutions (0.1 mM; pH 6). Initial quantum yields for the photodegradation at both 254 and 365 nm were 2×10^{-4}. Prolonged illumination of aerated solutions at 365 nm resulted in the formation of a photostable state, during which about 75% of the phosphonate groups of the original EDTMP had been released as orthophosphate. Illumination in the absence of air was slower, and the release of orthophosphate reached a plateau after 7 h, while about 50% of the phosphonate groups had been released. The primary photochemical step was proposed to be an oxidoreduction between Fe(III) and EDTMP, with stepwise release of two phosphonate groups and formation of Fe(II). In the presence of oxygen, a third phosphonate group was released, and the photostable product was *N*-methylaminomethylenephosphonic acid.[15]

$$[(HO)_2P(O)-CH_2]_2N-CH_2-CH_2-N[CH_2P(O)(OH)_2]_2 + 2\ Fe^{3+}$$

$$\downarrow h\nu$$

$$(HO)_2P(O)-CH_2-NH-CH_2-CH_2-NH-CH_2P(O)(OH)_2 + 2\ Fe^{2+} + 2HPO_4^{2-}$$

$$O_2 \downarrow h\nu$$

$$CH_3-NH-CH_2-P(O)(OH)_2 + HPO_4^{2-}$$

SCHEME VI.2

H. Sodium Dodecyl Bis(oxyethylene) Phosphate

See Chapter III, Section C.1 on the organophosphorus surfactant sodium dodecyl bis(oxyethylene) phosphate.

I. Aromatic Phosphates

1. Without Sensitizers

Ishikawa et al.[16] studied the photochemical behavior of five trialkyl phosphate esters — tributyl phosphate, tris(chloropropyl) phosphate, tris(2-chloroethyl) phosphate, trioctyl phosphate, and tris(dichloropropyl) phosphate — and of two triaryl phosphate esters, triphenyl phosphate and tricresyl phosphate.[16] In aqueous solutions at pH 3 and pH 10 under illumination with a 15-W low-pressure mercury lamp, pseudo first-order decay of the organophosphates was observed. Aryl phosphates degraded more rapidly, presumably because of their larger absorptivity in the range of 230 to 280 nm due to the aromatic ring. During the photodegradation of the aryl phosphates in alkaline media, phenol was produced as an intermediate, which was further degraded.[16]

Unsensitized photolysis of disodium phenyl phosphate occurs in the ultraviolet region and is due to its UV absorption band ($\lambda_{max} = 267$ nm).[17,18] The photoinduced hydrolysis of nitrophenyl phosphates in aqueous solutions by sunlight was described as a reaction of the photoexcited substrate with the nucleophilic solvent, in a mechanism of "heterolytic photosubstitution."[19,20]

a. Parathion, Edifenfos, and Fenitrothion

One of the most commonly used insecticides is the moderately toxic parathion, *O,O′*-diethyl *O*-4-nitrophenyl phosphorothioate. Its photochemical degradation in aqueous solutions has been the subject of several investigations.[21] Major photoproducts are the more toxic paraoxon (diethyl 4-nitrophenyl phosphate) and 4-nitrophenol. Méallier et al.[22] proposed a mechanism for the aqueous photolysis of parathion, involving both radical and ionic pathways.

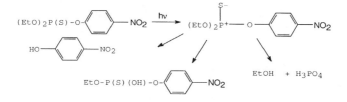

SCHEME VI.3

Gal et. al.[23] found that the photodegradation of parathion in homogeneous aqueous solution by illumination with a mercury lamp resulted in decomposition of the substrate by direct absorption in the 240 to 320 nm range. The "dark" reaction under these condition was negligible. At short irradiation times, the photodecomposition followed first-order kinetics.[23]

The organophosphorus fungicide edifenfos (*O*-ethyl *S,S*-diphenyl phosphorodithioate) is important for the control of blast disease in rice. UV illumination of its aqueous solution or of a thin film of the compound resulted mainly in stepwise photohydrolysis, to *O*-ethyl *S*-phenyl hydrogen phosphorothiolate, *S*-phenyl dihydrogen phosphorothiolate, ethyl dihydrogen phosphate, and eventually to orthophosphate. The sulfur atom ended up in benzene sulfonic acid and sulfuric acid.[24]

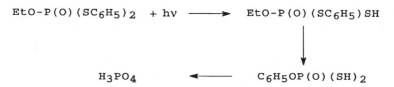

SCHEME VI.4

Fenitrothion (phenitrothion, *O,O*-dimethyl-*O*-4-nitro-*m*-tolyl phosphorothioate) is a cholinesterase inhibitor that is widely used as insecticide and acaricide. It is moderately toxic to mammals, and very toxic to fish and bees.[6] Its direct photolysis under illumination with simulated sunlight was very slow. Faster photolysis was attained by irradiation with a high-pressure Hg lamp. A solution of fenitrothion in water–methanol (5:1) after 7 h of illumination yielded a variety of products. Those identified included trimethyl phosphate, *O,O,S*-trimethyl phosphorothioate, and tetramethyl pyrophosphate, as well as the *S*-methyl isomer of fenitrothion. These products were thus formed by oxidation, solvolysis, and isomerization reactions.[14,25] See also Section II.C on the TiO_2 + O_3 photocatalyzed degradation of fenitrothion.

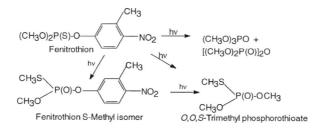

SCHEME VI.5

b. Tolclofos-Methyl

The nonsystemic contact fungicide tolclofos-methyl acts as phospholipid synthesis inhibitor. It is toxic to fish. In lake and river water it was photolytically degraded, with a half-life of 15 to 28 d.[6] Under illumination at 320 to 468 nm, in the presence of oxygen and water, tolclofos-methyl adsorbed on kaolinite or montmorillonite was photooxidized, presumably with hydroxyl and hydroperoxyl radicals as the active intermediates.[26]

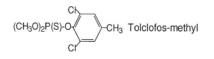

SCHEME VI.6

c. Pyridoxal 5'-Phosphate

The pyridine aldehyde derivative pyridoxal 5'-phosphate, which is an important coenzyme and enzyme inhibitor, was found in neutral aqueous oxygen-free solutions to undergo condensation to a benzoin-type dimer, 5,5'-bis(dihydroxyphenyl-oxymethyl)-3,3'-dihydroxy-2,2'-dimethyl 4,4'-pyridyl, while in oxygenated solutions the main product was 4-pyridoxic acid.[27] On the other hand, in alkaline oxygenated solutions the pyridine ring was ruptured, leading to furan derivatives.[28]

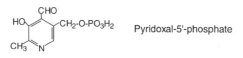

SCHEME VI.7

d.   Chlorpyrifos and Fenamiphos

Chlorpyrifos (3,5,6-trichloro-2-pyridinol) is a nonsystemic contact insecticide used for control of a wide range of insects, including flies, mosquitoes, and household pests. It is a cholinesterase inhibitor, of moderate toxicity to mammals, and it is also toxic to fish and to bees. Fenamiphos (ethyl 4-methylthio-*m*-tolyl-isopropylphosphoramidate) is a systemic nematicide, absorbed through leaves and roots. It is a cholinesterase inhibitor and is very toxic to mammals and to fish.[6]

The phototransformation of chlorpyrifos and fenamiphos in water containing 2 to 4% methanol was examined by illumination with simulated sunlight using a xenon arc. Chlorpyrifos reacted quickly, and only 15% remained after 30 min of illumination. The main product identified was 3,5,6-trichloro-2-pyridinol, which is more toxic than chlorpyrifos. Fenamiphos was even more labile to illumination, and after 30 min only 2% of the compound remained. Its main photolysis product was fenamiphos sulfoxide.[14,25]

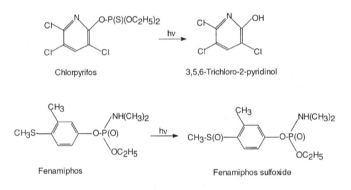

SCHEME VI.8

J.   Dye-Sensitized Photooxidation

Orthophosphate release from O_2-saturated aqueous disodium phenyl phosphate was strongly enhanced by methylene blue, rose bengal, and thionine. Much slower photolysis occurred under N_2. Experiments indicating the probable mechanism showed that the rate in D_2O solution was nine times faster than in H_2O. Also, the photolysis was considerably inhibited by the singlet oxygen quencher 1,4-diazabicyclo-(2.2.2)octane (DABCO). These results suggested a mechanism involving singlet oxygen.[29]

K.   Effect of Iron(III) Salts and H_2O_2

Inouye et al.[30] observed remarkably rapid mineralization of organophosphonates under illumination with visible light ($\lambda > 435$ nm) in the presence of iron(III) salts and hydrogen peroxide. Aqueous solutions of diethyl benzylphosphonate,

$Fe(NO_3)_3$ (10^{-2} M) and H_2O_2 (0.3 M) in the dark underwent slow and incomplete mineralization, accompanied by a temporary color change in the solution, which turned intensely black–brown. This deeply colored species ($\lambda_{max} = 380$ nm, with a broad shoulder at 550 nm) of unknown constitution decayed slowly in the dark with first-order kinetics (k = 0.1 min^{-1}), and very rapidly (k = 1 min^{-1}) under visible light illumination, yielding a yellow–brown colored species. With visible light illumination, 90% mineralization was achieved after 20 min, according to the stoichiometry

$$(C_2H_5O)_2(C_6H_5CH_2)P=O + 15O_2 \rightarrow 11CO_2 + H_3PO_4 + 7H_2O \quad (9)$$

In the absence of the ferric salt, no mineralization of diethyl benzylphosphonate occurred, even under illumination. The effect was explained by a peroxidase-like reaction of Fe(III) complexes and hydrogen peroxide, rather than by the conventional Haber–Weiss process, which involves the extremely short-lived $\cdot OH$ and $\cdot HO_2$ radicals. The mechanism proposed assumed the formation of two relatively long-lived intermediates between the organophosphonate, the Fe(III) ion and hydrogen peroxide, which at room temperature in the dark were stable for at least 40 h, while they decayed very rapidly in the light.

$$(EtO)_2(PhCH_2)P=O + Fe^{3+} + H_2O_2 \rightarrow$$
$$(EtO)_2(PhCH_2)P=O\cdots Fe^{3+}-OOH + H^+ \quad (10)$$

$$(EtO)_2(PhCH_2)P=O\cdots Fe^{3+}-OOH \rightarrow$$
$$(EtO)_2(PhCH_2)P=O\cdots Fe^{4+}-O + OH^- \quad (11)$$

The hypothetical high-valent iron intermediate, the ferryl species $(EtO)_2(PhCH_2)P=O...Fe^{4+}-O$ was assumed to have a high oxidation potential, enabling the light-promoted autodestruction of this intermediate complex to carbon dioxide and phosphoric acid.[30]

$$(EtO)_2P=O\cdots Fe^{4+}-O + h\nu \rightarrow CO_2 + H_2O + H_3PO_4 \quad (12)$$

II. HETEROGENEOUS PHOTODEGRADATION

Organic phosphates were shown to be selectively adsorbed on titanium dioxide, a property that had been applied to the use of titania microbeads (5 to 10 μm) as packing materials for high-performance liquid chromatography. Organic phosphates were selectively retained on such columns, while sugars, amino acids, and organic sulfates were not retained.[31] The high adsorption capacity of TiO_2 for organic phosphates facilitates the activity of this semiconductor as photocatalyst.

A. Trialkyl Phosphates

The photodecomposition of a variety of organic phosphates has been sensitized by illumination in the presence of suspended TiO_2. The photodegradation of trimethyl phosphate, trimethyl phosphite, and O,O-dimethyl ammonium phosphodithioate (DMPDT) was carried out in the presence of Pt-loaded (1.4 wt%) TiO_2, by illumination with either a 500-W super-high-pressure mercury lamp, or with sunlight. All these three compounds were found to be completely degraded under sunlight within 4 h, producing phosphoric acid, and in the case of DMPDT also sulfuric acid. An intermediate product observed in the photooxidation of DMPDT was formaldehyde. With "bare" TiO_2 (without Pt), the rates of degradation were 30 to 40% slower.[32]

While organophosphate pesticides were readily photodegraded in aqueous suspensions of TiO_2, this reaction was substantially retarded in the presence of fulvic acid. Presumably, fulvic acid competed with the organophosphates for the same active sites on the semiconductor particle surface, thus inhibiting the pesticide degradation.[33]

B. Dichlorvos, Trichlorfon, and Tetrachlorvinphos

In photocatalytic experiments with the insecticides dimethyl 2,2-dichlorovinyl phosphate [DDVP; dichlorvos; $(MeO)_2P(O)–O–CH=CCl_2$] and dimethyl 2,2,2-trichloro-1-hydroxyethylphosphonate [DEP; trichlorfon; $(MeO)_2P(O)–CH(OH)–CCl_3$], rates of photodegradation were increased about 4.5- and 6-fold for DDVP and DEP, respectively by loading the TiO_2 photocatalyst with Pt. Rates were also enhanced by addition of H_2O_2 to the reaction mixture.[34]

In a detailed study by Lu et al.[35] on the photocatalytic degradation of dichlorvos, the pesticide in oxygen-saturated aqueous solutions was continuously pumped through a glass tube coated on the inside with TiO_2 and illuminated with a 20-W blacklight UV fluorescent lamp. The disappearance of dichlorvos and release of chloride ions observed first-order kinetics, and the initial rate was quite rapid. Thus at an initial concentration of 0.1 mM, the half-life was only 11 min. The rate of phototransformation of dichlorvos as a function of its initial concentration could be fitted to a Langmuir–Hinshelwood relationship. The rate of photodegradation increased with the flow rate, reaching a plateau at a flow rate of about 500 ml min^{-1}. The rate also increased with temperature, yielding a linear Arrhenius plot. From this an activation energy of 28.4 kcal mol^{-1} was derived. The initial quantum yield for the degradation of dichlorvos was estimated to be about 2.7%. The rate of photodegradation was significantly inhibited by the presence of Cl^- ions, at concentrations of ≥ 0.01 M. Perchlorate ions had no effect. The complete mineralization, according to

$$(CH_3O)_2 P(O)-O-CH=CCl_2 + {}^9/_2 O_2 \rightarrow$$

$$H_3PO_4 + 2HCl + 4CO_2 + H_2O \tag{13}$$

was very much slower, and within 90 min about 90% of the total organic carbon (TOC) still remained in solution, indicating the formation of chlorine-free intermediates. These phosphorus-containing intermediates required much longer illumination times for complete mineralization.[35]

Tetrachlorvinphos, [Z-2-chloro-1-(2,4,5-trichlorophenyl) ethenyl dimethyl phosphate], is used for insecticidal control on fruits, tomatoes, and cotton. It is a cholinesterase inhibitor.[6] Both fenitrothion and tetrachlorvinphos in aerated aqueous suspensions of TiO_2 underwent photodegradation, with more than 75 to 80% of the phosphorus atoms released as orthophosphate. Reaction intermediates identified from tetrachlorvinphos included 2,4,5-trichlorobenzaldehyde, 2,4,5-trichlorobenzoic acid, and 2,4,5-trichlorophenol, which readily underwent further photocatalytic degradation. Other intermediates observed were trimethyl phosphate and dimethyl hydrogen phosphate.[36,37]

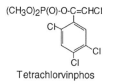

$(CH_3O)_2P(O)$-O-C=CHCl

Tetrachlorvinphos

SCHEME VI.9

C. O_3 and TiO_2

For some recalcitrant water pollutants, an ozone pretreatment considerably facilitated the photocatalytic degradation. Tanaka et al.[38] observed that dimethyl 2,2,2-trichloro-1-hydroxyethyl phosphonate (DEP) and O,O-dimethyl O-4-nitro-m-tolyl phosphorothioate (fenitrothion) in the photocatalytic reaction with aqueous suspensions of TiO_2 underwent very slow degradation, during which the rates of elimination of the TOC were equal to the rates of release of the equivalent amounts of carbon dioxide. This showed that in the photocatalysis with TiO_2 there was no appreciable accumulation of intermediate products. On the other hand, ozonation of the same compounds resulted in the formation of partially oxidized compounds, which still retained their phosphorus and chlorine atoms. However, photocatalytic treatment after ozonation caused much faster and more complete mineralization of these agrochemicals.[38] As noted earlier (Section I.I.a.) the direct photolysis of fenitrothion resulted in a slow transformation to a variety of organophosphorus compounds.

Krosley et al.[39] illuminated butylphosphonic acid, benzylphosphonic acid, and phenylphosphonic acid in aqueous suspensions of TiO_2. After prolonged irradiation of benzylphosphonic acid, complete mineralization was achieved. After shorter reaction times, the intermediate oxidation products benzyl alcohol, benzaldehyde, and bibenzyl were identified. The observation of the formation of bibenzyl suggested that benzyl radicals may be the primary unstable intermediate, formed by attack of surface holes (h^+) or hydroxyl radicals.

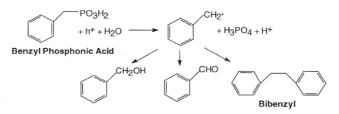

SCHEME VI.10

The kinetics were evaluated using the Langmuir–Hinshelwood model, by plotting the inverse rate vs. the inverse initial substrate concentration. In the linear plots obtained, the intercept represented the inverse ($1/k$) of the rate constants for the reaction of the substrate adsorbed on the TiO_2 surface, in which the slope may be considered to represent the inverse of this rate constant times the adsorption equilibrium constant ($1/kK$). The rate constants k for phenylphosphonic, butylphosphonic, and benzylphosphonic acids were 3.2, 9.4, and 13.5 μM min^{-1}, respectively.[39]

In a study of the photocatalyzed oxidation of 1-hydroxyethane-1,1-diphosphonate (HEDP) in aqueous suspensions of TiO_2, Sabin et al.[40] observed that the rate of reaction followed zero-order kinetics at all substrate concentrations. This was in contrast to the kinetic characteristics of halocarbon compounds, in which the rates followed zero-order kinetics only at high substrate concentrations (>3 mM), while at low concentrations (<1.4 mM), pseudo first-order kinetics were obeyed. The interpretation given on the basis of the Langmuir–Hinshelwood model is that at high substrate concentrations all the catalytic sites of the photocatalyst were occupied.[40]

4-Nitrophenyl diethyl phosphate (paraoxon) deposited on Nb-doped TiO_2 on a moist solid surface underwent photohydrolysis:

$$(EtO)_2 P(O) - O - C_6H_4 - NO_2 + H_2O \rightarrow$$
$$(EtO)_2 P(O) - OH + HO - C_6H_4 - NO_2 \tag{14}$$

This reaction may be considered a model for the photodegradation of active phosphates, possibly mediated by surface peroxy species.[41] In a further study of paraoxon and of several other organophosphorus insecticides in oxygenated aqueous solutions with suspensions of TiO_2, by illumination with near UV light ($\lambda > 350$ nm), all compounds were completely oxidized to carbon dioxide and phosphoric acid within 5 to 10 h. No reaction was found in an argon atmosphere. The stimulatory effects of several oxidant additives ($K_2S_2O_8$, H_2O_2, $KBrO_3$, and $NaIO_4$) were tested. Thus, for the widely used insecticide malathion [*O*,*O*-dimethyl 5-(1,2-dicarb ethoxyethyl) phosphoro dithioate], the half-times for CO_2 evolution in the presence of an air atmosphere, pure oxygen atmosphere, H_2O_2 additive, and $NaIO_4$ additive were 225, 73, 100, and 5 min. To account for the observed photocatalytic effects, and for the remarkable stimulatory effect of the periodate ions, the mechanism proposed involved a "dark" interaction of the organophosphorus compound adsorbed on the TiO_2 surface, possibly by hydrogen bonding between surface-bound hydroxyl groups with the phosphoryl moiety of the phosphate ester, with release of the phenolic moiety. The phenolic compounds underwent epoxidation by surface-bound O_2^- radicals, resulting in ring opening and further stepwise oxidation of the ring-opened intermediate. Illumination and band-gap excitation of the semiconductor resulted mainly in the formation of surface-bound hydroxyl radicals, which react by α-hydrogen abstraction from the $P–O–CH_2CH_3$ groups, leading to the release of aldehydes, which are further oxidized to CO_2.[42]

REFERENCES

1. CB Weapons Today, Vol. 2, in *The Problem of Chemical and Biological Warfare,* Stockholm International Peace Research Institute, Humanities Press, New York, 1973.
2. Halmann M., Photochemical and radiation-induced reactions of phosphorus compounds, *Topics Phosphorus Chem.,* 4, 49–84, 1967.
3. Halmann, M. and Platzner, I., The photochemistry of phosphorus compounds. Part II. Photolysis of ethyl dihydrogen phosphate in aqueous solution, *J. Chem. Soc.,* 5380–5385, 1965.
4. Benschop, H. and Halmann, M. Photochemistry of phosphorus compounds. Part IX. Photolysis of trimethyl phosphate in aqueous solution, *J. Chem. Soc.,* 1175–1180, 1974.
5. Mill, T., Gould, C. W., Epstein, J., and Schiff, L. J., Oxidative degradation of phosphorus esters, U.S. Patent No. 4108746, C-NO: 204–15R. 8 September 1977. *Chem. Abstr.,* 90, 55104.
6. Hartley, D. and Kidd, H. (Eds.), *The Agrochemicals Handbook,* Royal Society of Chemists, Nottingham, England, 1987.
7. Gignoux, J., Triantaphylides, C., and Peiffer, G., Photolysis of dimethyl vinyl phosphate in aqueous solution under an inert atmosphere, *Bull. Soc. Chim. France, Pt. 2,* 527–530, 1977.
8. Greenwald, J. and Halmann, M., Far-UV spectroscopy and photochemistry of glycerol-1-phosphate and of glycerol-2-phosphate in aqueous solutions, *J. Chem. Soc., Perkin Trans. II,* 1095–1101, 1972.

9. Triantaphylides, C. and Halmann, M., The photochemistry of phosphorus compounds. Part X. Photolysis of disodium α-D-glucose 6-phosphate in aqueous
   solution under nitrogen or oxygen, *J. Chem. Soc., Perkin Trans. II*, 34–40,
   1975.

10. Trachtman, M. and Halmann, M., Far-ultraviolet absorption spectra of D-
    glucose phosphates in aqueous solutions, *Carbohydr. Res.*, 19, 245–248, 1971.

11. Trachtman, M. and Halmann, M., The photochemistry of phosphorus com-
    pounds. Part XI. Photolysis of dipotassium α-D-glucose 1-phosphate in aqueous
    solution under argon and oxygen, *J. Chem. Soc., Perkin Trans. II*, 132–137,
    1977.

12. Triantaphylides, C. and Gerster, R., Photoreactions of fructose 6-phosphate in
    oxygenated and deoxygenated aqueous solutions, *J. Chem. Soc., Perkin Trans.
    II*, 1719–1724, 1977.

13. Goossen, J. T. H. and Kloosterboer, J. G., Photolysis and hydrolysis of adeno-
    sine 5′-phosphates, *Photochem. Photobiol.*, 27, 703–708, 1978.

14. Barceló, D., Durand, G., and De Bertrand, N., Photodegradation of the organo-
    phosphorus pesticides chlorpyrifos, fenamiphos and vamidithion in water,
    Toxicol. Environ. Chem., 38, 183–199, 1993.

15. Matthijs, E., De Oude, N. T., Bolte, M., and Lemaire, J., Photodegradation of
    ferric ethylenediamine-tetra(methylene-phosphonic acid) (EDTMP) in aqueous
    solution, *Water Res.*, 7, 845–851, 1989.

16. Ishikawa, S., Uchimura, Y., Baba, K., Eguchi, Y., and Kido, K., Photochemical
    behavior of organic phosphate esters in aqueous solutions irradiated with a
    mercury lamp, *Bull. Environ. Contam. Toxicol.*, 49, 368–374, 1992.

17. Getoff, N. and Solar, S., Photolysis of monophenyl phosphate and formation of
    e_{aq}^{-} in aqueous solutions, *Monatsh. Chem.*, 105, 241–253, 1974.

18. Köhler, G. and Getoff, N., Effect of molecular environment and of excitation
    energy on electron photoejection from monophenyl phosphate, *J. Chem. Soc.,
    Faraday Trans. I*, 74, 1029–1035, 1978.

19. Havinga, E., Some photochemical reactions, *Chimia*, 16, 145–151, 1962.

20. Havinga, E. and Cornelisse, J., Aromatic photosubstitution reactions, *Pure
    Appl. Chem.*, 47, 1–10, 1976.

21. Cavell, B. D., Methods used in the study of the photochemical degradation of
    pesticides, *Pestic. Sci.*, 10, 177–180, 1979.

22. Méallier, P., Nury, Y., Pouyet, B., Coste, C., and Bastide, J., Photodegradation
    of plant protectant molecules. II. Kinetics and mechanism of parathion
    photodegradation, *Chemosphere*, 12, 815–820, 1977.

23. Gal, E., Aires, P., Chamarro, E., and Esplugas, S., Photochemical degradation
    of parathion in aqueous solution, *Water Res.*, 26, 911–915, 1992.

24. Murai, T., Photodecomposition of *O*-ethyl *S,S*-diphenyl phosphorodithiolate
    (edifenphos), *Agric. Biol. Chem.*, 41, 71–77, 1977.

25. Barceló, D., Durand, G., Debertrand, N., and Albaigés, J., Determination of
    aquatic photodegradation products of selected pesticides by gas-chromatogra-
    phy mass-spectrometry and liquid-chromatography mass-spectrometry, *Sci.
    Total Environ.*, 132, 283–296, 1993.

26. Katagi, T., Photoinduced oxidation of the organophosphorus fungicide tolclofos-
    methyl on clay minerals, *J. Agric. Food*, 38, 1595–1600, 1990.

27. Morrison, A. L. and Long, R. F., The photolysis of pyridoxal phosphate, *J. Chem. Soc.*, 211–215, 1958.

28. Bazhulina, N. P., Kirpichnikov, M. P., Morozov, Y. V., Savin, F. A., Sinyavina, L. B., and Florentiev, V. L., Photochemistry of aldehyde forms of pyridoxal, pyridoxal 5′-phosphate, and their derivatives, *Mol. Photochem.*, 6, 367–396, 1974.

29. Halmann, M. and Levy, D., Dye-sensitized photolysis of disodium phenyl phosphate in aqueous solution under oxygen, *Photochem. Photobiol.*, 30, 143–146, 1979.

30. Inouye, Y., Jirousek, M., Grätzel, C. K., and Grätzel, M., Rapid mineralization of organophosphonates by visible light-enhanced peroxidase action of iron(III) complexes, *J. Mol. Catal.*, 68, L35–L40, 1991.

31. Matsuda, H., Nakamura, H., and Nakajima, T., New ceramic titania: selective adsorbent for organic phosphates, *Anal. Sci.*, 6, 911–912, 1993.

32. Harada, K., Hisanaga, T., and Tanaka, K., Photocatalytic degradation of organophosphorus compounds in semiconductor suspension, *N. J. Chem.*, 11, 597–600, 1987.

33. Hung, S. T. and Mak, M. K. S., Titanium dioxide photocatalysed degradation of organophosphate in a system simulating the natural aquatic environment, *Environ. Technol.*, 14, 265–269, 1993.

34. Harada, K., Hisanaga, T., and Tanaka, K., Photocatalytic degradation of organophosphorus insecticides in aqueous semiconductor suspensions, *Water Res.*, 24, 1415–1417, 1990.

35. Lu, M.-C., Roam, G.-D., Chen, J.-N., and Huang, C. P., Factors affecting the photocatalytic degradation of dichlorvos over titanium dioxide supported on glass, *J. Photochem. Photobiol. A: Chem.*, 76, 103–110, 1993.

36. Kerzhentsev, M., Guillard, C., Herrmann, J.-M., and Pichat, P., TiO_2-photosensitized degradation of the insecticide tetrachlorvinphos ((Z)-2-chloro-1(2,4,5-trichlorophenyl) ethenyl dimethyl phosphate), in: *Photocatalytic Purification and Treatment of Water and Air*, Ollis, D. F. and Al-Ekabi, H. (Eds.), Elsevier Science Publishers, Amsterdam, 1993, pp. 601–606.

37. Herrmann, J. M., Kerzhentsev, M., Tahiri, H., and Ait Ichou, Y., Photocatalytic degradation of some pesticides in aqueous medium, *Abstr. 2nd Int. Symp. New Trends in Photoelectrochemistry*, University of Tokyo, Japan, March 1994, p. 10.

38. Tanaka, K., Abe, K., Sheng, C. Y., and Hisanaga, T., Photocatalytic wastewater treatment combined with ozone pretreatment, *Environ. Sci. Technol.*, 26, 2534–2636, 1992.

39. Krosley, K. W., Collard, D. M., Adamson, J., and Fox, M. A., Degradation of organophosphonic acids catalyzed by irradiated titanium dioxide, *J. Photochem. Photobiol. A: Chem.*, 69, 357–360, 1993.

40. Sabin, F., Türk, T., and Vogler, A., Photooxidation of organic compounds in the presence of titanium dioxide: determination of the efficiency, *J. Photochem. Photobiol. A: Chem.*, 63, 99–106, 1992.

41. Grätzel, C. K., Jirousek M., and Grätzel M., Accelerated decomposition of active phosphates on titanium oxide surfaces, *J. Mol. Catal.*, 39, 347–353, 1987.

42. Grätzel, C. K., Jirousek, M., and Grätzel, M., Decomposition of organophosphorus compounds on photoactivated TiO_2 surfaces, *J. Mol. Catal.*, 60, 375–387, 1990.

Chapter VII_____
Organic Sulfur Compounds

Organic sulfur compounds include intermediates in the industrial synthesis of dyes, various pesticides, and components of crude petroleum. Many of these compounds are toxic or resistant to biodegradation. The photodegradation of sulfur-containing organophosphorus pesticides has been described in Chapter VI.

I. SULFONIC ACIDS

A. Benzenesulfonic Acid

The photolysis of aqueous solutions of benzenesulfonic acid yielded SO_2 and H_2SO_4 in a total yield of 39% (based on the consumed starting sulfonic acid). In acidic solutions, the quantum yield for the decomposition of benzenesulfonic acid was 0.26. In alkaline aqueous solutions the photolysis products included benzene (16%) and biphenyl (1%). In the proposed mechanism, the initial step was radical fission of the C–S bond, producing Ph· and ·SO_3H radicals.[1]

B. p-Toluenesulfonic Acid

The photodegradation of p-toluenesufonic acid (PTS) as a model of anionic surfactants was tested in the homogeneous H_2O_2/UV and the heterogeneous TiO_2/O_2/UV systems. The rates of formation of hydroxyl radicals in the two systems were determined by spin trapping with DMPO, and electron spin resonance (ESR) measurement of the radical-adduct ·DMPO-OH. In media buffered with a phosphate–borax mixture (pH 7), the dependences of the rates of formation of this radical-adduct on the initial concentrations of PTS were identical in the two systems. Since the ·DMPO-OH formation in the H_2O_2/UV system obviously occurred in the homogeneous phase, it was suggested that in the buffered media the reaction photocatalyzed by TiO_2 also occurred in the homogeneous phase, by radicals leaving the TiO_2 surface. With an initial concentration of 1 mM PTS in an unbuffered solution (pH 3) and dispersed TiO_2 (1 g/L), the initial decomposition rate was 23 μM/min. Much lower rates, of 5.6 μM/min, were obtained in solutions buffered with a mixture of phosphate and borax (pH 6–9). The rate decreases in the buffered media were explained by competitive adsorption of phosphate anions on the TiO_2 surface, thus preventing the adsorption of the PTS anions.[2]

C. Metanilic Acid

Aniline sulfonic acids are important intermediates in the manufacture of textile dyes. The direct photolysis of metanilic acid in deaerated aqueous solutions, by illumination at $\lambda = 254$ nm, led to the formation of aniline and to the isomerization products sulfanilic acid and orthanilic acid.[3]

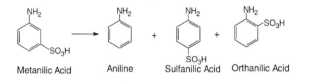

Metanilic Acid　　　　Aniline　　　Sulfanilic Acid　Orthanilic Acid

SCHEME VII.1

The distribution of the photoproducts depended on the pH. The yields after 1 min of irradiation at pH 4.2 and 11.2 were for aniline 46 and 63%, for sulfanilic acid 35 and 24%, and for orthanilic acid 13 and 8%, respectively. The absence of radical intermediates in the photoreaction of metanilic acid was shown by irradiation of its methanolic solution in the presence of *n*-butyl thiol. The yields of products and the product distribution were similar to that in the absence of this radical scavenger, indicating that the photoreactions of metanilic acid did not occur by a radical mechanism. In the presence of oxygen, a known triplet quencher, the photolysis of metanilic acid was inhibited. One of the proposed mechanisms involved excitation of metanilic acid to the first excited singlet state S_1, followed by intersystem crossing to a single triplet state T_1. Protonation of the T_1 state led to a σ complex of metanilic acid, which then decomposed to the photoproducts.[3]

D. Anthraquinone Sulfonic Acid

Anthraquinone derivatives, such as anthraquinone 2-sulfonic acid (ASS) and anthraquinone 2,6-disulfonic acid, are important intermediates in dye synthesis and are also widely used in the paper industry for the bleaching of pulp. These compounds are highly resistant to microbial biodegradation in conventional sewage treatment. Also, direct photolysis was very slow and caused only partial degradation. In photocatalytic oxidation of ASS in aerated aqueous suspensions of TiO_2 under illumination with simulated sunlight, 75 to 80% of the organic carbon was oxidized to CO_2 within 20 h, and mineralization was complete after about 50 h. The rate of photodegradation was enhanced threefold by addition of H_2O_2 (45 mM) at regular intervals during the irradiation. The optimal concentration of TiO_2 was about 2 g/L. The reaction involved degradation of the aromatic compound, presumably by hydroxylation, leading to formation of hydroxy-cyclohexadienyl radicals and subsequent ring open-ing. The photocatalyzed degradation of anthraquinone 2-sulfonate was also tested as a pretreatment prior to microbial treatment with waste water plant sludge bacteria. A 24-h photocatalytic pretreatment very markedly facilitated the biological oxidation.[4]

ASS also was oxidized in a homogeneous solution photo Fenton reaction using Fe^{3+} and H_2O_2 and illuminated at AM1 ($\lambda > 290$ nm). With initially 3 mM ASS, at 35 and 60°C, degradation was about 90% complete within 5 and 3 h, respectively. The rate was enhanced in the presence of oxygen. Maximal photoreactivity was obtained using 0.3 mM Fe^{3+}, 3 mM ASS, and 0.1 M H_2O_2, but significant photodegradation was observed even with Fe^{3+} concentrations as low as 10^{-5} M, indicating that Fe ions act as a photocatalyst. Complete mineralization required about 15 to 16 h. The primary excitation step was presumably light absorption by the $Fe(OH)^{2+}$ ion, with $\varepsilon = 2050$ M^{-1} cm^{-1} at 295 nm, followed by quite rapid dearomatization. Probably the temperature-dependent and slow step was the photooxidation of the fumaric, malic, and oxalic acids formed after ring opening. If Cu^{2+} was used instead of Fe^{3+}, similar reactions occurred, but at a slower rate. The photo Fenton reaction for 5 h was useful as a pretreatment, prior to biodegradation, cutting down the biological treatment time from 8 to 3 d in order to achieve 95% biodegrada-tion. The photo Fenton reaction destroyed the aromatic rings and facilitated the biotreatment.[5]

E.   Thiolcarbamates

Molinate, *S*-ethyl-*N,N*-hexamethylene thiocarbamate, is used for herbicidal control of broad-leaf and grassy weeds.[6] The photooxidation of molinate under sunlight was initiated in the presence of dilute hydrogen peroxide (5 to 100 μM).[7]

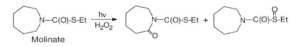

SCHEME VII.2

F.   Asulam

Asulam (sodium *N*-methoxycarbonylsulfanilamide) is an important systemic herbicide that is active in both pre- and postemergence treatment against many broad-leaf and grass weeds.[6] Ozone pretreatment prior to photocatalytic reaction with suspended TiO_2 considerably enhanced the rate and complete-ness of mineralization of Asulam.[8]

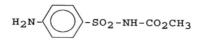

SCHEME VII.3

See Chapter V, Section II.F, on the photodegradation of the pesticide aldicarb.

In the rose bengal-catalyzed photooxidation of thioglycolic acid in aqueous solution, the isolated product was dithiodiglycolic acid. The proposed mechanism involved formation of singlet oxygen, and the intermediate production of thiyl radicals.[9]

With eosin-Y, methylene blue, or rose bengal as photosensitizers, the oxidation of D,L-, L-, and D-methionine resulted in formation of methionine sulfoxide. Singlet oxygen involvement was indicated by the inhibitive effects of known singlet oxygen quenchers.[10]

G. Methylene Blue

Matthews[11] observed the TiO_2-photocatalyzed degradation of methylene blue in aqueous solution.

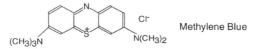

SCHEME VII.4

The reaction effected complete mineralization, according to

$$C_{16}H_{18}N_3SCl + 5.5O_2 \rightarrow 16CO_2 + 6H_2O + 3HNO_3 + H_2SO_4 + HCl \quad (1)$$

Under illumination with a 20-W blacklight fluorescent lamp, with TiO_2 coated as a thin film inside a spiral glass reactor surrounding the lamp, with initially 0.01 mM methylene blue flowing through the reactor, the half-time for bleaching of the color was 11.8 min. The reaction rate observed apparent first-order kinetics and was in agreement with a Langmuir–Hinshelwood model. Increasing flow rates resulted in increased reaction rates, which may indicate mass flow limitations. At high flow rates, a maximal quantum yield was estimated to be 0.0092. The photodegradation of methylene blue was also successful with sunlight, by mounting the reactor glass spiral in the focus of a parabolic trough solar concentrator.[11]

A revised value of the quantum yield for the photobleaching of methylene blue by illumination at 320 nm was 0.056, as described in detail in Chapter I, Section IV.H.[12]

The bleaching of methylene blue has been developed by Nogueira and Jardim[13] as a student exercise, teaching in one experiment some basic principles of semiconductors as photocatalysts, chemical kinetics, and the amelioration of environmental pollution.[12] Methylene blue (0.1 mM) with suspended TiO$_2$ (1.0 g/L) in 25 ml of water was placed in open petri dishes and was exposed to sunlight. The disappearance of the methylene blue was observed by the decrease of its absorbance at 660 nm ($\varepsilon = 66,700$ L mol^{-1} cm^{-1}). The reaction followed first-order kinetics, with a half-life of about 9 min.[13]

H. Thioacetamide

Davis and Huang[14] tested CdS from different sources on the photocatalytic degradation of thioacetamide. The most reactive CdS photocatalysts were those with the largest surface area per unit weight. However, on a surface area basis, the most effective CdS samples were those of highest purity, such as the electronic-grade materials. The presence of sulfur-containing organic compounds decreased the photocorrosion of CdS.[14]

I. Sethoxydim and Clethodim

The cyclohexane 1,3-dione herbicides sethoxydim and clethodim are used for control of annual and perennial grasses. Dicotyledonous plants are tolerant of these herbicides.[6] Sethoxydim in aqueous solution under UV illumination was within 1 h almost completely (>80%) transformed to six major products. One of these was identified by mass spectrometry as desethoxysethoxydim.[15] Clethodim in aqueous solutions was subject both to acid (dark) catalyzed and to photolytic transformation. Thus under UV light, its half-lives at pH 5, 6, and 7 were 2.4, 2,6, and 3.2 h, respectively. The four photoproducts isolated by high-performance liquid chromatography (HPLC) were not identified.[16]

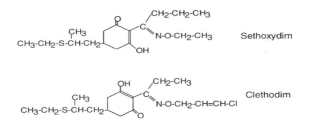

SCHEME VII.5

J. Benzo[b]thiophene and Dibenzothiophene

Polycyclic aromatic sulfur heterocyclic compounds are major components of most crude oils, often constituting 10 to 30% of the polycyclic aromatic compounds in the crude petroleum. Crude oil spills in the oceans result in a variety of processes, including the partial evaporation of volatile components, dispersion into the water column, chemical and biodegradation, and adsorption. Following a major oil spill, the analysis of the weathering products in oil-slick samples indicated the formation of sulfoxides of dibenzothiophenes. Such compounds may increase the toxicity of petroleum. Benzo[b]thiophene (thianaphthene) is sometimes found in crude oils, while its alkylated derivatives are common components of crudes.[17]

See Chapter III, Section II.C, on the photocatalytic destruction of oil spills.

Benzo[b]thiophene and dibenzothiophene in aqueous solutions under illumination at 313 nm or by sunlight underwent photolysis. The addition of humic acid had little effect on the rate of photodegradation.[18] After illuminating dispersions of benzo[b]thiophene for 1 h in a two-phase system of water and tetradecane (simulating an oil spill) with a 250-W medium-pressure Hg lamp ($\lambda > 300$ nm), two products could be identified, benzo[b]thiophene-2,3-quinone and the tetracyclic benzonaphtho[2,1-d]thiophene (1,2-benzodiphenylene sulfide). The sulfone of this compound, which may be a plausible oxidation product under environmental conditions, is known as a potent mutagen to bacteria. UV illumination of benzo[b]thiophene in a water–methanol mixture also yielded benzo[b]thiophene 2,3-quinone, and in addition a strong acid, which was identified as 2-sulfobenzoic acid.[19]

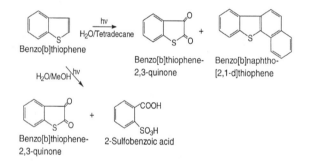

SCHEME VII.6

In a further simulation of the photochemical effects during an oil spill, several polycyclic aromatic and thioaromatic compounds in a tetradecane solution were floated on water in an open vessel. The relative losses of phenanthrene, dibenzothiophene, and of several mono- and dimethyl derivatives of these compounds under sunlight illumination during 27 h were determined.[20] The order of decreasing losses was: phenanthrene >1-methyl-phenanthrene = dibenzothiophene >2-methylphenanthrene >4-methyldibenzo-thiophene >1-methyldibenzothiophene >2- and 3-methyl-dibenzothiophene >3,4-dimethyldibenzothiophene >1,7-dimethyldibenzothiophene >3,7-dimethyldibenzothiophene. Thus, the sulfur heterocycles were more stable than the analogous homocyclic compounds, and the stability increased with the larger number of methyl substituents. These results suggest that aromatic sulfur heterocycles should be environmentally relatively stable and therefore useful as persistent markers of oil pollution.[20]

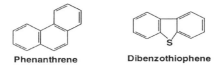

Phenanthrene   **Dibenzothiophene**

SCHEME VII.7

K. 4-Thiomethyl-*N*-methylphenylcarbamate

Thiocarbamates, as well as carbamates, are widely used as pesticides. In order to identify the photochemical transformations of thiocarbamates in the environment, the photolysis of a model compound, 4-thiomethyl-*N*-methyl-phenylcarbamate, was studied in ethanol solution by illumination with a medium-pressure Hg lamp ($\lambda > 290$ nm). The photoproducts identified were 4-(methylmercapto)phenol and 3-*N*-methylamide-4-hydroxythioanisole. The proposed mechanism involved excitation of the starting compound ($\lambda_{max} = 254$ nm), presumably due to an n–π* transition, resulting in C–O bond cleavage with formation of an *N*-methylacyl radical, ·CONHCH$_3$. This radical then may migrate to a position ortho to the hydroxyl group, resulting in a photo Fries rearrangement product.[21]

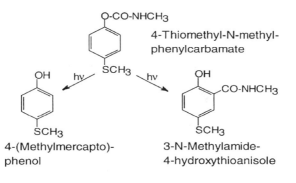

SCHEME VII.8

L. Endosulfan

Endosulfan (thionex) is widely used as an insecticide and acaricide. It is very toxic to mammals and extremely toxic to fish.[6] The photocatalytic degradation of this water-insoluble pesticide was achieved using a dispersion of TiO$_2$ (3 g/L).[22]

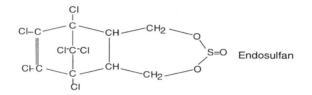

Endosulfan

SCHEME VII.9

M. Fenitrothion

See Chapter VI for the phosphorothioate pesticide fenitrothion (phenitrothion).

REFERENCES

1. Ogata, Y., Takagi, K., and Yamada, S., Photolysis of benzenesulfonic acid, *Bull. Chem. Soc. Jpn.,* 50, 2205–2206, 1977.
2. Brezová, V., Stasko, A., Biskupic, S., Blazková, A., and Havlínová, B., Kinetics of hydroxyl radical spin trapping in photoactivated homogeneous (H_2O_2) and heterogeneous (TiO_2, O_2) aqueous systems, *J. Phys. Chem.,* 98, 8977–8984, 1994.
3. Lally, J. M. and Spillane, W. J., Photoisomerization and photodegradation of metanilic acid, *J. Chem. Soc., Perkin Trans. 2,* 333–338, 1991.
4. Kiwi, J., Pulgarin, C., Peringer, P., and Grätzel, M., Beneficial effects of heterogeneous photocatalysis on the biodegradation of anthraquinone sulfonate observed in water treatment, *N. J. Chem.,* 17, 487–494, 1993.
5. Kiwi, J., Pulgarin, C., Peringer, P., and Grätzel, M., Beneficial effects of homogeneous photo-Fenton pretreatment upon the biodegradation of anthraquinone sulfonate in waste water treatment, *Appl. Catal. B: Environ.,* 3, 85–99, 1993.
6. Hartley, D. and Kidd, H. (Eds.), *The Agrochemicals Handbook,* Royal Society of Chemists, Nottingham, England, 1987.
7. Draper, W. M. and Crosby, D. G., Solar photooxidation in dilute hydrogen peroxide, *J. Agric. Food Chem.,* 32, 231–237, 1984.
8. Tanaka, K., Abe, K., Sheng, C. Y., and Hisanaga, T., Photocatalytic wastewater treatment combined with ozone pretreatment, *Environ. Sci. Technol.,* 26, 2534–2636, 1992.
9. Gandhi, P., Dubey, R., Ameta, S., and Sharma, T. C., Dye-sensitized photooxygenation of thioglycolic acid by singlet oxygen, *Natl. Acad. Sci. Lett. (India),* 8, 181, 1985.
10. Gandhi, P., Dubey, R., Ameta, S., and Sharma, T. C., Photosensitized oxidation of D,L-, L- and D-methionine by singlet oxygen, *Natl. Acad. Sci. Lett. (India),* 7, 331, 1984.
11. Matthews, R. W., Photocatalytic oxidation and adsorption of methylene blue on thin films of near ultraviolet illuminated TiO_2, *J. Chem. Soc., Farad. Trans. I,* 85, 1291–1302, 1989.

12. Valladares, J. E. and Bolton, J. R., A method for determination of quantum yields in heterogeneous systems: the TiO_2 photocatalyzed bleaching of methylene blue, in: *Photocatalytic Purification and Treatment of Water and Air,* Ollis, D. F. and Al-Ekabi, H. (Eds.), Elsevier Science Publishers, Amsterdam, 1993, pp. 111–120.
13. Nogueira, R. F. P. and Jardim, W. F., Photodegradation of methylene blue — using solar light and semiconductor (TiO_2), *J. Chem. Educ.,* 70, 861–862, 1993.
14. Davis, A. P. and Huang, C. P., Effects of cadmium sulfide characteristics on the photocatalytic oxidation of thioacetamide, *Langmuir,* 7, 709–713, 1991.
15. Campbell, J. R. and Penner, D., Abiotic transformation of sethoxydim, *Weed Sci.,* 33, 435–439, 1985.
16. Falb, L. N., Bridges, D. C., and Smith, A. E., Jr., Effects of pH and adjuvants on clethodim photodegradation, *J. Agric. Food Chem.,* 38, 875–878, 1990.
17. Patel, J. R., Overton, E. B., and Laseter, J. L., Environmental photooxidation of dibenzothiophenes following the Amoco Cadiz oil spill, *Chemosphere,* 8, 557–561, 1979.
18. Mill, T., Mabey, W. R., Lan, B. Y., and Baraze, A., Photolysis of polycyclic aromatic hydrocarbons in water, *Chemosphere,* 10, 1281–1290, 1981.
19. Andersson, J. T. and Bobinger, S., Polycyclic aromatic sulfur heterocycles. II. Photochemical oxidation of benzo[b]thiophene in aqueous solution, *Chemosphere, 24,* 383–389, 1992.
20. Andersson, J. T., Polycyclic aromatic sulfur heterocycles. 3. Photochemical stability of the potential oil pollution markers phenanthrenes and dibenzothiophenes, *Chemosphere,* 27, 2097–2102, 1993.
21. Addison, J. and Cote, K. A., Structure–photodegradation of 4-thiomethyl-*N*-methylphenyl carbamate, *Chemosphere,* 24, 181–188, 1992.
22. Porter, J. F. and Yue, P. L., Simple supported photocatalysts, in: *Photocatalytic Purification and Treatment of Water and Air,* Ollis, D. F. and Al-Ekabi, H. (Eds.), Elsevier Science Publishers, 1993, pp. 759–764.

Chapter VIII

Natural and Waste Waters

I. NATURAL TRANSFORMATIONS IN FRESHWATER AND OCEANS

The phototransformation of organic pollutants in natural waters is usually a combination of direct photolysis, in which the pollutant molecules undergo direct light absorption and photoexcitation, and indirect photolysis, through electron transfer from other excited compounds or by reaction with naturally produced reactive species such as ·OH and organic peroxy radicals, or $^1O_2$ molecules.[1,2] From the observed first-order photolysis rate constants in the natural waters and the rates in buffered solutions in distilled water, the relative contributions of the direct and indirect photolysis in the specific natural waters could be derived. These measurements were applied to one eutrophic lake and one oligotrophic lake in Japan, and to coastal seawater of the Sea of Japan, using added 2-chlorophenol as a test compound. In the eutrophic lake, the contribution of indirect photolysis was substantial in spring, summer, and fall, and amounted on average to about 17% of the overall photolysis. In the summer, decomposition through the action of singlet oxygen contributed about 10% of the indirect photolysis. In the oligotrophic lake, the indirect photolysis was noted only in the summer. In the seawater, indirect photolysis was observed during the spring, summer, and fall, and contributed about 15% of the total photolysis.[3] The photolysis rates of 2-chlorophenol both in humic acid solutions and in waters from a wastewater treatment plant were determined by illumination at 313 nm. The overall rate constants for all these water samples were proportional to the concentration of dissolved organic carbon (DOC). Also, the relative contribution of the indirect photolysis to the overall photolysis increased with increasing DOC.[4]

Photodegradation of organic compounds results in a variety of products. Complete mineralization requires oxidation of all the organic carbon to carbon dioxide. In natural waters, a substantial alternative process is the formation of carbon monoxide, the photoproduction of which may be involved in the turnover of CO of up to 0.5% C/d (at 40° N). This reaction may be an important process in the cycling of refractory DOC in natural waters. The photochemical emission of CO from both marine and terrestrial sources (particularly from wetlands and near-coastal areas) is partly accompanied by its biological conversion.[5]

A. Humic Substances and Singlet Oxygen

In natural waters, the predominant light-absorbing species are usually the humic substances, which are involved in the formation of reactive oxidants such as singlet oxygen and hydrogen peroxide.[6-10] Humic substances are high-molecular-weight materials resulting from the oxidative decomposition of plant and animal residues. Singlet oxygen photoproduction in natural waters containing humic substances (HS) may be described by the reactions[1]

$$^1HS + h\nu \rightarrow \,^1HS^* \rightarrow \,^3HS^* \tag{1}$$

$$^3HS^* + \,^3O_2 \rightarrow \,^1HS + \,^1O_2 \tag{2}$$

The photochemical formation of singlet oxygen results in the indirect photolysis of organic compounds in natural waters. The steady-state concentration of singlet oxygen is controlled by its production under sunlight and its decay by quenching with water and with various substances in the water. In sunlight-exposed near-surface water this steady-state concentration reached 2×10^{-13} M.[11] The singlet oxygen concentration in natural water may be assayed by trapping with furans.[1]

B. Fulvic Acid Sensitization of Semiconductors

The naturally occurring fulvic acid has a light absorption increasing gradually from the visible range into the UV region, and a broad fluorescence spectrum ($\lambda_{max} = 475$ nm). In the presence of the large-band-gap semiconductor colloidal ZnO (particle diameter 20Å), which absorbs light below 330 nm, the fluorescence quantum yield of fulvic acid decreased. This decrease in the fluorescence emission intensity was due to the quenching of the excited singlet state of fulvic acid by the ZnO. The very strong adsorption of fulvic acid on the colloidal ZnO particles enabled efficient electron transfer from the excited fulvic acid to the semiconductor. Laser flash photolysis of aqueous solutions of fulvic acid at 532 nm revealed the transient formation of the hydrated electron, e_{aq}^- (with $\lambda_{max} = 600$ to 700 nm), presumably due to photoionization of the fulvic acid, as well as a transient absorption at 480 nm attributed to the cation radical of fulvic acid. In the presence of colloidal ZnO, the yields of these transients are much higher, indicating an efficient charge transfer from the excited fulvic acid, and leading to an electron trapped on the ZnO surface.[12] In a similar study, naturally occurring humic acid was used to sensitize colloidal TiO_2 particles (approximate diameter 30 nm), resulting in electron transfer from the photoexcited humic acid to the semiconductor. Thus, in the presence of colloidal TiO_2, the fluorescence of the humic acid was quenched. The charge injected into the semiconductor was applied to the reduction and bleaching of the oxazine dyes oxazine 725 (*N,N,N′,N′*-tetraethyloxonine) and Nile Blue A.[13]

The sensitization of the large-band-gap semiconductors TiO_2 and ZnO may possibly lead to enhanced photocatalytic oxidation of organic pollutants.

C. Singlet Oxygen Generation by Photosensitized Soil

Near-UV-illuminated soil in contact with molecular oxygen sensitized the formation of singlet oxygen. This reaction was applied to the photodegradation of the pyrethroid insecticide resmethrin (bioresmethrin). It was proposed that such an action of the electrophilic singlet oxygen may be a primary pathway for the photodegradation of electron-rich xenobiotics in contact with soil.[14] Resmethrin is used for the control of crawling and flying insects.[15]

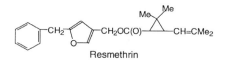

Resmethrin

SCHEME VIII.1

D. Hydrated Electrons

Jensen-Korte et al.[16] observed increased rates of photodegradation of pesticides such as organophosphorus compounds, carbamates, triazole derivatives, and pyrethroids in the presence of humic substances. Flash photolysis experiments indicated the production of solvated electrons as primary transient species in the illumination of humic acid.[17] The primary quantum yields for electron ejection by illumination at 355 nm from the dissolved organic matter (DOM) was determined by laser flash photolysis, both for a river in the United States (Suwannee River, Georgia) and a eutrophic Swiss Lake (Greifensee). These quantum yields, measured in deaerated waters, were in the range of $\phi = 0.005$ to 0.008. The hydrated electron was identified by its absorption spectrum, with a broad $\lambda_{max} = 700$ to 750 nm.[18]

E. Superoxide Ion-Radicals (O_2^-)

Hydrated electrons are very rapidly trapped by molecular oxygen ($k = 1.6 \times 10^{10}\ M^{-1}\ s^{-1}$), forming superoxide radical ions,

$$e_{aq}^- + O_2 \rightarrow O_2^- \qquad (3)$$

The radical ions then undergo disproportionation (autoredox dismutation) to hydrogen peroxide,

$$2O_2^- + 2H^+ \rightarrow H_2O_2 + O_2 \qquad (4)$$

The estimated rate of production of hydrogen peroxide in natural waters by the reaction of hydrated electrons with oxygen was much lower than measured values for H_2O_2 production in such waters. It was therefore concluded that most of the hydrogen peroxide formed may be due to some other photochemical process.[18]

In order to calibrate the peroxy radical activity of humic substances, 2,4,6-trimethylphenol (TMP) has been proposed as a probe compound.[2] In this test, it is essential to use TMP concentrations lower than about 6×10^{-8} M (~8 µg/L), because only at such low concentrations does the phototransformation of TMP follow first-order kinetics. At higher initial concentrations of TMP, more complicated kinetics were observed.[19]

The one-electron reduction of oxygen to the superoxide ion-radical is often the rate-detemining step in photocatalytic reactions involving semiconductors such as TiO_2, ZnO, and CdS. The production and the kinetics of decay of this ion-radical could be observed by pulse radiolysis of aerated water and measuring the absorption spectrum of the transient formed at 250 nm. The absorption maxima of the radical $\cdot HO_2$ and of the ion-radical O_2^- are 225 nm and 245 nm, respectively.[20] For the acid–base equilibrium

$$O_2^- + H^+ \rightarrow \cdot HO_2 \tag{5}$$

the value of pK_a in seawater is 4.6 ± 0.15,[21] which is similar to the value in pure water, $pK_a = 4.75 \pm 0.08$.[20]

The major pathway for the decay of the superoxide ion-radical is the already described autoredox dismutation,

$$2O_2^- + 2H^+ \rightarrow H_2O_2 + O_2 \tag{4}$$

Above pH 6, the decay of O_2^- was found to be second order and dependent on pH:

$$-d\left[O_2^-\right]/dt = 2k_2\left[O_2^-\right]^2\left[H^+\right] \tag{6}$$

In seawater (S = 30 to 36‰), $k_2 = (5 \pm 1) \times 10^{12}[H^+]$ M^{-1} s^{-1}.[21]

The superoxide dismutation reaction was proposed to be the predominant source of hydrogen peroxide in the ocean. In midday surface water at pH 8.3, the H_2O_2 concentration thus formed was estimated to reach 80 nM, and the O_2^- turnover time was about 1 h.[21]

F.   Nitrate-Induced Photodegradation

In natural waters, nitrate and nitrite photolysis is an additional important factor in the production of hydroxyl radicals (see also Chapter II, Section C), and thus a contributor to the natural abiogenic degradation of organic compounds in these waters.[22,23] Mean annual values for the rate of photoproduction of ·OH radicals derived from nitrate in near-surface waters (averaged over both night and day) were estimated to decrease from 0.0076 d^{-1} for latitude 0° (at the equator) to 0.0057 and 0.0034 d^{-1} for latitudes 30° and 50° (N). Nitrate-induced photooxidation of organic micropollutants may be significant for clear water bodies with high nitrate concentrations (e.g., 1 mM), such as small shallow lakes and rivers. For a typical pollutant, such as chlorobenzene, the environmental half-life due to photodecomposition was estimated to be about 1 month, while polychlorinated halocarbons, such as tetrachloroethylene, required much longer times.[22]

G. Fe(III)/Fe(II) Reactions

One important pathway for the oxidation of organic compounds in natural waters could be light-induced redox reactions with the hydrous amorphous Fe(III) phase, according to the schematic reaction,

$$Fe(OH)_3 + \text{reduced organics} = Fe(III) + \text{oxidized organics} \qquad (7)$$

Common organic constituents of natural waters such as phenols and tannic acid are oxidized by Fe(III). It was proposed that the amorphous Fe(III)(hydr) oxides may in oxic aquatic environments be reductively dissolved through photochemical processes involving aquo Fe–ligand complexes. As noted by Sulzberger,[24,25] such redox processes of iron may possibly be of major importance in providing available iron to organisms in the natural environment.

Many Fe(III) complexes absorb light in the 300 to 400 nm region, at which the sunlight intensity is substantial, with a flux of about 1 einstein m^{-2} h^{-1}. The photochemical reactions of these Fe(III) complexes have been suggested to be the driving force for many abiotic geochemical processes, such as the oxidation of SO_2 to H_2SO_4, the oxidation of organic compounds to CO_2 and HCO_3^- ions, the acidification of natural waters, and the dissolution of phosphate.[26]

H. Hematite–Oxalate Photolysis

Iron(III) (hydr)oxides such as hematite in the presence of oxalate undergo photoinduced dissolution under near-UV irradiation ($\lambda < 400$ nm). In the absence of oxygen, oxalate is oxidized, and Fe(II) is released into solution — a process that is crucial in the geochemical cycling of iron in natural waters. The overall stoichiometry is

$$2Fe(III)C_2O_4^+ + h\nu \rightarrow 2Fe(II) + 2CO_2 + C_2O_4^{2-} \qquad (8)$$

This reaction presumably occurred by formation of a surface bidentate complex of oxalate with Fe(III) in an inner-sphere coordination complex. Since the

Fe(II) produced catalyzed the thermal dissolution of hematite, this reaction was autocatalytic, yielding dissolved Fe(III)–oxalate complexes, which also readily underwent photolysis by near-UV light. In the presence of oxygen, the reductive dissolution of hematite under illumination was inhibited, but the photooxidation of oxalate was accelerated. Thus, hematite catalyzed the photooxidation of oxalate by oxygen.[27]

Ferric oxide (Fe_2O_3, hematite) occurs widely in the suspended matter of natural waters. Since α-Fe_2O_3 is a semiconductor, it has been considered as a potential photocatalyst. However, it was shown that only relatively strong reducing agents were oxidized in the presence of ferric oxide, and that it is a very poor photocatalyst relative to TiO_2 and ZnO.[28] Solubilization and photoreduction of Fe(III) is, however, possible in the presence of complexing ligands. A detailed kinetic study was made by Zepp et al.[29] on the rates of production of hydroxyl radicals formed by the photo Fenton reaction on Fe(III) complexes. In this process, the Fe(III) complexes with ligands such as oxalate or citrate were photoreduced to Fe(II), which reacted with hydrogen peroxide to form hydroxyl radicals. Aerated aqueous solutions of Fe(III)–oxalate (100 μM) in the pH range 3.1 to 4.9 or Fe(III)–citrate in the pH range 4.5 to 8.0 and H_2O_2 (100 μM) contained 30 μM 1-octanol as a substrate ·OH scavenger, and either anisole (1.5 μM) or nitrobenzene (1.5 μM) as competitive probe ·OH scavengers. These solutions were illuminated continuously at λ = 436 nm. With the steady-state assumption of equal rates of production and disappearance of hydroxyl radicals, and from the measured pseudo first-order rates of the disappearance of the substrate scavenger and the probe scavenger, and using the previously known second-order rate constants of the hydroxyl radicals with the probe molecules, it was possible to derive the rates of production of the ·OH radicals. Also, in similar experiments, but with argon-saturated solutions, and in the absence of added H_2O_2, the rates of the production of Fe(II) were determined. Throughout the pH range studied, the rates of formation of ·OH and of Fe(II) were practically equal. A model of the photo Fenton reaction could thus be proposed with the equations

$$Fe(III) - ligand + h\nu \rightarrow Fe(II) + ligand \text{ - } radical \qquad (9)$$

$$Fe(II) + H_2O_2 \rightarrow \cdot OH + Fe(III) \qquad (10)$$

$$\cdot OH + scavenger \rightarrow oxidized \ scavenger \qquad (11)$$

$$\cdot OH + probe \rightarrow oxidized \ probe \qquad (12)$$

Thus Fe(II), produced photochemically from Fe(III) organic complexes, may react with hydrogen peroxide to form hydroxyl radicals, providing an important pathway for the oxidation of organic compounds in the environment.[29]

Fe(III) complexes of polycarboxylate ions such as oxalate, citrate, and malonate are common in atmospheric and surface waters and provide a soluble form of Fe(III). Sunlight exposure of argon-saturated solutions of such complexes resulted in rapid photodecomposition of these complexes, within minutes, with formation of Fe(II). Quantum yields determined at 436 nm for Fe(II) production from Fe(OH) (citrate)$^-$, Fe(oxalate)$_2^-$, and Fe(oxalate)$_3^{3-}$ were 0.28, 1.0, and 0.6, respectively, and at 366 nm from Fe(malonate)$_2^-$ was 0.027. In the presence of air, the quantum yields for Fe(II) formation were much lower. The proposed mechanism assumes reaction of O_2 with the photogenerated organic radical-anions, producing the $\cdot O_2^-/\cdot HO_2$ couple, which then generated H_2O_2 and $\cdot OH$ radicals, reoxidizing Fe(II) to Fe(III).[30] The decarboxylation of the carboxylate radical anions thus produced was extremely rapid. For the decarboxylation of $C_2O_4^-\cdot$, the half-life was determined by pulse radiolysis to be 0.3 μs.[26,31]

In seawater, iron is often a limiting factor to phytoplankton growth. The stable form of iron in oxic water is the insoluble Fe(III). Solubilization of iron is possible by complexation with hydroxycarboxylic acids, such as sugar acids, which may be excreted from the phytoplankton. In sunlight experiments in natural seawater containing a culture of a diatom and adjusted to pH 8, Fe(III) (5 μM) was within 90 min reduced to Fe(II) in the presence tartaric, gluconic, and glucaric acids and glucaric acid 1,4-lactone — even in low concentrations of such hydroxycarboxylic acids (1 μM). Among the hydroxycarboxylic acids, glucaric acid gave the highest reduced Fe(II) concentration.[32]

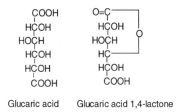

Glucaric acid   Glucaric acid 1,4-lactone

SCHEME VIII.2

Among the amino acids, only cystine was effective in the reduction of Fe(III). The Fe(II) concentration reached using glucaric acid was higher at 5°C than that at 20°C. This was proposed to be due to the slower rate of reoxidation of the Fe(II) produced at the lower temperature. These laboratory experiments of the photoreduction of Fe(III) to Fe(II) in seawater containing hydroxycarboxylic acids may explain the field observations during several years in the spring algal blooms in Funka Bay (Japan). In these observations, 0.02 to 0.04 μM concentrations of dissolved Fe(II) were measured in oxic surface seawater.[32]

The two-carbon compounds glycolaldehyde, glyoxal, glycolic acid, and glyoxylic acid in aqueous solutions were photooxidized with molecular oxygen in the presence of ferric ions, with overall quantum yields of 0.3 to 1.2. The reaction was observed gas-volumetrically by the consumption of oxygen. The light-absorbing species were Fe(III)–substrate complexes, absorbing strongly in the 300 to 400 nm region. Experiments of intermittent UV illumination and dark periods (a few minutes each of irradiated and dark periods) indicated a strong postirradiation effect with glyoxylic acid. The oxygen consumption continued for more than 6 min even during the dark periods. In the proposed mechanism, the photochemical reaction resulted in formation of a catalyst containing Fe(II), which was then reoxidized by molecular oxygen in a dark thermal reaction.[33]

I.   Fog, Cloud, and Rain Waters

Photocatalytic oxidation may also contribute to the formation of acid rain. Rain water contains nitric and sulfuric acids, which may be formed by oxidation of nitrous and sulfurous acids. Nitrogen and sulfur oxides, NO and NO_2, and SO_2 and SO_3, dissolve in water to form mixtures of nitrous, nitric, sulfurous and sulfuric acids. The oxidation of nitrous to nitric and of sulfurous to sulfuric acids may presumably occur as a photocatalytic process. Exposure of rain and fog water to 313 nm illumination resulted in the formation of H_2O_2 and the oxidation of sulfite and dissolved organic carbon (DOC). Atmospheric water droplets contain both Fe(II) and Fe(III), H_2O_2, oxalate, and various low-molecular-weight acids and aldehydes, sulfite, and sulfate. The proposed mechanism of formation of H_2O_2 is the photolysis of Fe(III) complexes with organic ligands such as oxalate, glyoxalate, and pyruvate by near-UV and visible light, with production of Fe(II) and a ligand-radical. The ligand-radical is oxidized by electron transfer to molecular oxygen, forming the superoxide radical-ion O_2^- (and its conjugate acid HO_2), which disproportionates to hydrogen peroxide and oxygen:[34]

$$Fe(III) - L + h\nu \rightarrow Fe(II) + \cdot L \tag{13}$$

$$\cdot L + O_2 \rightarrow \text{oxidized } L + O_2^- \tag{14}$$

$$2O_2^- + 2H^+ \rightarrow H_2O_2 + O_2 \tag{15}$$

Typical concentrations of H_2O_2 in rain and cloud water are 10^{-7} to 10^{-4} M. H_2O_2 is both a sink for $\cdot HO_2$ and a source of $\cdot OH$ radicals. In a search for a plausible source of H_2O_2 in atmospheric water, Fe(III)–oxalate complexes were proposed to play a most important role. Oxalate forms stable complexes with Fe(III), which absorb strongly in the 290 to 570 nm region. Photoabsorption by a Fe(III)–oxalate complex results in electron transfer from the oxalate ligand to the ferric ion, producing an Fe(II) ion and an oxalate radical anion. In the presence of oxygen, the oxalate radical anion transfers an electron to produce the superoxide ion, $\cdot O_2^-$, or its conjugate acid, the hydroperoxyl radical, $\cdot HO_2$. The organic radical dissociates to two molecules of CO_2, while the superoxide ion undergoes rapid disproportionation to H_2O_2 and O_2.[38] The rate of hydrogen peroxide production was found to depend on pH, sunlight intensity, and the concentrations of oxalate and Fe(III) in solution. A subsequent dark process (at night) by the Fenton reaction, $H_2O_2 + Fe(II)$, may then be a major source of $\cdot OH$ radicals. These hydroxyl radicals in atmospheric waters may be the true culprit in the damage to forests by "acid rain".[38]

In acidic fog water (pH < 5), the oxidation of SO_2 to H_2SO_4 was shown to be predominantly due to H_2O_2, while above pH 5, ozone was the major oxidant. The Fe(II) may be recycled to Fe(III) by the Fenton reaction with H_2O_2, releasing $\cdot OH$ radicals as most active oxidants.[34] Quantum yields for the formation of H_2O_2 at 313 nm observed in authentic cloud and fog waters reached the remarkably high values of 6×10^{-4} to 5×10^{-3}. Since the oxidation of sulfite or SO_2 by H_2O_2 is the dominant source of H_2SO_4 in the troposphere, H_2O_2 was proposed to be the limiting reagent for the production of sulfuric acid, and thus of acid rain.[35] In cloud and fog waters, the rate of photoproduction of $\cdot OH$ (observed by adding benzene as $\cdot OH$ scavenger and measuring the production of phenol) reached 0.32 to 3.0 μM h^{-1} under midday equinox solar illumination. The quantum yield at 313 nm for the photoproduction of the hydroxyl radical in cloud water ranged from 4.6×10^{-4} to 0.01.[36]

An alternative proposed mechanism of photooxidation of S(IV) to S(VI) leading to acid rain involved suspended particles and was supported by model experiments with TiO_2 particle dispersions.[37]

Acetaldehyde (ethanal) is one of the important organic components of atmospheric waters. Its phototransformation in the presence of solid mineral particles is of geochemical interest. The photodegradation of acetaldehyde in aqueous slurries of TiO_2 at pH 5.5 caused rapid mineralization, with intermediate formation and decay of acetic acid.[39]

J. Surface Waters

In most surface waters, both amorphous ferric hydroxide, $Fe(OH)_3$, and crystalline forms of ferric oxyhydroxide, FeOOH, are major components of the suspended particulate matter. In order to understand the light-induced interactions of Fe(III) with organic compounds occurring in natural waters, the photolysis of ethylene glycol was measured in aerated aqueous dispersions of goethite (α-FeOOH), under illumination at 300 to 400 nm. Ethylene glycol is widely used as an automobile antifreeze and in industrial coolants. Ethylene glycol and polysaccharides are important constituents of the organic pollutant load in the natural aquatic environment. The products of the goethite photocatalyzed degradation were formaldehyde and glycolaldehyde. Also, immediately after photolysis, Fe^{2+} was detected, but this was rapidly reoxidized to Fe^{3+}. In oxygen-free media, only formaldehyde was observed as the organic reaction product. The proposed mechanism involved electron transfer in an "encounter complex" from a surface-adsorbed ethylene glycol molecule to an excited state of Fe(III), forming Fe^{2+} and an ethylene glycol cation, which then decomposed to formaldehyde and glycolaldehyde.[40]

Photolysis with UV light of aqueous suspensions of the Fe(III) oxide minerals lepidocrite and hematite, as well as of amorphous iron oxides, led to the formation of Fe(II) and of molecular oxygen. This was explained by an electron transfer from a hydroxide ion bound in an inner coordination sphere to a photoexcited surface Fe(III) lattice atom [designated >Fe(III)]:

$$>Fe(III)(:OH^-) + h\nu \rightarrow Fe(III)^*(:OH^-) \rightarrow >Fe(II) + \cdot OH \qquad (16)$$

An analogous mechanism was proposed for the observed photodecomposition of benzoate, oxalate, and succinate ions adsorbed on crystalline goethite, α-FeOOH, in aqueous suspensions. The rates of formation of ·OH radicals were measured by their reaction with benzoic acid, forming salicylic acid, while the Fe(II) produced in N_2-purged media was measured colorimetrically. The mechanism involved as primary step the photooxidation of surface complexes of Fe(III) with the $RCOO^-$ ligands, forming $Fe(II)_{aq}$ and $RCOO\cdot$ radicals,

$$>Fe(III)(:OH^-) + RCOOH \leftrightarrow Fe(III)(:RCOO^-) + H_2O \qquad (17)$$

$$>Fe(III)(:RCOO^-) + h\nu \rightarrow >Fe(II) + RCOO\cdot \qquad (18)$$

In oxygenated solutions, the Haber–Weiss mechanism of oxidation of Fe(II) led to hydroxyl radicals via hydrogen peroxide:

$$Fe_{aq}^{2+} + O_2 \rightarrow Fe^{3+} + O_2^- \qquad (19)$$

$$2O_2^- + 2H^+ \rightarrow H_2O_2 + O_2 \qquad (20)$$

$$Fe_{aq}^{2+} + H_2O_2 \rightarrow Fe^{3+} + \cdot OH + OH^- \qquad (21)$$

Thus, it was concluded that most of the hydroxyl radicals were produced by this thermal oxidation of Fe_{aq}^{2+}.[41]

Goldberg et al.[42] compared the aquatic photolysis at 300 to 400 nm of a large variety of adsorbed mono-, hydroxy-, and dicarboxylic acids on goethite. This mineral is the most insoluble and thermodynamically stable of the iron oxyhydroxides, which are ubiquitous components of suspended matter and of most sediments. Such photolysis causes the oxidation of the carboxylic acids and the solubilization of Fe(III) as Fe(II). The reactivity of the carboxylic acids depended strongly both on their structure and on the pH of the medium. At pH

6.5, the relative rates of reaction were tartrate > citrate > oxalate > glycolate > maleate > succinate > formate > fumarate > malonate > glutarate > benzoate = butanoate. This sequence showed that carboxylic acids with α-hydroxyl groups were more reactive than unsubstituted carboxylic acids. Also, dicarboxylic acids with an even number of carbon atoms in the central chain were more reactive than those with an odd number of carbon atoms. The most photoreactive were polycarboxylic acids containing an α-hydroxyl group. Presumably, the anions of such acids were most strongly adsorbed to the Fe(III) lattice atoms, facilitating the electron transfer from these anions to the Fe(III) atoms upon photoexcitation.[42]

K. Cr(VI) and Cr(III)

In marine and shallow estuarine waters and in freshwater lakes in the southeastern United States, a diurnal cycle was observed in the oxidation state of chromium, between Cr(VI) and Cr(III). The proposed mechanism involved the photoreduction of ferric oxyhydroxide to Fe(II) by organic compounds, but only in the presence of particulate matter. This sunlight-driven process occurred even in the presence of oxygen, and was followed by rapid oxidation of Fe(II) by Cr(VI). An alternate mechanism could be the direct photodegradation of some organic Cr(VI) complex. The concentrations measured for Cr(III) and Cr(VI) in fresh waters were in the range of 0.03 to 0.17 nM and 0.003 to 0.01 nM, respectively, and in seawater 0.1 to 0.2 nM and 0.01 to 0.2 nM, respectively. Thus, Cr(III) was the predominant form of chromium in several natural waters in this study. Since Cr(VI) is very toxic, while Cr(III) is quite nontoxic to aquatic organisms, this photochemical detoxification may be an important process in natural waters.[43,44] See also Chapter II, Section VI.B on the photoreduction of Cr(VI) to Cr(III).

L. Dissolved Gaseous Mercury

Sunlight was found to induce the reduction in lake waters of Hg^{2+} to $Hg°$, dissolved gaseous mercury (DGM). This reaction, which could be due to either photochemical or photobiological processes, was mainly induced by UV_A light (320 to 400 nm), and less than 25% by UV_B light (280 to 320 nm). The diurnal variation in surface-water DGM concentrations closely followed the incident light flux, with highest concentrations observed at noon. The formation of $Hg°$ competes with the formation of methylmercury in fresh water and contributes to the volatilization of mercury.[45]

M. Eutrophic Waters

Eutrophication and excessive growth of plankton lead to the production of substances imparting a musty odor and taste to fresh water. Thus the algae *Phormdium tenue* and *Oscillatoria tenue* release 2-methylisoborneol, while *Anabaena macrospora* causes the formation of geosmin. Photocatalytic degradation with TiO_2 has been applied successfully to eliminate these musty odor substances (see Chapter III, Section II.B.13, for details).[46]

N. Seawater

In seawater, dissolved organic matter (DOM), which has a broad shallow absorption in the ultraviolet B region (280 to 320 nm), seems to sensitize the photoproduction of ·OH radicals, possibly due to hydroquinolic and phenolic moieties within humic substances. Another potentially important group of photosensitizers in seawater is that of the flavins, which include riboflavin and its oxidation products, lumiflavin and lumichrome. The total concentration of the flavins in seawater is in the range of 0.1 to 2 nM. Due to the high absorptivity of the flavins in the visible light region, and to the high efficiencies of intersystem crossing from singlet to triplet states, these compounds are effective photosynthesizers in seawater. The photodegradation of riboflavin in deoxygenated seawater led to the isoalloxazine products lumichrome, lumiflavin, and formylmethylflavin. Lumichrome is the most photostable of these compounds and was found in highest concentrations in seawater,[47,48] where its presence was proven by high-performance liquid chromatography (HPLC) separation and identification.[48] Flavins in natural waters sensitized the photooxidation of substituted phenols.[49]

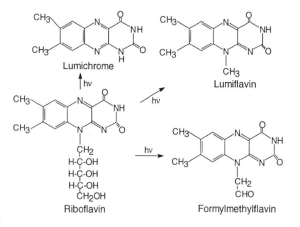

SCHEME VIII.3

O. Anthraquinone Photosensitization

Among the components of DOM in seawater, photosensitizing properties of hydrogen-atom abstraction have been attributed to anthraquinone derivatives.[50–52] The branched-chain hydrocarbon pristane (2,4,10,14-tetramethylpentadecane) is often found in seawater. In the presence of anthraquinone as sensitizer, a surface film of pristane on synthetic seawater under sunlight illumination underwent photooxidation to a variety of degradation products, such as ketones, tertiary alcohols, and alkanes.[53] In similar experiments, the chlorophyll biodegradation product phytol in synthetic seawater in the presence of anthraquinone was rapidly photooxidized under sunlight to several products, such as to 2,6,10-trimethyltridecane and 6,10,14-trimethyl-2-pentadecanone. The latter compound has often been detected in marine sediments.[54] By sunlight illumination of anthraquinone solutions of seawater containing dispersions of cholesterol, Rontani et al.[55] observed the formation of several hydrogen-abstraction products of cholesterol. Sterols such as cholesterol are chemically rather stable compounds that occur in relatively large concentrations in surface seawater and can be considered as indicators of biological activity. The products of photodegradation were both nonacidic and acidic compounds. The nonacidic products identified by gas chromatography/ mass spectrometry presumably originated from an initial hydrogen-atom abstraction reaction at the carbon-7 atom of cholesterol (a), via a carbon-7 radical intermediate (b) to the two epimers, cholest-5-ene-3β,7β-diol (c) and cholest-5-ene-3β,7α-diol (d). Further anthraquinone-sensitized photooxidation of these two epimeric diols led to the ketonic product 3β-hydroxy-cholest-5-en-7-one (e), which underwent dehydration and oxidation to the dienone cholesta-3,5-dien-7-one (f). Compound (f) underwent cleavage reactions to the indene derivative 2,3,3aα,6,7,7aα-hexahydro-7a-methyl-1β-(1,5-dimethylhexyl)indene (g). After 15 d of sunlight illumination of anthraquinone sensitized cholesterol in seawater, the relative percentages of unreacted cholesterol and of the products (b), (c), (d), (e), and (f) were 82, 2.8, 4.6, 5.8, and 0.9, respectively, while product (g) appeared only in traces. The formation of the acidic product 2,6-dimethylheptanoic acid was proposed to occur by an initial oxidation of cholesterol at the carbon 17 position (see Scheme VIII.4). Such photosensitized oxidation reactions involving radical intermediates may contribute to the nonbiological degradation of sterols in the euphotic layer of seawater.[55,56]

P. Photodegradation in Marine Surface Microlayers

The ocean surface is often an accumulation zone of organic materials, including oil slicks and various organic compounds. Among these, UV-light-absorbing compounds such as phenols and humic and fulvic acids may be photochemically active. The remarkably high photoactivity of marine surface microlayers was demonstrated in experiments on the photodegradation of phloroglucinol (1,3,5-trihydroxybenzene), TMP (2,4,6-trimethylphenol), and cresol (3-methylphenol). Quartz tubes containing tracer amounts of these phenols added either to surface waters containing oil slicks, or to surface water free of oil slicks, or to bulk (subsurface) seawater were floated on seawater and exposed to sunlight. Photodegradation of all three phenols was much faster in the oil-slicked samples than in the nonslicked surface water or in the bulk water. Very little degradation occurred in the dark, indicating that microbial

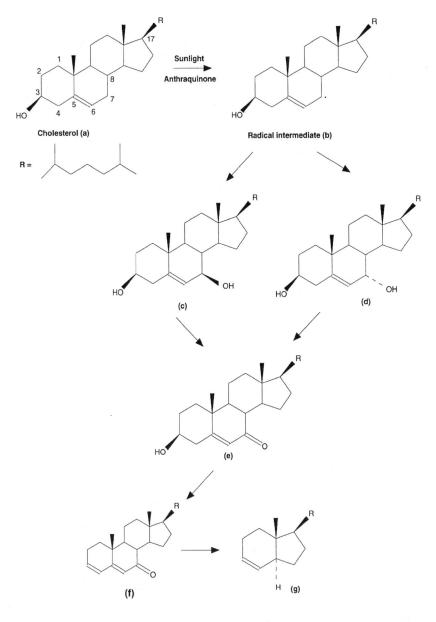

SCHEME VIII.4

degradation was much slower than photodegradation. The relative order of photodegradation of these phenols was phloroglucinol > TMP ≫ cresol. The rates in the nonslicked surface waters were similar to those in the bulk sea-water. In order to test if the higher photoactivity in the slicked surface water was due to the action of singlet oxygen or of peroxides, experiments were made with added NaN$_3$ (50 μM), a known quencher of singlet oxygen. The photodegradation of phloroglucinol in the surface slick decreased by 25%, while that of TMP and cresol was unaffected by the presence of azide. These result seem to indicate that the photodegradation of phloroglucinol was at least partly by a singlet oxygen mechanism, while the disappearance of TMP and cresol may be mainly by a peroxide pathway.[57]

Interest in bromide photooxidation has been particularly with respect to the attenuation of light energy in seawater. The most reactive species formed by UV-light excitation of natural waters are ·OH radicals. Hydroxyl radicals have been shown to be consumed in ocean water by several reactions, the mechanisms of which were clarified in flash photolysis studies.[56-59] A predominant trap for hydroxyl radicals was found to be the bromide ion.

In acidified Dead Sea water, which contains 5.3 g/L bromide, illumination with natural or simulated sunlight caused a marked oxidation of bromide to bromine, when oxygen or air was passed through the solution.[60] The yield of this oxidation increased with rising acid concentration. Since bromide salts dissolved in dilute aqueous solutions are not susceptible to photooxidation under these conditions, it must be concluded that the Dead Sea brine contains a natural photosensitizer. As noted earlier, such sensitizers had previously been identified in ocean water. In order to obtain more effective photosensitization of bromide oxidation in Dead Sea water, various organic sensitizers were tested. Very effective sensitization was obtained with anthraquinone sulfonates and with some tetrasulfonated phthalocyanines.[61]

II. TREATMENT OF POLLUTED GROUNDWATER

Photocatalyzed reactions are extremely attractive for the decomposition of organic pollutants in groundwater. The pollutants include aliphatic and aromatic hydrocarbons, their halogenated derivatives, and various pesticides. The conventional methods of water purification, such as chlorination, often cause an increase in the concentration of chlorinated organic products. The degradation of pollutants in drinking or waste waters may be achieved by homogeneous processes, using reagents such as ozone or hydrogen peroxide, which often can be substantially enhanced by UV irradiation.[62,63] The photooxidation of organic contaminants in water containing hydrogen peroxide has been extensively studied.[64-66]

Some pollutants are recalcitrant to such treatments, but as described in detail in Chapters I to VII, may be degraded by illumination with UV or visible light in the presence of semiconductor particles serving as heterogeneous catalysts.[67-69]

See Section V on the detoxification of polluted groundwater with concentrated sunlight.

III. TREATMENT OF WASTEWATER

Municipal wastewater treatment plants often also receive wastewater from various industries. In one study of effluent samples of such a treatment plant receiving a heavy loading from textile dying operations, the environmental effects of solar irradiation were investigated. Modern textile dyes are quite resistant to photodegradation since they have been designed to release the absorbed photon energy to neighboring molecules. Using kinetic competition techniques (trapping with furfuryl alcohol and with azide, and rate enhancement in D_2O vs. H_2O solutions), the enhanced production of singlet oxygen $^1O_2$ in the textile dye-containing effluent was proven. Also, the rate of photooxidation under sunlight of the phenol derivatives 2,4,6-trimethylphenol, 4-cresol, and 4-chlorophenol was very considerably enhanced in the dye-containing effluent, compared with that in dye-free effluents. However, with pentachlorophenol, the rates of photodegradation in the dye-containing and dye-free effluents were equal, suggesting that pentachlorophenol underwent predominantly direct photolysis.[70]

See also Chapter V, Section X on TiO_2-photocatalyzed degradation of textile dyes, and this chapter on photodynamic sterilization (Section IV) and on photodegradation with concentrated sunlight (Section V).

A. Treatment of Landfill Leachates

Photocatalytic detoxification was carried out on a biologically pretreated landfill effluent, which still had a very high pollutant content. After 1:1 dilution with water and adjustment to pH 3, in the presence of suspended TiO_2 (Degussa P25, 1 g/L) and H_2O_2 (32 mM), with O_2 saturation, under illumination with a 500-W Xe lamp ($\lambda > 320$ nm), the total organic carbon (TOC) content decreased to 25% of its original value within 3 h. Also, the original dark-brown color of the effluent disappeared and the water was almost transparent. Slower photodegradation occurred at pH 5, and practically none at pH 11, presumably because of scavenging of the hydroxyl radicals at the more alkaline pH. At pH 3, in the absence of either O_2, or H_2O_2, or TiO_2, the decreases in TOC were much slower.[71] In a further study, the removal of nonbiodegradable hazardous contaminants in wastewater from landfill leachates was again performed after biological treatment. This biological treatment had decreased the chemical oxygen demand (COD) by about 60%. Photochemical degradation of the residual organics was optimal at pH 3. Rapid mineralization of the organic contaminants was achieved, at almost equal rates, with $TiO_2/O_2/H_2O_2/UV$ and $O_3/H_2O_2/UV$. Under illumination with a 700-W high-pressure Hg lamp, about 80% of the residual organics were removed within about 2.5 h. The treatment with $TiO_2/O_2/H_2O_2/UV$ was proposed to be considerably more economic than that using O_3.[72]

IV. PHOTODYNAMIC STERILIZATION

A. Dye Sensitization

For highly polluted wastewater, photodynamic disinfection was applied, using dye sensitizers and molecular oxygen. Photodegradation was shown to involve singlet oxygen as an intermediate. This method enabled oxygenation and partial degradation of some organic compounds, and by photodynamic action also caused sterilization of microorganisms, including bacteria, algae, and viruses.[73-75] A photochemical disinfection of municipal wastewater effluents was developed in a pilot plant study, taking as input the effluent from an activated sludge treatment plant. The main purpose of this work was to disinfect this effluent, which still contained high concentrations of microorganisms, and to make it available for agricultural irrigation. The pilot plant consisted of a cascade of 10 shallow open troughs (each $6 \times 2 \times 0.3$ m^3) connected in series and exposed to sunlight and aeration. Using methylene blue (0.6 to 0.8 g/m^3) as a sensitizer, maintaining the dissolved oxygen (DO) concentration at more than 6 g/m^3, and adding calcium hydroxide (33 g/m^3) for pH correction (pH 8.7 to 8.9), the pilot plant was operated at an effluent flow rate of about 33 m^3/h and an effluent retention time of about 35 min. Calcium hydroxide was preferred for pH control, rather than sodium hydroxide, since Ca^{2+} ions had no detrimental effect on the soil properties. In order to maintain a fairly constant concentration of methylene blue along the cascade of photoreactors, the sensitizer was pumped in continuously in four portions, into the first (50%), third (17%), fifth (17%), and seventh (17%) reactor. Decreases of two to four orders of magnitude were achieved in the counts of coliforms, enterococci, and polioviruses. The effluent after this photochemical treatment could be infiltrated into a sand dune, without forming an impermeable soil crust (in contrast to the original effluent, after only the activated sludge treatment, which caused formation of such a crust). The solar photosensitized disinfection of domestic effluents was proposed as an economical and environmentally safe method, particularly useful for arid and semiarid regions.[76-78]

A major advantage of the dye-sensitized sterilization of water is the possibility of using visible light, including sunlight, instead of the energy-intensive germicidal lamps required for direct disinfection by UV irradiation.[79]

B. Bactericidal Activity of TiO₂

Irradiation of oxygen-saturated suspensions of *Escherichia coli* (10^6 cells/ml) in the presence of TiO$_2$ (Degussa P25; 1 g/L) by near-UV light (>380 nm) or by sunlight resulted in killing of the bacteria within 30 min. Practically no cell death occurred in the absence of TiO$_2$, oxygen, or illumination. The rate of cell inactivation was proportional to the light intensity in the range of 180 to 1600 μE s^{-1} m^{-2}, and was proportional to the square root of the concentration of TiO$_2$.

The role of oxygen was proposed to be both that of an acceptor of photogenerated electrons on the TiO_2 surface, forming the superoxide radical anion O_2^-, and of an inhibitor of electron–hole recombination, thus freeing holes for the production of $\cdot OH$ radicals.[80]

The bactericidal effect of illumination with a fluorescent lamp (290 to 390 nm) of suspensions of TiO_2 (P25; anatase form) on physiological salt solutions containing *Streptococcus mutans* (causing dental caries) was observed by Onoda et al.[81] Optimal concentrations of TiO_2 were between 0.01 and 0.1 wt%, resulting in essentially complete disappearance of viable cells within a few minutes.[81]

With dechlorinated tapwater flowing through a photocatalytic reactor containing anatase TiO_2 firmly bound on a sleeve formed of fiberglass mesh and surrounding a UV lamp (300 to 400 nm), rapid cell death occurred of both *Escherichia coli* and of the natural microbial population of such water. At a flow rate of 2 L/min, the concentration of viable organisms of *E. coli* decreased by 6 min of exposure from the initial 2×10^9 colony-forming units (CFU)/100 ml by 7 orders of magnitude. After 9 min total exposure the *E. coli* counts were below the detection limit. The process was proposed to depend on the biocidal and disinfecting activity of the hydroxyl radicals. In natural waters containing high concentrations of hole or hydroxyl radical scavengers such as bicarbonate, improved photocatalytic disinfection by TiO_2 may be achieved by adding hydrogen peroxide as an irreversible electron acceptor, thus preventing the semiconductor electron–hole recombination process and producing additional hydroxyl radicals by the reaction

$$H_2O_2 + e_{cb}^- \rightarrow \cdot OH + OH^- \tag{22}$$

The combined action of TiO_2 and H_2O_2 (6.5 mM) was applied to photoinactivate the heterotrophic bacteria in a pond.[82]

Solar disinfection of microorganisms in water was tested by illuminating suspensions of TiO_2 (Degussa P25; 0.2 g/L) in water containing *Eschericia coli*. Within 23 min of sunshine irradiation, more than 99% of the *E. coli* was inactivated.[83]

Ultrafine UV-illuminated TiO_2 particles were found to kill cancer cells. These experiments were carried out *in vitro*, with HeLa cells attached to TiO_2 particles (P25, average diameter 30 nm). In the dark, the TiO_2 particles alone showed little cytotoxicity, while UV illumination alone caused slow cell killing. With both TiO_2 and UV illumination, rapid and effective cell death was observed. The tumor growth-suppressing activity of illuminated TiO_2 was observed also *in vivo*, on HeLa cells transplanted into nude mice. Most effective tumor killing was achieved by modifying the TiO_2 with a hematoporphyrin derivative that is known to be selectively absorbed into tumors.[84,85]

V. CONCENTRATED SUNLIGHT

In an effort to apply the semiconductor-promoted degradation of organic compounds to the practical detoxification of polluted water, experiments at Sandia National Laboratories (Albuquerque, NM) were made with concentrated sunlight, either from parabolic trough mirrors on a glass pipe reactor, or from a heliostat to a falling film reactor. Using TiO_2 and H_2O_2 as a catalyst system, rapid degradation rates of salicylic acid and of trichloroethylene were obtained. Destruction rates appeared to be linearly related to light intensity.[86,87]

In experiments performed at the Plataforma Solar de Almería (PSA) (Almería Spain), a comparison was made on the solar photodegradation of dichloroacetic acid in an aqueous suspension of TiO_2 (Degussa P25, 0.25 g/L) illuminated in a parabolic trough reactor with that in a fixed-bed reactor, in which a thin film of the polluted water (~100 μm) rinsed the TiO_2 photocatalyst immobilized on a plate. Both solar reactors were effective. Although the parabolic trough reactor provided a 30 to 50 times concentration of the ultraviolet part of the solar spectrum, nevertheless the efficiency of the fixed-bed reactor was reported to have been superior. With initially 1 mM dichloroacetic acid, pH ~3, and a flow rate of 1.6 L/h, nearly 100% degradation was achieved.[71,78,89]

A pilot-plant-scale solar photocatalytic system was tested to remediate groundwater at the Lawrence Livermore National Laboratory Superfund Site in California, which was heavily polluted with chlorinated solvents, principally trichloroethylene, but also tetrachloroethylene, 1,1-dichloroethylene, carbon tetrachloride, and chloroform.[90] The system used two single-axis parabolic troughs covered with a highly reflective aluminized film, with a total aperture area of 155 m^2. The photoreactors consisted of 2-in.-diameter Pyrex® glass pipes in the focus of the troughs and connected in series. These troughs provided about 20 times concentration of normal incident solar radiation. Water from a polluted well was pumped into a storage tank, for adding TiO_2 (Degussa P25) and adjusting the pH. In batch mode, the water after passing through the photoreactors was recirculated to this tank, while in single-pass (plug-flow) mode the treated water was released to an effluent (settling) tank. In a single-pass experiment, at a flow rate of 15 L/min (corresponding to a residence time of 10 min), with the natural pH of 7.2, the trichloroethylene concentration was reduced from the initial high of 80 ppb (μg/L) to 10 ppb. A considerable enhancement in the degradation rate was achieved by adjusting the initial pH to 5.6, thus achieving in a single-pass operation a reduction to less than 0.5 ppb trichloroethylene. This concentration was below that required by the U.S. Environmental Protection Agency (EPA) drinking water standard. The higher photodegradation rate attained by acidification was explained to be due to the conversion of the bicarbonate ion (380 mg/L in the groundwater) into carbon dioxide. The HCO_3^- ion is an efficient scavenger of hydroxyl radicals, competing with the oxidation of the organic compounds, while the neutral CO_2 is much less reactive toward ·OH radicals. The kinetics of the photodegradation of the chlorocarbon compounds were studied by the batch-mode recirculation operation, varying the flow velocity, pH, catalyst loading, and high or low solar intensity. The apparent rate constant was strongly dependent on flow rate, increasing with higher flow rate, and on pH, increasing with lower pH. Increasing the TiO_2 catalyst loading from 0.3 to 1.0 g/L only slightly enhanced the reaction rate. A comparison of the rate of degradation by illumination with "one sun" (obtained by moving the troughs out of focus, so that the photoreactors collected only the global UV radiation and not the concentrated light), versus the degradation with the concentrated light (20 suns) revealed that the efficiency per light collection area was about eight times higher in the "one sun" experiment. The explanation given was that at the higher light intensity, the losses of hydroxyl radicals due to electron–hole recombination resulted in lower oxidation rates. Also, with the solar troughs, only the direct light was concentrated, while in the "one sun" mode the diffuse indirect UV light could also be absorbed by the photocatalyst. The conclusion was that nonconcentrating solar collectors will be more efficient for photocatalytic water detoxification.[90] Such a nonconcentrating collector could be designed in what has been named the "Lazy River" concept, in which the polluted water, mixed with the photocatalyst, was made to flow by gravity through a long serpentine channel, providing a residence of up to several days in order to assure complete detoxification. Additional mixing could preferably be achieved by bubbling air or oxygen at several points along the channel.[91]

The solar degradation of tetrachloroethylene was also performed by the combined effect of illumination and heat. Samples of tetrachloroethylene containing TiO_2 in aqueous suspension in sealed Pyrex® glass ampules were illuminated with concentrated sunlight from one or several heliostats at the Weizmann Institute central receiver facility (solar tower). Reaction rates as a function of solar concentration are presented in Figure VIII.1.[92]

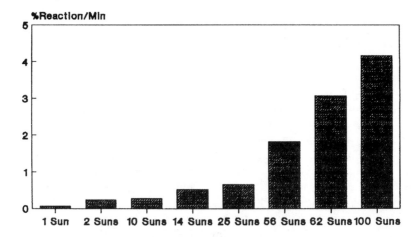

FIGURE VIII.1

Trichloroethylene (TCE) mixed with water vapor and oxygen was rapidly oxidized in a gas-phase photocatalytic reaction over TiO_2 coated on a reticulated alumina foam, under illumination with an ultraviolet argon-ion laser (20 W) emitting mainly at 330 and 360 nm. This reaction was performed in a reactor system incorporating rapid observation of the reaction products, using either a molecular-beam mass spectrometer or a Fourier-transform infrared spectrometer. The reaction products identified were HCl, CO, CO_2, Cl_2, $COCl_2$, and dichloroacetyl chloride. The dichloroacetaldehyde was decomposed by longer residence time in the reactor. These products differ from those of the photodegradation of TCE in aqueous dispersions of TiO_2, in which dichloroacetic acid and trichloroacetaldehyde were the main intermediate products. The apparent quantum yield for the destruction of the TCE (number of molecules oxidized divided by number of photons impinging on the reactor) was very large, in the range of 0.5 to 0.8. These values were an order of magnitude higher than the apparent quantum yields observed for the photodegradation of TCE and other halocarbons in aqueous slurries of TiO_2. Slightly slower destruction rates were achieved in the vapor-phase photocatalytic oxidation of 1,1-dichloroethylene, 1,2-dichloroethylene, and perchloroethylene.[93] This vapor-phase photodegradation may be a model for the destruction of hazardous organic wastes by intense concentrated sunlight.

In a simulation of reaction in a solar furnace, chloronaphthalene vapor mixed with oxygen and helium was introduced into an electrically heated furnace at temperatures of up to 772°C, with and without illumination with an intense argon-ion laser. The destruction of the chloronaphthalene was monitored by a molecular-beam mass spectrometer. The residence time in the reactor was about 17 s. The addition of ultraviolet photons considerably enhanced the rate of destruction of the chloronaphthalene. The reaction observed first-order kinetics, and the apparent first-order rate constants were linearly related to the light intensity, measured up to a flux equivalent to 800 suns. The high flux of ultraviolet photons caused a shift in the destruction rate to the rate obtained in a purely thermal reactor at a higher temperature.[94] The mechanism of this photothermal oxidation is not yet clear.

A comparison was made on the photochemical degradation of pentachlorophenol in a small laboratory batch photoreactor illuminated with a 1500-W xenon lamp, and in a large solar pilot-plant system consisting of parabolic trough solar concentrators.[95] The solar plant, operated at the Plataforma Solar de Almería (PSA) in southern Spain, used 12 heliostat modules in series, each supporting four parabolic collectors, in the focus of which were placed the glass absorber tubes. The collectors had a reflecting surface of an aluminized polymeric film (aluminum is best for UV reflectivity). The total effective aperture of the loop was 384 m^2, and the total useful length of the glass reactor tubes was about 200 m, with an internal diameter of 56 mm. The intensity of UV light available inside the absorber tube was estimated at about 6 suns. Flow conditions in the absorber loop were those of a plug flow reactor. With an initial concentration of 10 mg/L of sodium pentachlorophenolate (pH adjusted to 4) and 0.1 g/L of TiO_2, with injection of oxygen, the disappearance of pentachlorophenol observed first-order kinetics and led to degradation of pentachlorophenol to parts per billion (ppb) concentrations. No reaction occurred in the absence of oxygen. Intermediates identified were tetrachloro-1,4-benzoquinone, tetrachloro-1,2-benzoquinone, and tetrachlorohydroquinone, which were mineralized completely with longer residence times. The rate constants of photodegradation observed with the solar plant agreed closely with those obtained in the laboratory experiment with simulated sunlight. Considerable enhancement in the rate of photocatalytic degradation was obtained by addition of peroxydisulfate. Thus with TiO_2 (0.1 g/L), $S_2O_8^{2-}$ (0.01 M), and pentachlorophenol (initially 10 mg/L), at a flow rate of 2500 L h^{-1}, the pentachlorophenol concentration was brought down to ppb levels in a few minutes. In the presence of peroxydisulfate it was not necessary to inject oxygen into the system. This should be advantageous in large plant operation, as it avoids problems of bubble formation. The peroxydisulfate was reduced to harmless sulfate ions. The mechanism of action of peroxydisulfate may be both in the scavenging of conduction-band electrons and in the production of hydroxyl radicals:[95]

$$S_2O_8^{2-} + e_{cb}^- \rightarrow SO_4^{2-} + \cdot SO_4^- \tag{23}$$

$$\cdot SO_4^- + H_2O \rightarrow SO_4^{2-} + \cdot OH^- \tag{24}$$

The addition of persulfate increased the reaction rate by an order of magnitude.[96] With peroxydisulfate and TiO_2, but without illumination, the degradation of pentachlorophenol was extremely slow. Even higher rates were achieved in this photocatalytic system for the photodegradation of dichloroacetic and acetic acids.[97]

The PSA pilot plant flow reactor just discussed was applied to the solar photodestruction of real industrial wastewater from a resins factory. This effluent water contained a mixture of many aliphatic and aromatic compounds, with an original TOC concentration of about 600 ppm. This water was diluted to about 55 ppm TOC and was treated with TiO_2(0.1 g/L) and $S_2O_8^{2-}$ (7 mM), as well as with intermittant injection of oxygen. At a solar UV radiation of 36 W m^{-2} and a residence time of 44 min, the TOC degradation was 100% complete. The rate of destruction of the TOC amounted to 1.25 mg L^{-1} min^{-1}. In the absence of the persulfate, the rate was only 0.14 mg L^{-1} min^{-1}. With a more concentrated effluent from the same resins factory, at an initial TOC concentration of 160 ppm, and using the same concentrations of TiO_2 and $S_2O_8^{2-}$, a 75% decrease in TOC (to 40 ppm) was achieved, with a residence time of 50 min.[97]

The methylene blue-sensitized transformation of bromacil (see Chapter V, Section II.G) was studied with concentrated solar light, using both the Weizmann

Institute central receiver for illumination intensities of up to 200 kW/m², and the solar furnace (consisting of a 100-m² heliostat illuminating a 7-m-diameter spherical focussing mirror) for illuminations of up to 800 kW/cm².[98] The presence of O_2 was required for the reaction, and transformation yields were higher in solutions saturated with oxygen than with air. At a light flux of about 160 kW/cm², in the presence of 2 ppm methylene blue, complete disappearance of bromacil was attained within 5 s. These illuminations were made in batch experiments, in Pyrex® glass tubes. The only photoproduct was 3-*sec*-butyl-5-acetyl-5-hydroxyhydantoin, as in the experiments with unconcentrated sunlight. At the much higher light fluxes in the solar furnace, a flow system was used, providing a light exposure time of 62 ms. In the range of light flux of up to 800 kW/m², the rate of disappearance of bromacil was proportional to the light flux and to the initial bromacil concentration. With these very high light fluxes (>200 kW/m²), the irradiation of air-saturated solutions of bromacil and methylene blue resulted in complete mineralization, without formation of intermediate products. A mechanism of multistep photooxidation of the substrate was therefore proposed to occur with such high light fluxes.[98]

REFERENCES

1. Zepp, R. G., Wolfe, N. L., Baughman, G. L., and Hollis, R. C., Singlet oxygen in natural waters, *Nature,* 267, 421–423, 1977.
2. Faust, B. C. and Hoigné, J., Sensitized photooxidation of phenols by fulvic acid and in natural waters, *Environ. Sci. Technol.,* 21, 957–964, 1987.
3. Kawaguchi, H., Photolysis of 2-chlorophenol in natural waters, *J. Contam. Hydrol.,* 9, 105–114, 1992.
4. Kawaguchi, H., Determination of direct and indirect photolysis rates of 2-chlorophenol in humic acid solution and natural waters, *Chemosphere,* 25, 635–641, 1992.
5. Valentine, R. L. and Zepp, R. G., Formation of carbon monoxide from the photodegradation of terrestrial dissolved organic carbon in natural waters, *Environ. Sci. Technol.,* 27, 409–412, 1993.
6. Zafiriou, O. C., Blough, N. V., Micinski, E., Dister, B., Kieber, D., and Moffett, J., Molecular probe systems for reactive transients in natural waters, *Marine Chem.,* 30, 45–70, 1990.
7. Zepp, R. G., Baughman, G. L., and Schlotzhauer, P. F., Comparison of the photochemical behavior of various humic substances in water: I. Sunlight induced reactions of aquatic pollutants photosensitized by humic substances, *Chemosphere,* 10, 109–117, 1981.
8. Hoigné, J., Faust, B. C., Haag, W. R., Scully, F. E., and Zepp, R. G., Aquatic humic substances as sources and sinks of photochemically produced transient reactants, in: *Aquatic Humic Substances,* Suffet, I. H. and Mac Carthy, P. (Eds.), *Adv. Chem. Ser.,* Chapter 23, American Chemical Society, 219, 363–381, 1989.
9. Haag, W. R., Measurement of sunlight-induced transient species in surface waters, *J. Res. NBS,* 93, 285–288, 1988.

10. Haag, W. R. and Hoigné, J., Photosensitized oxidation in natural water via hydroxyl radicals, *Chemosphere,* 14, 1659–1671, 1985.

11. Haag, W. R. and Hoigné, J., Singlet oxygen in surface waters. 3. Photochemical formation and steady state concentrations in various types of waters, *Environ. Sci. Technol.,* 20, 341–348, 1986.

12. Vinodgopal, K. and Kamat, P. V., Environmental photochemistry on surfaces. Charge injection from excited fulvic acid into semiconductor colloids, *Environ. Sci. Technol.,* 26, 1963–1966, 1992.

13. Vinodgopal, K., Environmental photochemistry: Electron transfer from excited humic acids to TiO_2 colloids and semiconductor mediated reduction of oxazine dyes by humic acids, *Res. Chem. Intermed.,* 20, 825–833, 1994.

14. Clements, P. and Wells, C. H. J., Soil sensitized generation of singlet oxygen in the photodegradation of bioresmethrin, *Pestic. Sci.,* 34, 163–166, 1992.

15. Hartley, D. and Kidd, H. (Eds.), *The Agrochemicals Handbook,* Royal Society of Chemists, Nottingham, England, 1987.

16. Jensen-Korte, U., Anderson, C., and Spiteller, M., Photodegradation of pesticides in the presence of humic substances, *Sci. Total Environ.,* 62, 335–340, 1987.

17. Langford, C. H., Bruccoleri, A., Arbour, C., and Power, J. F., Primary photochemical events in natural waters, *XIII IUPAC Symp. Photochemistry,* University of Warwick, England, Abstract P340, 1990.

18. Zepp, R. G., Braun, A. M., Hoigné, J., and Leenheer, J. A., Photoproduction of hydrated electrons from natural organic solutes in aquatic environments, *Environ. Sci. Technol.,* 21, 485–490, 1987.

19. Kawaguchi, H., Rates of photosensitized photo-oxidation of 2,4,6-trimethylphenol by humic acid, *Chemosphere,* 27, 2177–2182, 1993.

20. Bielski, B. H. J., Cabelli, D. E., Arudi, R. L., and Ross, A. B., Reactivity of HO_2/O_2^- radicals in aqueous solution, *J. Phys. Chem. Ref. Data,* 14, 1041–1100, 1985.

21. Zafiriou, O. C., Chemistry of superoxide ion-radical (O_2^-) in seawater. I. pK_{asw}^* (HOO) and uncatalyzed dismutation kinetics studied by pulse radiolysis, *Mar. Chem.,* 30, 31–43, 1990.

22. Zepp, R. G., Hoigné, J., and Bader, H., Nitrate-induced photooxidation of trace organic chemicals in water, *Environ. Sci. Technol.,* 21, 443–450, 1987.

23. Mopper, K. and Zhou, X., Hydroxyl radical photoproduction in the sea and its potential impact on marine processes, *Science,* 250, 661–664, 1990.

24. Sulzberger, B., Suter, D., Siffert, C., Banwart, S., and Stumm, W., Dissolution of Fe(III) (hydr) oxides in natural waters; laboratory assessment on the kinetics controlled by surface coordination, *Mar. Chem.,* 28, 127–144, 1989.

25. Sulzberger, B., Photoredox reactions at hydrous metal oxide surfaces: a surface coordination chemistry approach, in: *Aquatic Chemical Kinetics,* Stumm, W. (Ed.), Wiley & Sons, New York, 1990, pp. 401–429.

26. Faust, B. C., A review of the photochemical redox reactions of iron(III) species in atmospheric, oceanic, and surface waters: Influences on geochemical cycles and oxidant formation, in: *Aquatic and Surface Photochemistry;* Helz, G. R., Zepp, R. G., and Crosby, D. G. (Eds.), Lewis Publishers, Boca Raton, FL, 1993, pp. 3–37.

27. Siffert, C. and Sulzberger, B., Light-induced dissolution of hematite in the presence of oxalate: a case study, *Langmuir,* 7, 1627–1634, 1991.

28. Kormann, C., Bahnemann, D. W., and Hoffmann, M. R., Environmental photochemistry: is iron oxide (hematite) an active photocatalyst? A comparative study: α-Fe_2O_3, ZnO, TiO_2, *J. Photochem. Photobiol.,* 48, 161–169, 1989.

29. Zepp, R. G., Faust, B. C., and Hoigné, J., Hydroxyl radical formation in aqueous reactions (pH 3–8) of iron (II) with hydrogen peroxide. The photo-Fenton reaction, *Environ. Sci. Technol.,* 26, 313–319, 1992.

30. Faust, B. C. and Zepp, R. G., Photochemistry of iron(III)-polycarboxylate complexes: roles in the chemistry of atmospheric and surface waters, *Environ. Sci. Technol.,* 27, 2517–2522, 1993.

31. Mulazzani, Q. G., d'Angelantonio, M., Venturi, M., Hoffman, M. Z., and Rodgers, M. A. J., Interaction of formate and oxalate with radiation-generated radicals in aqueous solution. Methylviologen as a mechanistic probe, *J. Phys. Chem.,* 90, 5347–5352, 1986.

32. Kuma, K., Nakabayashi, S., Suzuki, Y., Kudo, I., and Matsunaga, K., Photoreduction of Fe(III) by dissolved organic substances and existence of Fe(II) in seawater during spring blooms, *Mar. Chem.,* 37, 15–27, 1992.

33. Klementová, S. and Wagnerová, D. M., Photocatalytic effect of Fe(III) on oxidation of 2-carbon substrates related to natural waters, *Collect. Czech. Chem. Commun.,* 59, 1066–1076, 1994.

34. Zuo, Y. and Hoigné, J., Evidence for photochemical formation of H_2O_2 and oxidation of SO_2 in authentic fog water, *Science,* 260, 71–73, 1993.

35. Faust, B. C., Anastasio, C., Allen, J. M., and Arakaki, T., Aqueous phase photochemical formation of peroxides in authentic cloud and fog water, *Science,* 260, 73–75, 1993.

36. Faust, B. C. and Allen, J. M., Aqueous phase photochemical formation of hydroxyl radical in authentic cloudwaters and fog-waters, *Environ. Sci. Technol.,* 27, 1221–1224, 1993.

37. Hori, Y., Bandoh, A., and Nakatsu, A., Electrochemical investigation of photocatalytic oxidation of NO_2^- at TiO_2 (anatase) in the presence of O_2, *J. Electrochem. Soc.,* 137, 1155–1161, 1990.

38. Zou, Y. and Hoigné, J., Formation of hydrogen peroxide and depletion of oxalic acid in atmospheric water by photolysis of iron(III)-oxalato complexes, *Environ. Sci. Technol.,* 26, 1014–1022, 1992.

39. Guillard, C., Amalric, L., D'Oliveira, J.-C., Delprat, H., Hoang-Van, C., and Pichat, P., Heterogeneous photocatalysis: use in water treatment and involvement in atmospheric chemistry, in: *Aquatic and Surface Photochemistry,* Helz, G. R., Zepp, R. G., and Crosby, D. G. (Eds.), Lewis Publishers, Boca Raton, FL, 1994, pp. 369–386.

40. Cunningham, K. M., Goldberg, M. C., and Weiner, E. R., The aqueous photolysis of ethylene glycol adsorbed on goethite, *J. Photochem. Photobiol.,* 41, 409–416, 1985.

41. Cunningham, K. M., Goldberg, M. C., and Weiner, E. R., Mechanisms for aqueous photolysis of adsorbed benzoate, oxalate, and succinate on iron oxyhydroxide (geothite) surfaces, *Environ. Sci. Technol.,* 22, 1090–1097, 1988.

42. Goldberg, M. C., Cunningham, K. M., and Weiner, E. R., Aquatic photolysis. Photolytic redox reactions between goethite and adsorbed organic acids in aqueous solutions, *J. Photochem. Photobiol. A: Chem.,* 73, 105–120, 1993.

43. Kieber, R. J. and Helz, G. R., Indirect photoreduction of aqueous chromium(VI), *Environ. Sci. Technol.,* 26, 307–312, 1992.

44. Kaczynski, S. E., and Kieber, R. J., Aqueous trivalent chromium photoproduction in natural waters, *Environ. Sci. Technol.,* 27, 1572–1576, 1993.

45. Amyot, M., Mierle, G., Lean, D. R. S., and McQueen, D. J., Sunlight-induced formation of dissolved gaseous mercury in lake waters, *Environ. Sci. Technol.,* 28, 2366–2371, 1994.

46. Ishibai, Y., Suita, T., Murakami, H., and Murasawa, S., Purification of water in Lake Biwa using TiO_2 photocatalyst, *Abstr. 2nd Int. Symp. New Trends in Photoelectrochemistry,* University of Tokyo, March 1994, p. 57.

47. Momzikoff, A., Santus, R., and Giraud, M., A study of the photosensitizing properties of seawater, *Mar. Chem.,* 12, 1–14, 1983.

48. Mopper, K. and Zika, R. G., Natural photosensitizers in seawater: riboflavin and its breakdown products, in: *Photochemistry of Environmental Aquatic Systems,* Zika, R. G. and Cooper, W. J. (Eds.), *Am. Chem. Soc., Symp. Ser.,* 32, 174–190, 1987.

49. Tatsumi, K., Ichikawa, I., and Wada, S., Flavin-sensitized photooxidation of substituted phenols in natural water, *J. Contam. Hydrol.,* 9, 207, 1992.

50. Ehrhardt, M., Marine Gelbstoff, in: *Handbook of Environmental Chemistry,* Part C, Hutzinger, O. (Ed.), Springer Verlag, Berlin, 1984, pp. 63–77.

51. Ehrhardt, M. and Petrick, G., On the sensitized photo-oxidation of alkylbenzenes in seawater, *Mar. Chem.,* 15, 47–58, 1984.

52. Ehrhardt, M. and Petrick, G., The sensitized photo-oxidation of n-pentadecane as a model for abiotic decomposition of aliphatic hydrocarbons in seawater, *Mar. Chem.,* 16, 227–238, 1985.

53. Rontani, J.-F. and Giusti, G., Photosensitized oxidation of pristane in seawater: Effect of photochemical reactions on tertiary carbons, *J. Photochem. Photobiol. A: Chem.,* 40, 107–120, 1987.

54. Rontani, J.-F. and Giusti, G., Photosensitized oxidation of phytol in seawater, *J. Photochem. Photobiol. A: Chem.,* 42, 347–355, 1988.

55. Rontani, J.-F., Raphel, D., and Aubert, C., Photochemical degradation of cholesterol induced by anthraquinone, *J. Photochem. Photobiol. A: Chem.,* 72, 189–193, 1993.

56. Rontani, J. F., Identification by GC/MS of acidic compounds produced during the photosensitized oxidation of normal and isoprenoid alkanes in seawater, *Int. J. Env. A.,* 45, 1–9, 1991.

57. Lin, K. J. and Carlson, D. J., Photo-induced degradation of tracer phenols added to marine surface microlayers, *Mar. Chem.,* 33, 9–22, 1991.

58. Zafiriou, O. C., True, M. B., and Hayon, E., Consequences of OH radical reaction in seawater: formation and decay of Br_2^- radical, in: *Photochemistry of Environmental Aquatic Systems,* Zika, R. G. and Cooper, W. J. (Eds.), *ACS Symp. Ser.,* 327, 89–105, 1987.

59. True, M. and Zafiriou, O. C., Reaction of Br_2^- produced by flash photolysis of seawater with components of the dissolved carbonate system, in: *Photochemistry of Environmental Aquatic Systems,* Zika, R. G. and Cooper, W. J. (Eds.), *ACS Symp. Ser.,* 327, 106–115, 1987.

60. Halmann, M. and Porat, Z., Photooxidation of bromide to bromine in Dead Sea water, *Solar Energy,* 41, 417–421, 1988.

61. Grätzel, M. and Halmann, M., Photosensitized oxidation of bromide in Dead Sea water, *Mar. Chem.,* 29, 169–182, 1990.
62. Oliver, B. G. and Carey, J. H., Photodegradation of wastes and pollutants in aquatic environment, *NATO ASI Ser., Ser. C.,* 174, 629–650, 1986.
63. Clark, R. M., Fronk, C. A., and Lykins, B. W., Jr., Removing organic contaminants from groundwater, *Environ. Sci. Technol.,* 22, 1126–1130, 1988.
64. Malaiyandi, M., Sadar, M. H., Lee, P., and O'Grady, R., Removal of organics in water using hydrogen peroxide in presence of ultraviolet light, *Water Res.,* 14, 1131–1135, 1980.
65. Leitis, E., Zelf, J. D., Smith, M. M., and Crosby, D. G., The chemistry of ozone and ozone/UV light for water reuse, *Proc. Water Reuse Symp.,* 1121–1144, 1982.
66. Meulemans, C. C. E., The basic principles of UV disinfection of water, *Ozone: Sci. Eng.,* 9, 299–313, 1987.
67. Ollis, D. F., Contaminant degradation in water, *Environ. Sci. Technol.,* 19, 480–484, 1985.
68. Zepp, R. G., Factors affecting the photochemical treatment of hazardous waste, *Environ. Sci. Technol.,* 22, 256–257, 1988.
69. Matthews, R. W., Photooxidation of organic impurities in water using thin films of titanium oxide, *J. Phys. Chem.,* 91, 3328–3333, 1987.
70. Tratnyek, P. G., Elovitz, M. S., and Colverson, P., Photoeffects of textile dye wastewaters. Sensitization of singlet oxygen formation, oxidation of phenols and toxicity to bacteria, *Environ. Toxicol. Chem.,* 13, 27–33, 1994.
71. Bahnemann, D. W., Bockelmann, D., Goslich, R., Hilgendorff, M., and Weichgrebe, D., Photocatalytic detoxification: novel catalysts, mechanisms and solar applications, in: *Photocatalytic Purification and Treatment of Water and Air,* Ollis, D. F. and Al-Ekabi, H. (Eds.), Elsevier Science Publishers, Amsterdam, 1993, pp. 301–319.
72. Weichgrebe, D., Vogelpohl, A., Bockelmann, D., and Bahnemann, D., Treatment of landfill leachates by photocatalytic oxidation using TiO_2: a comparison with alternative photochemical technologies, in: *Photocatalytic Purification and Treatment of Water and Air,* Ollis, D. F. and Al-Ekabi, H. (Eds.), Elsevier Science Publishers, Amsterdam, 1993, pp. 575–584.
73. Acher, A. J. and Rosenthal, I., Dye-sensitized photooxidation — a new approach to the treatment of organic matter in sewage effluents, *Water Res.,* 11, 557–562, 1977.
74. Acher, A. J. and Juven, B. J., Destruction of coliforms in water and sewage water by dye-sensitized photooxidation, *Appl. Environ. Microbiol.,* 33, 1019–1022, 1977.
75. Lewis, A. L., Wellings, F. M., and Martin, D. F., Photoinactivation of three viruses by rose bengal, *J. Environ. Sci. Health,* A23, 127–137, 1988.
76. Acher, A. J. and Saltzman, S., Photochemical inactivation of organic pollutants from water, *Ecol. Stud.,* 73, 302–319, 1989.
77. Acher, A. J., Fischer, E., Zellingher, R., and Manor, Y., Photochemical disinfection of effluents. — Pilot plant studies, *Water Res.,* 24, 837–843, 1990.
78. Acher, A. J., Fischer, E., and Manor, Y., Sunlight disinfection of domestic effluents for agricultural use, *Water Res.,* 28, 1153–1160, 1994.
79. Meulemans, C. C. E., The basic principles of UV-disinfection of water, *Ozone: Sci. Eng.,* 9, 299–314, 1987.

80. Wei, C., Lin, W. Y., Zainal, Z., Williams, N. E., Zhu, K., Kruzic, A. P., Smith, R. L., and Rajeshwar, K., Bactericidal activity of TiO$_2$ photocatalyst in aqueous media. Toward a solar-assisted water disinfection system, *Environ. Sci. Technol.,* 28, 934–938, 1994.

81. Onada, K., Watanabe, J., Nakagawa, Y., and Izumi, I., Photocatalytic bacteriocidal effect of powdered TiO$_2$ on *Streptococcus mutans, Denki Kagaku,* 56, 1108–1109, 1988.

82. Ireland, J. C., Klostermann, P., Rice, E. W., and Clark, R. M., Inactivation of *Escherichia coli* by titanium dioxide photocatalytic oxidation, *Appl. Environ. Microbiol.,* 1668–1670, 1993.

83. Zhang, P. C., Scrudato, R. J., and Germano, G., Solar catalytic inactivation of *Escherichia coli* in aqueous solutions using TiO$_2$ as catalyst, *Chemosphere,* 28, 607, 1994.

84. Cai, R., Hashimoto, K., Itoh, K., Kubota, Y., and Fujishima, A., Photokilling of malignant cells with ultrafine TiO$_2$ powder, *Bull. Chem. Soc. Jpn.,* 64, 1268, 1991.

85. Fujishima, A., Cai, R., Hashimoto, K., Sakai, H., and Kubota, Y., Biochemical application of TiO$_2$ photocatalysts, in: *Photocatalytic Purification and Treatment of Water and Air,* Ollis, D. F. and Al-Ekabi, H. (Eds.), Elsevier Science Publishers, Amsterdam, 1993, pp. 193–205.

86. Pacheo, J. E. and Tyner, C. E., Enhancement of processes for solar photocatalytic detoxification of water, *Proc. ASME, Solar Energy Div., Internat. Solar Energy Conf.,* April 1–4, 1990, Miami, FL.

87. Pacheco, J. E. and Yellowhorse, L., Summary of Engineering-Scale Experiments for the Solar Detoxification of Water Project, *SAND* 92–0385, March 1992.

88. Bockelmann, D., Goslich, R., Weichgrebe, D., and Bahnemann, D., Solar detoxification of polluted water: Comparing the efficiencies of a parabolic trough reactor and a novel thin-film-fixed-bed reactor, in: *Photocatalytic Purification and Treatment of Water and Air,* Ollis, D. F. and Al-Ekabi, H. (Eds.), Elsevier Science Publishers, 1993, pp. 771–776.

89. Dillert, R. and Bahnemann, D., Photocatalytic degradation of organic pollutants: mechanisms and solar applications, *EPA Newslett.,* 52, 33–52, 1994.

90. Pacheco, J. E., Mehos, M., Turchi, C., and Link, H., Operation of a solar photocatalytic water treatment system at a superfund site, in: *Photocatalytic Purification of Water and Air,* Ollis, D. F. and Al-Ekabi, H. (Eds.), Elsevier Science Publishers, Amsterdam, 1993, pp. 547–556.

91. Turchi, C., Mehos, M., and Pacheco, J., Design issues for solar-driven photocatalytic systems, in: *Photocatalytic Purification of Water and Air,* Ollis, D. F. and Al-Ekabi, H. (Eds.), Elsevier Science Publishers, Amsterdam, 1993, pp. 789–794.

92. Halmann, M., Hunt, A. J., and Spath, D., Photodegradation of dichloromethane, tetrachloroethylene and 1,2-dibromo-3-chloropropane in aqueous suspensions of TiO$_2$ with natural, concentrated and simulated sunlight, *Solar Energy Mater. Solar Cells,* 26, 1–16, 1992.

93. Nimlos, M. R., Jacoby, W. A., Blake, D. M., and Milne, T. A., Direct mass spectrometric studies of the destruction of hazardous wastes. 2. Gas-phase photocatalytic oxidation of trichloroethylene over TiO$_2$: products and mechanisms, *Environ. Sci. Technol.,* 27, 732–740, 1993.

94. Nimlos, M. R., Milne, T. A., and McKinnon, J. T., Photothermal oxidative destruction of chloronaphthalene, *Environ. Sci. Technol.,* 28, 816–822, 1994.
95. Minero, C., Pelizzetti, E., Malato, S., and Blanco, J., Large solar plant photocatalytic water decontamination: degradation of pentachlorophenol, *Chemosphere,* 26, 2103–2119, 1993.
96. Blanco, G. B. and Malato, R. S., Influence of solar irradiation over pentachlorophenol solar photocatalytic decomposition, in: *Photocatalytic Purification and Treatment of Water and Air,* Ollis, D. F. and Al-Ekabi, H. (Eds.), Elsevier Science Publishers, Amsterdam, 1993, pp. 639–644.
97. Blanco, G. B. and Malato, R. S., Solar photocatalytic mineralization of real hazardous waste water at pre-industrial level, in: *Joint Solar Eng. Conf.,* March 27–30, 1994, San Francisco, pp. 103–109.
98. Turnheim-Ashkenazi, R., Use of concentrated sunlight for photochemical inactivation of organic pollutants in water, M. Sc. Thesis, Weizmann Institute of Science, Rehovot, Israel, 1993.

Chapter IX

Evaluation and Future Trends

I. PHOTODEGRADATION COMPARED WITH OTHER METHODS

A. Ozonation

Ozonation is effective for the degradation of many organic compounds. However, the installation of ozonation requires considerable capital costs and electric energy consumption. Also the excess of unused ozone must be removed, to prevent the escape of this toxic gas to the atmosphere.[1] Ozonation by itself is ineffective for the oxidation of halomethanes, such as trichloroethylene and carboxylic acids (e.g., formic acid), and also of the triazine herbicides, such as atrazine. Ozone in combination with H_2O_2 or UV irradiation provides an effective supply of ·OH radicals for complete degradation and mineralization of most pollutants.[2-5] In natural waters containing a high concentration of bromide ions, which may occur due to the intrusion of saltwater into fresh-water supplies, ozonation causes formation of hypobromous acid, HOBr. This highly reactive brominating agent converts organic precursors in the water into brominated disinfection by-products (DBPs). Such brominated organic compounds include the carcinogenic and lachrymatory bromoform, bromoacetone, and various four- and five-carbon bromohydrins such as $(CH_3)_2C(OH)-CH(Br)-CH_3$ and dibromides, which probably are more toxic than the organic precursors.[6]

See also Chapter IV on trihalomethane formation by chlorination of drinking water.

B. Biodegradation

Biodegradation was shown to be hindered in the presence of certain xenobiotic compounds. Thus, methyl vinyl ketone, pentachlorophenol, and 2,4-dichlorophenol inhibited the respiratory activity of activated sludge. These compounds could be detoxified by sunlight exposure in the presence of TiO_2 (1 g/L). Complete mineralization was necessary, since the intermediates formed by partial degradation were even more toxic than the starting compounds.[7] For the biodegradation of xenobiotic substances, an important phase is the acclimation of microorganisms. The degradation of recalcitrant compounds may often be facilitated by the use of mixed populations of bacteria.[8-10] Photocatalytic oxidation of bioresistant substances has been proposed as a useful pre- or post-treatment for activated sludge treatment.[11-13]

C. Radiolysis

A comparison was made of the γ-radiolysis of several chlorocarbon compounds in aerated aqueous solutions with the direct photolysis of these compounds under vacuum-UV illumination (184.9 nm). In order to make such a comparison, a "photochemical G-value," that is, absorbed energy/quantum (eV hv^{-1}) was calculated for each compound, in analogy to the radiolytic G-value. The compounds tested were CH_2Cl_2, $ClHC=CHCl$, $ClHC=CCl_2$, and $Cl_2C=CCl_2$. The yields (G-values) of Cl^- release were in the case of the vacuum-UV irradiation about one order of magnitude larger than those due to radiation-induced degradation. The enhanced decomposition at 184.9 nm was explained by the combined effects of the direct excitation of the halocarbons at this wavelengths, in addition to the photolysis of water.[14,15]

A pulsed electric discharge was used to decompose a variety of organic compounds dissolved in water, including benzene, toluene, xylene, styrene, phenol, *p*-benzoquinone, hydroquinone, and catechol. The decomposition was enhanced in the presence of H_2O_2.[16]

D. Ultrasonics

Ultrasonic irradiation of water by sound waves at frequencies higher than 15 kHz causes the formation and collapse of microbubbles as a result of the acoustic cavitation created by the expansion and compression waves. In the collapsing cavitation bubbles, temperatures of several thousand degrees Kelvin and pressures of several hundred atmospheres effect the thermal dissociation of water,

$$H_2O \rightarrow \cdot H + \cdot OH \qquad (1)$$

The resulting hydroxyl radicals are thus available for oxidation reactions.

The sonolysis of parathion was performed in a water-jacketed stainless steel cell operated at 20 kHz and ~75 W/cm^2, thermostatted at 30°C. The initial sonolytic reaction was breakage of the P=S and P-nitrophenylate bonds, with the intermediate formation of 4-nitrophenol, sulfate, and diethyl phosphate. At a slower rate, 4-nitrophenol degraded to hydroquinone, quinone, and 4-nitrocatechol, and eventually to nitrite and nitrate, oxalate, acetate, and formate, while diethyl phosphate hydrolyzed to ethanol and orthophosphate.[17,18]

Up to 99.9% destruction of chlorinated C1 and C2 halocarbons in aqueous solutions at ambient temperature and pressure was achieved by ultrasonic irradiation. Sonication was performed at 20 kHz and 0.1 kW/L. Essentially, complete degradation and mineralization within 1 h was obtained with methylene chloride, chloroform, carbon tetrachloride, 1,2-dichloroethane, 1,1,1-trichloroethane, trichloroethylene, and tetrachloroethylene.[19]

E. Corona Discharge

Water decomposition similar to that by electron bombardment and γ-radiolysis, with formation of hydroxyl and hydrogen radicals, hydrated electrons, and hydrogen peroxide, may be achieved by a pulsed streamer corona discharge. In an oxygenated aqueous solution of an anthraquinone dye, rapid discolorization was achieved by such a discharge.[20] This method was applied to the degradation of phenol in aqueous solution, using a rotating spark gap power supply (peak voltage 25 to 40 kV; pulse width 500 to 1000 ns; repetition frequency 60 Hz). The discharge occurred between the tip of a hollow hypodermic needle (through which nitrogen or oxygen was bubbled) and a stainless steel round wide counterelectrode. The degradation of the phenol by the corona discharge was considerably accelerated by the presence of oxygen. This was attributed to the formation of ozone in the gas phase, followed by the diffusion of the ozone into the aqueous phase and its decomposition, with formation of additional ·OH radicals. Further enhancement of the rate of degradation of phenol was attained by addition of $FeSO_4$ (~1 mM), and breakdown then was complete in less that 15 min. The effect of Fe^{2+} was explained by a Fenton reaction on the H_2O_2 produced in the discharge, releasing ·OH radicals, which were the active oxidant species.[21]

F. Electrochemical Oxidation

For the electrochemical oxidation of organic compounds in water it is necessary to choose low-resistivity anode materials with a high overvoltage for oxygen gas evolution. Among the known materials with high oxygen evolution overvoltage, PbO_2 and graphite are unstable under prolonged electrolysis, while Pt/Ti is expensive. A stable and highly efficient SnO_2 electrode doped with Sb was developed, using a spray hydrolysis method (spraying $SnCl_4$ and $SbCl_3$ in aqueous ethanol on a Ti plate at 500°C). SnO_2 is an n-type semiconductor with a direct band gap of about 3.5 eV, and doping is necessary to increase its conductivity. The best Sb-doped SnO_2 had a resistivity of 10^{-3} ohm cm. With such an anode and a platinized titanium counterelectrode, in a simple single-compartment beaker cell, with galvanostatic control at 30 mA/cm^2, an initially 1000 ppm phenol solution (and 0.5 N Na_2SO_4 at pH 12), the phenol disappeared after 1 Ah, while the total organic carbon (TOC) was destroyed in about 8 Ah. With the SnO_2 electrode, no quinone intermediates were observed during the oxidation of phenol. Presumably, such intermediates, if formed, were very rapidly mineralized. This electrode was also applied to the oxidation of a large variety of aliphatic and aromatic compounds, including biorefractory substances such as sulfonic acid derivatives of benzene, naphthalene, and anthraquinone. The SnO_2 electrode enabled about five times faster rates of oxidation of organic compounds than Pt electrodes. A pilot-scale undivided flow-through plate-and-frame electrochemical reactor was designed, with a SnO_2 anode and platinized titanium cathode. At an operating current density of 30 mA/cm^2, the oxidation of organic compounds was not mass transfer limited, enabling a space-time yield of 6.4 kg COD h^{-1} m^{-3}. The power consumption was 40 to 50 kWh for the removal of 1 kg COD. The process was proposed to be competitive with other advanced oxidation processes for the removal of biorefractory pollutants.[22,23]

G. Oxidation in Supercritical Water

The rapid and complete oxidation of organic compounds in supercritical water, in the temperature range of 380 to 390°C and pressures of 230 to 235 bar, was achieved in the presence of a commercial catalyst (12.1 wt% CuO and 22.7 wt% ZnO supported on a porous cement, pretreated for 2 h at 860°C in an O_2 stream). The reaction was performed by preheating (at 430°C) and then mixing aqueous solutions of these compounds with aqueous solutions of H_2O_2 (which decomposed to O_2 and water), and then passing the mixed solution through the reactor in a 1 m long (3.2 mm OD) stainless steel tube. A residence time of 25 s was sufficient to cause reductions of TOC for 2-propanol, *tert*-butanol, acetic acid, 1-methyl-2-pyrrolidone, benzoic acid, and phenol of 98, 87, 98, 51, 73, and 100%, respectively. Without the catalyst, the rates of degradation were very much smaller.[24] The catalytic oxidation in supercritical water may be attractive as a high-temperature solar-thermal process using concentrated sunlight.

With even higher temperatures, in the range of 700 to 1000°C, steam reforming of organic wastes was accomplished in a gas–solid reaction over a rhodium catalyst supported on a porous ceramic absorber, causing conversion of organic compounds to CO, CO_2, and H_2, and of chlorocarbons also to HCl.[25]

II. COST ESTIMATES, ENERGETICS, AND CONCLUSIONS

A. Cost Estimates

The economics of photocatalytic treatment of water have been discussed by Matthews,[26] by comparing photocatalytic destruction vs the removal with granular activated carbon (GAC). Phenol was chosen as a model pollutant. The cost of mineralization of 1000 L of initially 100 ppm phenol by 90%, down to 10 ppm phenol, using electrically powered UV lamps and 0.1% TiO_2, was estimated to be about $7.70 (U.S.), of which about 80% are those of the electric power. The cost of removal of 90% of the phenol by adsorption in GAC would cost only about $4.60 (U.S.). However, this requires that the spent GAC will continue to be permitted to be disposed of by landfill or high-temperature incineration. Another calculation was for the sunlight treatment of 1 million L of more dilute phenol, initially 10 ppm, with 0.1% TiO_2, in a shallow lagoon.

The major running cost in the operation of such a lagoon would the replacement of the catalyst, estimated for 20 reuses at $1060 (U.S.)/million L. To this must be added the cost of the land and of building the lagoon. With GAC adsorption, for 20 reuses, the cost estimate would be only $770 (US)/million L, but to this must again be added the disposal costs of the spent GAC, which may rise markedly if landfill will not be permitted.[26] More reliable values for the economics of photocatalytic oxidation of water pollutants will require pilot-plant and large-scale experiments. Some efforts in this direction have been carried out using TiO_2 coated on beach sand for the solar oxidation of phenol in open ponds,[8] described in Chapter III, Section II.B.5, as well as in the methylene blue photosensitized disinfection of the effluent of an activated sludge treatment plant, described in Chapter VIII, Section IV.

In the pilot-plant study of solar photodegradation with TiO_2 and peroxydisulfate of wastewater from a resins factory (Chapter VIII, Section V), an estimate was made of the cost for industrial operation. Assuming treatment of 10 m^3/d (2250 m^3/year) at 225 working days per year, at 65% rate of sunny days and average 6 sunny hours/d, with average 20 W m^{-2} UV radiation, with 600 m^2 of parabolic trough collectors (25% overdesigned), the total capital required is estimated at $186,000 (U.S.). Not considering the land cost, the estimated annual costs are: for the capital, 13% of the total capital required, or $24,180; for consumables, mainly peroxydisulfate and TiO_2, $10,000; and for operation and maintenance, $9,300, totaling $43,480, or $19.30/m^3. This cost estimate of about $20/m^3 for highly polluted toxic industrial waste water is considered competitive with other advanced oxidation processes (AOPs).[27]

Another cost estimate was based on the previously described pilot-plant-scale experiments performed at the Lawrence Livermore National Laboratory, using parabolic trough collectors with UV reflectors, and TiO_2 fixed on fiber-glass cloth packed in a glass receiver tube (see Chapter VIII, Section V). A conceptual design for solar detoxification was proposed to treat contaminated groundwater at the Rocky Flats Plant in Colorado. The groundwater at this site has a pH of 7.6 and a total dissolved solids concentration (mainly calcium bicarbonate) of 817 ppm. In order to reach drinking water quality, this hardness should be removed in a cation exchange pretreatment step, to yield water with pH 6 to 7, and maximally 400 ppm dissolved solids. Assuming direct normal insolation of 960 W/m^2, of which 3.47% are UV photons, the UV flux would be 1.82×10^{18} photons/W-s. Using a parabolic trough collector area of 1145 m^2, with a trough aperture width of 2.1 m and receiver inside diameter of 5.1 cm, the geometric concentration ratio will be 42:1. The plant was designed to be operated at a flow rate proportional to the light flux, in order to attain an approximately constant rate of photodegradation. Hydrogen peroxide and oxygen should be added to the water to enhance the rate of degradation. For an annual treated volume of 32,000 m^3 (8,500,000 gal), at an annual capacity factor of 16%, and assuming 355 d/year of operation, the cost estimate (in 1991 U.S. dollars), and taking as the installed costs (U.S.) of the collector and the receiver tube (including the catalyst) $215/m^2 and $8.15/m, the levelized cost of the treated water was about $10.63/m^3 ($40/1000 gal). Of this cost, 63% was for capital recovery (assuming a levelized capital carrying charge of 12.5%) and 37% for operating expenses (mainly labor, maintenance, chemicals, and waste disposal). This cost is prohibitive compared with present water treatment techniques, such as air stripping for the removal of volatile organic compounds.[28]

B. Energy Requirement Evaluation

Matthews and McEvoy[29] carefully tested the energetics of the photooxidation of phenol in oxygen- or air-saturated aqueous dispersions of TiO_2. With a blacklight fluorescent lamp (15 W NEC, mainly emitting between 300 and 425 nm) as light source, in initially 1 mM phenol at pH 4.0 and 0.1% TiO_2 (Degussa P25), 50% of the phenol was oxidized within 87 min. This corresponded to about 2400 ppm L^{-1} kWh^{-1} phenol, equivalent to 820 ppm total organic carbon oxidized L^{-1} kWh^{-1}. Even higher energy efficiency was obtained using a germicidal lamp (15 W, mainly emitting at 254 nm), but such lamps are more expensive because of the requirement of quartz shields between the lamps and the solutions.[29]

C. Future Trends

The application of photochemical methods for water purification will have to be carried out in the context of existing water treatment technologies. For highly contaminated waters, especially those also containing "normal" organic pollutants, such as those from domestic wastes or from food industries, certainly the best approach will be to continue with the conventional activated sludge treatment, with anaerobic and aerobic stages. This treatment may also decrease the concentration of some toxic compounds by several orders of magnitude. Many xenobiotic organic compounds, such as chlorophenols, herbicides, textile dyes, and surfactants, are resistant to such treatment. In those cases, in which the effluent of the biological treatment still contains unacceptably high levels of contaminants, photocatalytic degradation is becoming highly attractive as an additional stage of treatment. For highly concentrated toxic industrial effluents, the photocatalytic method may be the only effective option.

An important aspect of photocatalytic degradation is the advantage of combining the mineralization of organic pollutants with the removal both of harmful heavy metals and of the microbial contamination.

A limitation on the wide-scale application of solar photocatalytic methods of water decontamination based on semiconductors such as TiO_2 is the small flux of photons available for the photodegradation reactions. This may be overcome by concentration of sunlight, but this requires considerable land areas and may be economic only in desert regions. While the sun is "free," solar installations, in particular solar concentration, require considerable capital investment. Since the rates of photocatalytic reactions on semiconductors such as TiO_2 at high light intensities are often proportional to the square root of the light intensity, the use of very highly concentrated sunlight (more than 10 suns) is less likely to be economic for the degradation of water pollutants. Also, the degradation rates have been found to be very slow if the TOC (total organic carbon) content was higher than about 100 mg/L.[30] However, as noted in Chapter VIII (Section E), higher TOC concentrations (160 ppm) were rapidly treated in a solar plant using a combination of TiO_2 and peroxydisulfate.[27] Solar photodegradation is obviously possible only in sites of high insolation and unused land, which excludes most of the industrial regions. The development of efficient and economic UV and visible light sources will facilitate the application of photochemical detoxification also to these regions.

REFERENCES

1. Davis, A. P. and Huang, C. P., The photocatalytic oxidation of toxic organic compounds, in: *Physicochemical and Biological Detoxification of Hazardous Wastes,* Wu, Y. C. (Ed.), Vol. 1, Technomic, Lancaster, 1989, pp. 337–352. *Chem. Abstr.*, 112:104225Z.
2. Hoigné, J., The chemistry of ozone in water, in: *Process Technologies for Water Treatment,* Stucki, S. (Ed.), Plenum, New York, 1988, pp. 121–143.
3. Masten, S. J. and Hoigné, J., Comparison of ozone and hydroxyl-induced oxidation of chlorinated hydrocarbons in water, *Ozone Sci. Eng.*, 14, 197–214, 1992.
4. Masten, S. J. and Davies, S. H. R., The use of ozonation to degrade organic contaminants in wastewater, *Environ. Sci. Technol,* 28, 180A–185A, 1994.
5. Prado, J., Arantegui, J., Chamarro, E., and Esplugas, S., Degradation of 2,4-D by ozone and light, *Ozone Sci. Eng.,* 16, 235–245, 1994.
6. Collette, T. W., Richardson, S. D., and Thruston, A. D., Jr., Identification of bromohydrins in ozonated waters, *Appl. Spectrosc.,* 48, 1181–1192, 1994.
7. Manilal, V. B., Haridas, A., Alexander, R., and Surender, G. D., Photocatalytic treatment of toxic organics in wastewater. Toxicity of photodegradation products, *Water Res.,* 26, 1035–1038, 1992.
8. Matthews, R. W. and McEvoy, S. R., Destruction of phenol in water with sun, sand and photocatalysis, *Solar Energy,* 49, 507–513, 1992.
9. Gottschalk, G. and Knackmuss, H.-J., Bacteria and the biodegradation of chemicals achieved naturally, by combination, or by construction, *Angew. Chem. Int. Ed.,* 32, 1398–1408, 1993.
10. Mandelbaum, R. T., Wackett, L. P., and Allan, D. L., Mineralization of the *s*-triazine ring of atrazine by stable bacterial mixed cultures, *Appl. Environ. Microbiol.,* 59, 1695–1701, 1993.
11. Buitrón, G., Koefoed, A., and Capville, B., Control of phenol biodegradation by using CO_2 evolution rate as an activity indicator, *Environ. Technol.,* 14, 227–236, 1993.

12. Kiwi, J., Pulgarin, C., Peringer, P., and Grätzel, M., Beneficial effects of homogeneous photo-Fenton pretreatment upon the biodegradation of anthraquinone sulfonate in waste water treatment, *Appl. Catal. B: Environ.*, 3, 85–99, 1993.

13. Tanaka, S. and Ichikawa, T., Effects of photolytic pretreatment on biodegradation and detoxification of surfactants in anaerobic digestion, *Water Sci. Technol.*, 28, 103, 1993.

14. Getoff, N., Radiation-degradation and photoinduced degradation of pollutants in water — a comparative study, *Radiat. Phys. Chem. — Int. J. Rad. Ap.*, 37, 673–680, 1991.

15. Getoff, N., Purification of drinking water by irradiation. A review, *Proc. Indian Acad. Sci. — Chem. Sci.*, 105, 373–391, 1993.

16. Shvedchikov, A. P., Belousova, E. V., Polyalova, A. V., Ponizovskii, A. Z., and Goncharov, V. A., Removal of organic impurities in aqueous solutions by a pulse discharge, *Khim. Vysok. Energ.*, 27, 63–66, 1993. *Chem. Abstr.*, 118, 244366v.

17. Kotronarou, A., Mills, G., and Hoffmann, M. R., Ultrasonic irradiation of *p*-nitrophenol in aqueous solution, *J. Phys. Chem.*, 95, 3630–3638, 1991.

18. Kotronarou, A., Mills, G., and Hoffmann, M. R., Decomposition of parathion in aqueous solution by ultrasonic irradiation, *Environ. Sci. Technol.*, 26, 1460–1462, 1992.

19. Bhatnagar, A. and Cheung, H. M., Sonochemical destruction of chlorinated C1 and C2 volatile organic compounds in dilute aqueous solution, *Environ. Sci. Technol.*, 28, 1481–1486, 1994.

20. Clements, J. S., Sato, M., and Davis, R. H., Preliminary investigation of breakdown phenomena and chemical reactions using a pulsed high-voltage discharge in water, *IEEE Trans. Ind. Appl.*, IA-23, 224–235, 1987.

21. Sharma, A. K., Locke, B. R., Arce, P., and Finney, W. C., A preliminary study of pulsed streamer corona discharge for the degradation of phenol in aqueous solution, *Haz. Waste Haz. Mater.*, 10, 209, 1993.

22. Kötz, R., Stucki, S., and Carcer, B., Electrochemical waste water treatment using high overvoltage anodes. Part I. Physical and electrochem. properties of SnO_2 anodes, *J. Appl. Electrochem.*, 21, 14–20, 1991.

23. Stucki, S., Kötz, R., Carcer, B., and Suter, W., Electrochemical waste water treatment using high overvoltage anodes. Part II. Anode performance and applications, *J. Appl. Electrochem.*, 21, 99–104, 1991.

24. Krajnc, M. and Levec, J., Catalytic oxidation of toxic organics in supercritical water, *Appl. Catal. B: Environ.*, 3, L101–L107, 1994.

25. Tyner, C. E., Application of solar thermal technology to the destruction of hazardous wastes, *Solar Energy Mater.*, 21, 113–129, 1990.

26. Matthews, R. W., Photocatalysis in water purification: Possibilities, problems and prospects, in: *Photocatalytic Purification and Treatment of Water and Air*, Ollis, D. F. and Al-Ekabi, H. (Eds.), Elsevier Science Publishing, Amsterdam, 1993, pp. 121–138.

27. Blanco, G. B. and Malato, R. S., Solar photocatalytic mineralization of real hazardous waste water at pre-industrial level, in: *Joint Solar Eng. Conf.*, March 27–30, 1994, San Francisco, pp. 103–109.

28. Kelly, B. D. and De Laquil, P., Conceptual design of a photocatalytic wastewater treatment plant, *Proc. 6th Int. Symp. on Solar Thermal Concentrating Technologies,* CIEMAT, Madrid, 1992, pp. 1123–1131.

29. Matthews, R. W. and McEvoy, S. R., A comparison of 254 nm and 350 nm excitation of TiO_2 in simple photocatalytic reactors, *J. Photochem. Photobiol. A: Chem.,* 66, 355–366, 1992.

30. Pelizzetti, E., Minero, C., Hidaka, H., and Serpone, N., Photocatalytic processes for surfactant degradation, in: *Photocatalytic Purification and Treatment of Water and Air,* Ollis, D. F. and Al-Ekabi, H. (Eds.), Elsevier Science Publishing, Amsterdam, 1993, pp. 261–273.

Index

A

Acenaphthene, 83
Acetaldehyde, in atmospheric waters, 261
Acetic acid, 70–72, 274
Acetophenone, 82
Acid Orange 7, 214
Acid rain, 260–261
Activated sludge treatment, 283, 288
Adenosine triphosphate (ATP), 230
Adrenaline, 213
Advanced Oxidation Processes (AOPs), 1
Alachlor, 198–200
Alcohols, secondary, 12
Aldicarb, 195–196
Aliphatic diamines, 202
Aliphatic halocarbons, 114
Alkyl phosphates, 227
Alkylbenzenesulfonates, 92
Alkylphenol polyethoxylates, 92
Ametryn, 184
Amides, 188
Amines, 188
p-Aminophenol, 201
Aminopolycarboxylates, 190
Ammonia to nitrate, 48
Ammonium persulfate, see Peroxydisulfate
Anatase, see Titanium dioxide
Anilines, 189
Anionic sensitizers, 136
Anionic surfactants, 92
Anthracene, 83
Anthraquinone 2, 6-disulfonic acid, 244
Anthraquinone photosensitization, 264, 267
Anthraquinone 2-sulfonic acid, 244–245
Aroclor^R, see Polychlorinated biphenyls (PCBs)
Aromatic aminoacids, 99
Aromatic orientation, 7, 85, 148, 154, 203, 205, 206
Aromatic phosphates, 231
Asulam, 245–246
Atmospheric water, 260–261
Atrazine, 181–188
Azide ions, 50, 77
Azo dyes, 213–217

B

Baygon, 195–196
Beach sand, 91, 285
Bentazone, 188
Benzamide, 182, 192
Benzene, 17
Benzenesulfonic acid, 243
Benzo[b]thiophene, 248
Benzo[f]quinoline, 218
Benzoic acid, 17, 81, 83
Benzophenones, as sensitizers, 68
Benzoquinone, 89
Benzyl dodecyl dimethyl ammonium chloride, 92
Benzylphosphonic acid, 238
Benzyl trimethylammonium chloride, 94
Biodegradation, 97, 283
Biphenyl, 102
1, 2-Bis(2-chloroethoxy)ethane, 124
Bis(tributyl)tin oxide, 59
Bis-2-ethylhexyl phthalate, 97
Br_2^- radical anion, 43
Bromacil, 197, 274–275
Bromide ion photooxidation, 43, 267
Brominated disinfection-byproducts (DBPs), 283
Bromoacetic acid, 17
Bromocarbons, 127
4-Bromo-3-chloroaniline, 189
4-Bromodiphenyl ether, 138
2-Bromoethyl 4-bromophenylether, 137
2-Bromoethyl phenyl ether, 137
4-Bromophenyl ethyl ether (4-bromophenetole), 137
Bromoform, 1, 113, 125
Bromoxynil, 208–210
Butylphosphonic acid, 237

C

Carbamates, 189
Carbaryl, 194–196
Carbazole, 218
Carbetamide, 193–195
Carbofuran, 195–196
Carbon monoxide formation, 253
Carbon tetrachloride, 113, 114, 125, 127, 129, 271–272

Carboxylic acids, 70
Catalyst preparations, 16
Catecholamines, 213
Catechols, 41
Cationic surfactants, 92
CdS as photocatalyst, 52, 54, 132, 154,
    156, 247, 256
Cellulose, 98
Charge transfer to solvent (CTTS)
    transitions, 41, 43, 49, 51
Chloral hydrate, 114
Chlordane, 166
Chlorinated ethanes, 117
Chlorinated ethenes (ethylenes), 120
Chlorinated solvents, 271
Chlorine, 1
    bleaching of wood pulp, 161
Chlorine dioxide treatment, 97, 113
1, 2-Chlorine shift in carbon-centered
    radicals, 120
Chloroacetic acid, 17, 114, 122
Chloroacetic acids, 71, 122
Chloroalkyl ethers, 123
4-Chloroaniline, 77, 189
Chlorobenzenes, 132
Chlorobenzoic acids, 136
Chlorocarbon compounds, 113
Chlorocarbon mixtures, 115
4-Chlorocatechol, 16, 157
2-Chlorodibenzo-p-dioxin, 161
2-Chloroethyl ether, 123
Chloroform, 23, 113–116, 129, 158,
    271
Chloronaphthalene, 273
Chlorophenols, 138, 144, 148
    dye-sensitized photooxidation, 142,
        143
2-Chlorophenol, 22, 81, 146, 148, 253
3-Chlorophenol, 139, 148
4-Chlorophenol, by excimer irradiation,
        28, 143
    by photo-Fenton reaction, 76, 144,
        145
    by photoelectrolysis, 157
    by photohydrolysis, 139
    by UV/ozone, 140
    dye sensitized, 268
    TiO$_2$ photocatalyzed, 10, 11, 16, 22, 26,
        81, 149–152
Chlorophenoxy herbicides, 146
3-Chlorosalicylic acid, 12, 14, 135
Chloroxynil, 208–210

Chlorpyrifos, 234
Cholesterol, 265
Cholinesterase inhibitors, 194, 227,
    237
Chromium(III) and (VI), 54–55, 57
    in natural waters, 263
Clethodim, 247
Cobalt, 55
Cobalt (II) tetrasulfo phthalocyanine-TiO$_2$,
    42
Concentrated sunlight, 29, 271
Conduction-band electrons, 6
Contaminants in groundwater, 3
Continuous flow reactor, 23
Copper, Cu(II) to Cu(0), 56–57
Copper(II) nitrilotriacetate, 191
Corona Discharge, 285
Cost estimates, 286–288
Cresols, 86–87, 265, 268
Cyanate ion, 48
Cyanide ions, 48
Cyanuric acid, 186
Cyclic acetals, 91
Cyclohexanol, 77
Cyclohexanone, 77
Cyclophosphamide, 181
Cystine, 259

D

DABCO, 77, 142, 234
DDT, 165
Dead Sea brine, 267
Dependency of reactions rates on light
    intensity, 30
Dequalinium chloride, 200
Detergents, 92
Deuterium oxide solvent isotope effect, 12,
    77, 234, 268
Diaminocarboxylic acids, 202
Dibenzothiophene, 248
1,2-Dibromo-3-chloropropane (DBCP),
    127–129
1,2-Dibromoethane (ethylene dibromide,
    EDB), 114, 126–127
1,2-Dibromopropane, 127
Dibutyltin oxide, 59
Dichloroacetaldehyde, 114
Dichloroacetic acid, 114, 115, 122–123,
    271, 274
Dichlorobenzenes, 133
Di-n-butyl phthalate, 98

1,2-Dichloroethane, 114
1,1-Dichloroethylene, 271, 273
1,2-Dichloroethylene, 273
2,6-Dichloroindophenol, 208
Dichloromethane, 114, 116
Dichlorophenols, 11, 144
2,4-Dichlorophenol, 115, 134, 141,
  153
2,6-Dichlorophenol, 154
3,5-Dichlorophenol, 141
2,4-Dichlorophenoxyacetic acid (2, 4-D),
  146, 152
1,2-Dichloropropane, 115
1,3-Dichloropropane, 115
Dichlorvos, 228, 236
Diethyl benzylphosphonate, 234–235
Diethyl phthalate, 98
Diethylenetriamine-pentaacetate, 190
1,1-Difluoro-1, 2, 2-trichloroethane
  (FC 122), 129
1,1-Difluoro-1, 2-dichloroethane (FC132b),
  129
3,5-Dihydroxybenzoic acid, 41
Dimethoxybenzenes, 88
Dimethyl 2, 2, 2-trichloro-1-hydroxyethyl
  phosphonate, 237
Dimethyl ammonium phosphodithioate,
  236
5,5'-Dimethyl-1-pyrroline-N-oxide (DMPO),
  3–4, 45, 87, 131–132,
  243
Dimethyl vinyl phosphate, 228
Dimethylamine, 99
2,4-Dinitrophenol, 81, 205
2,4-Dinitrotoluene, 203
Di-*iso*-octyl phthalate, 97
Dioxins, see Polychlorinated dioxins
Diphenylmercury, 60
Direct and indirect photodegradation, 14,
  253
Direct Blue, 1, 214
Direct electron transfer, 6
Direct photolysis of water, 41
Disodium phenyl phosphate, 231, 234
DMPO spin adducts, 4, 45, 87, 131–132,
  243
Dodecyl benzene sulfonate, 92
Dodecylpyridinium chloride, 93
Dopamine, 213
Dyes, 202, 213–217
  sensitized sterilization of water, 269
  in municipal wastewater, 216–217

E

Economics of photocatalytic treatment,
  286–288
Edifenfos, 231–232
Effluents from a pulp factory, 97
Electrochemical oxidation, 285
Electron capture reactions, 126
Electron spin resonance (EPR), 3, 73
Electron transfer by organic anions, 136
Electron-hole recombination, 6, 30
Endosulfan, 250
Energy requirements, 288
Eosin, sensitization by, 142, 205, 246
Ethanol, 73
2-Ethoxytetrahydropyran, 91
Ethyl dihydrogen phosphate, 227–228
Ethylbenzene, 82
Ethylene glycol, 261
Ethylenediamine tetraacetate (EDTA),
  190
Ethylenediaminetetra(methylenephosphonic
  acid) (EDTMP), 230
European Community directives, 82, 181
Eutrophication, 50, 253, 263
Excimer light sources, 28, 130, 187
Experimental techniques, 21
Explosives, 202, 207

F

Falling film reactor, 27, 271
Fe(III)-complexes of polycarboxylate ions,
  184, 196, 259
Fe(III)-EDTA, 190
Fe(III)+H_2O_2, 234–235
Fe(III)/Fe(II) reactions, 257
Fe(malonate)$^{2-}$, 259
Fe(OH)(citrate)$^-$, 259
Fe(oxalate)$^{2-}$, 259
Fenamiphos, 234
Fenarimol, 212
Fenitrothion, 231–233, 237
Fenton reaction, 144, 205, 215, 261
Ferro- and Ferricyanide, 48–50
Ferrous to ferric ions, 52
Ferric oxide (a-Fe_2O_3, hematite), 51, 54,
  72, 258, 262
  microcrystallites, 72
  colloidal, 51, 85
Fixed-bed photocatalyst, 23
Flavins, 213, 264

Flow-through photoreactor, 55, 135,
        186
Fluorene, 83
Fluorescein, 22, 81
Fluoroalkenes, 130
Fluorocarbons, 129–130
Fluorophenols, 152
Fluorotrichloromethane (FC 11 or
        Freon 11), 129
5-Fluorouracil, 181
Fog, cloud and rain waters, 260
Formaldehyde, 29
Formic acid, 30, 68, 69, 72
Fructose 6-phosphate, 229
Fulvic acid, 254
Furfuryl alcohol, 89

G

Gallic acid, 41
Geosmin, 90, 263
Glass fiber cloth coated with TiO_2, 23,
        135
Glass pipe reactor, 271, 287
Glucaric acid, 259
Glucaric acid-1,4-lactone, 259
Glucose 1-phosphate, 229
Glucose 6-phosphate, 229
Glycerophosphates, 229
Glycolaldehyde, 259
Glycolic acid, 259
Glyoxalic acid, 259–260
Goethite, 261–262
Gold, 57
Granulated activated carbon (GAC), 27,
        113, 130, 286–287
Groundwater remediation, 128–129, 130,
        159, 188, 209, 267, 268, 271–272

H

Haber-Weiss mechanism of Fe(II)
        oxidation, 262
Haloaromatic ethers, 137
Haloaromatics, 131
Halobenzenes, reaction with ·OH radicals,
        131
m-Halobenzene derivatives, 132
Halocarbon pesticides, 164
Halocarbons, 113
Halogenated peroxy radicals, 126
Halogenophenols, 138

2-Halogenophenols, sensitization by
        hydroquinone, 141
Halomethanes, 116
Hammett correlation, 152, 154
Heavy metal ions, 52
Hematite, see Ferric oxide
Hematite-oxalate photolysis, 257
Heterogeneous photodegradation, 81, 147,
        185
Hexachloroplatinate, 58
Hexacyano-ferrate(II), 48–50
Hexacyano-ferrate(III), 48–50
Hexadecyl trimethylammonium bromide,
        94
Humic substances, 253–254, 264
Hydrated electrons, 3, 51, 255
Hydrocarbon derivatives, 67
Hydrogen peroxide, 88, 114, 134, 141,
        146, 184, 271, 287
    in natural waters, 256
    in rain- and cloud-water, 260
    photolysis of, 42, 68–69
    production of, 41
Hydroquinone, 16, 76, 80, 141
4-Hydroxyazobenzene, 214
Hydroxycarboxylic acids, 259
Hydroxyethane-1,1-diphosphonate (HEDP),
        238
Hydroxyl radicals, 3, 89
    production of, 257
    reaction rates with chlorophenols, 141
    reaction rates with halobenzenes,
        131–132
    reaction rates with *s*-triazines, 183

I

Iminodiacetic acid, 190–191
Industrial waste-water, 29
    from pulp factory, 97, 155
    from resins factory, 274
    from textile dying, 216–217, 268
Inhibitory effects of chloride, sulfate and
        phosphate, 9
Inorganic ions and molecules, 9, 41
Intersystem crossing (ISC), 5, 244
Iron(III) and hydrogen peroxide, 234–235
Isoelectric point (pI), 15
Isopropyl methylphosphonofluoridate,
        228
Isoproturon, 198
Isoxaben, 192

K

Kinetic models, 8
Kolbe reaction, 70–72
Kraft wastewater, 95, 97

L

Landfill leachates, treatment of, 268
Langmuir adsorption coefficient, 8, 10
Langmuir-Hinshelwood model, 8–11, 84,
    86, 93, 94, 115, 133, 135, 150, 152,
    207, 216, 236, 238
Laser flash photolysis, 28
Lawrence Livermore National Laboratory,
    271, 287
Lead, 55
Lepidocrite, 262
Light-dark cycling, 29
Light intensity, effect on reaction rates, 30,
    123, 289
Lignin, 95
Lignin sulfonates, 95
Long-chain compounds, 92
Lumichrome, 264
Lumiflavin, 264

M

Malathion, 239
Maleic anhydride, 78, 90
Manganese, 55
Marine surface microlayers, 265
Mass transfer limitations, 26–27
Mechanisms, 3
Mercurochrome, 58
Mercury, 55–58
    dissolved gaseous, 263
    salts as sensitizers, 67
Metalaxyl, 201
Metanilic acid, 244
Metazachlor, 185
Methanol, 73
Methionine, 246
1-Methoxy-1-methyl-3-phenylurea
    (MMPU), 199
Methyl ethyl ketone, 115
2-Methylisoborneol, 90, 263
4-Methyl-2-nitrophenol, 206
Methyl orange, 215
Methyl vinyl ketone, 75
Methyl violet, 217

Methyl viologen, 217
Methylene blue, bleaching of, 20
    sensitization by, 142, 143, 197, 205,
        234, 246, 269, 274–275
    photocatalytic degradation, 246–247
Metobromuron, 199
Mirex, 166–167
Molecular oxygen, 3, 5
Molecular probes for reactive transients, 4
Molinate, 245
Monochloroacetic acid, see Chloroacetic
    acid
Monochlorophenols, 10, 143
Monophenylmercury chloride, 60
Monuron, 196–197
Mucondialdehyde, 81
Municipal wastewater treatment, 216–217,
    268
Musty odor and taste in freshwater, 90,
    263
Mutagen formation, 99–103, 124

N

Nanocrystalline titanium dioxide, 25, 157
Naphthalene, 22, 81, 83
2-Naphthol, 22, 81, 101
Naphthol blue black B, 29
Natural transformations in freshwater and
    oceans, 253
Nerve gases, 227
Nitrate ions, 44
Nitrate to nitrite, hydroxylamine and
    ammonia, 46
Nitrate-induced photodegradation, 256–257
Nitration, 99
Nitrilotriacetic acid, 190
Nitrite ion, 44
Nitrite to ammonia, 47
Nitrite to nitrate, 45
3-Nitroaniline, 77
4-Nitroaniline, 77, 189
Nitrobenzene, 182, 202, 205–206
Nitrobenzyl derivatives, 203
4-Nitrocatechol, 181
Nitrogen compounds, inorganic, 44
    organic, 181
2-Nitrophenol, 205–207
3-Nitrophenol, 204, 207
4-Nitrophenol, 203, 205–208
Nitrophenols, 202–208
Nitrophenyl phosphates, 231

Nitropolychlorodibenzo-*p*-dioxins, 163
Nitrosation, 99
Nitrosodimethylamine, 99
Nitrotoluenes, 202
Nonyl phenyl poly (oxyethylene), 92
Nonylphenol, 67, 78, 92, 93
Nonylphenol polyethoxylates, 78, 92
Noradrenaline, 213

O

Octachloro-dibenzo-*p*-dioxin (OCDD), 155,
    158–160
n-Octane, 91
Off-flavor in freshwater, 90, 263
Oil spills and oil slicks, 91, 248–249, 265
Open ponds, 29, 85
Open-flow continuous operation, 22
Orange G, 29, 214
Orange I, 29
Organic nitrogen compounds, 181
Organic phosphorus compounds, 227
Organic sulfur compounds, 243
Organo-metallic compounds, 58
Oxalic acid, 73, 260
Oxazine dyes, 254
Oxidation in supercritical water, 286
Oxone^R, see Potassium peroxymonosulfate
Ozonation, 1, 98, 182, 217, 283
Ozone pretreatment, 237–238, 245–246

P

PCBs, see Polychlorinated biphenyls
Paper production, 80
Parabolic trough solar concentrator, 22, 55,
    271, 273, 287
Paraoxon, 4-nitrophenyl diethyl phosphate,
    238
Parathion, 231
    sonolysis of, 284
Pendimethalin, 198–200
2,3,4,7,8-Pentachlorodibenzofuran,
    163
Pentachlorophenol, 143, 155–156, 158,
    268, 273, 283
Perchloroethylene, see Tetrachloroethylene
Periodate, see Sodium periodate
Permethrin, 165–166
Peroxy-cyclohexadienyl derivatives, 81
Peroxydisulfate, persulfate, 6, 115,
    134–135, 153, 239, 274, 287, 289

Phenol, by corona discharge, 285
    direct photolysis, 75–76
    electrooxidation, 285
    nitrosation of, 100,
    photocatalysis of, 12, 27, 81, 83–85,
        286, 288
Phenols, 257
p-Phenols, Hammett correlation, 152
Phenoxyacetic acid, 87
a-Phenoxylacetophenone, 95
p-Phenylenediamines, 189
Phenylethanamide, 182
Phenylmercury compounds, 60
2-Phenylphenol, 100
Phenylphosphonic acid, 237
Phloroglucinol, 265
Phophonofluoridates, 228
Phosphate ions inhibited photodegradation,
    15
Phosphomolybdates, 67
Phosphorus oxyanions, 50
Phosphotungstates, 67
Photocatalytic membranes, 24
Photocatalytic reactions, 2
Photocatalytic turnover number, 20
Photodechlorination, 135
Photodynamic sterilization, 269
Photoelectrochemical oxidation, 30, 72,
    157–158
Photo-Fenton reaction, 76, 88, 98, 129,
    144–145, 245, 258
Photographic processing, 80
Photo-Kolbe reaction, 70–72, 74, 122, 152
Photoreduction of metal salts, 53
Photosensitized reactions, 5
Photosensitized soil, 255
Phthalate esters, 97
Phthalic anhydride, 78, 90
Phytol, 265
Pilot plant scale solar photocatalysis, 271,
    273, 287
a-Pinene, 79
Plataforma Solar de Almeria, 55, 271,
    273
Platinum, 57
Point of zero charge, 15
Polluted groundwater, 267
Polycaproamide, 98
Polymethylmethacrylate, 98
Polynuclear aromatic hydrocarbons, 82
Polynuclear aromatic nitrogen heterocycles,
    218

Polynuclear aromatic sulfur heterocycles, 248
Polychlorinated biphenyls (PCBs), 158–159, 162
Polychlorinated dibenzo-*p*-dioxins (PCDDs), 155, 158–161
Polychlorinated dibenzofurans (PCDFs), 158, 163
Polychlorobenzenes in surfactant micelles, 135
Polyethylene, 98
Polyethylene glycols, 98
Polyethylene oxide, 98
Polymers, 98
Polyvinyl alcohol, 75, 98
Polyvinyl chloride (PVC), 98
Potassium bromate, 134, 239
Potassium peroxydisulfate, $K_2S_2O_8$, see Peroxydisulfate
Potassium peroxymonosulfate (Oxone[R]), 115, 134, 135
Pristane, 264
Promazine, 182
Prometon, 184, 186
Prometryn, 184, 186
Propachlor, 198–199
Propanal, 74
1-Propanol, 74
2-Propanol, 70
Propazine, 183
Propyzamide, 200–201
Pyrene, 103
Pyridine, 93, 182
Pyridoxal-5'-phosphate, 233
Pyrogallol, 41
Pyruvic acid, 260

Q

Q-sized ZnO, 42
Quantitative structure-activity relationship (QSAR)
    of $^1O_2$ reaction with phenols, 77
    for photohydrolysis of *m*-substituted halobenzenes, 132
Quantum yields, 19
Quinoline, 218

R

Radiolysis, 117–120, 150–151, 284
Rate determining process (RDP), 12

Reactive Black 5, 216
Reactivity constant, 8
Reactor design, 22
Reductive mechanism, 125
Regeneration of adsorbents, 27
Resmethrin (bioresmethrin), 255
Rhodamin B, by excimer irradiation, 29
Rhodamine 6G, 202
Riboflavin, 77, 78, 142, 184, 197, 213, 264
Role of oxygen, 13
Rose bengal, 77, 89, 142, 156, 187, 205, 234, 246
Rutile, see Titanium dioxide

S

Salicylic acid, 26, 81–82, 115, 271
Sandia National Laboratories, 271
Scavenging of the hydroxyl radical, 4
Seawater, dissolved organic matter (DOM), 264
    iron limitation in, 259
Sethoxydim, 247
Shallow lagoons and ponds, 29, 85
Silver, 57
Simazine, 182, 186
Singlet oxygen, 3, 77, 89, 142, 206, 213, 254–255, 268
Singlet oxygen inhibitors, 142
Slurry reactor, 21–22
Sodium azide, 142
Sodium benzenesulfonate, 94
Sodium dodecyl bis(oxyethylene) phosphate, 94
Sodium dodecyl sulfate, 94
Sodium dodecylbenzene sulfonate, 93
Sodium periodate, 121, 153, 239
Sodium 12-phenyldodecyl sulfonate, 93
Solar disinfection, 270
Solar furnace, reaction in, 273–274
Solar parabolic trough, 22, 271, 273, 287
Solar pilot plant, 271, 273
Solvent Red 1, 213–214
Spin trapping of hydroxyl radicals, 3
Sterols, 265
Strontium titanate, 54
Sugar acids, 259
Sugar phosphates, 229–230
Sulfide ions, 51–52
Sulfite ions, 51–52

Sulfonic acids, 243
Sulfur oxyanions, 51–52
Supercritical water, reaction in, 286
Superfund Site, 271
Superoxide dismutation reaction, 256
Superoxide radical anion, 3, 88, 128,
    255–256
Surface density of OH groups on titanium
    dioxide, 18
Surface properties, 15
Surface waters, 261
Surfactants, 92–94

T

Tannery dyes, 216
Tannic acid, 41, 257
Terbacil, 197–198
Terbutylazine, 187
Terpene compounds, 79
Tetrachloroethylene (perchloroethylene),
    17, 23, 114, 120–122, 271–273
2,3,7,8-Tetrachlorodibenzo-*p*-dioxin
    (TCDD), 159, 164
Tetrachloromethane, see Carbon
    tetrachloride
Tetrachlorvinphos, 236–237
Tetrasulfonated phthalocyanines, 267
Textile dying effluents, 213, 216–217, 268
Thallium, 55
Thioacetamide, 247
Thiocyanate, 50
Thioglycolic acid, 246
Thiolcarbamates, 245–246
4-Thiomethyl-N-methylphenylcarbamate,
    249
Thionex, 250
Thionone, 234
Thioridazine, 181
Thymine, 210–211
Tin oxide electrode doped with Sb, 285
Ti^{3+} ions, 14
Titanium dioxide, TiO$_2$, 5
    aerogels, 26
    anatase, 4, 16, 17, 20, 22, 23, 25, 27,
        30, 54, 83
    anchored on porous alumina, 24
    and peroxydisulfate, 115, 134–135, 274,
        287, 289
    aqueous slurries, 9
    as single crystal electrode, 158
    as stationary phase, 22–27

bactericidal activity of, 269–270
by sol-gel method, 72, 151, 186
ceramic membranes of, 55, 72
coated on reticulated alumina foam, 273
coated on oleophilic glass microbubbles,
    91
coated on fiberglass mesh, 23, 135
colloids and microbeads, 6, 57, 73, 74,
    186, 215, 235, 254
dye sensitized with rose bengal, 187
fixed on fiberglass cloth, 23, 117, 287
immobilized on beach-sand, 85
isoelectric point, 15
microcrystalline, 74
nanocrystalline, 25, 157
Nb-doped, 238
on cellulose, 24
on conductive glass, 30, 72
on glass fibers, 23
on photoelectrodes, 157, 158
on silica gel, 23
on optical fibers, 25, 151–152
on polyester membranes, 24
optically transparent electrode
    (OTE), 157
particles, HeLa cells attached to, 270
particles, ultrafine, 270
platinized, 16, 114, 115, 123, 236
P25, 21–22, 123, 133
rutile, 16, 17, 20, 22, 25, 83, 158
supported on glass, 72, 135, 157
thin-film electrodes, 30
Tolclofos-methyl, 233
Toluene, 115
Toluenes, 75–76, 88
p-Toluenesulfonic acid, 243
p-Toluidine, 189
Toxic industrial effluents, 288
Trialkyl phosphates, 236
s-Triazines, 182
Tributyl phosphate, 231
Tributyltin chloride, 59
Trichlorfon, 236
Trichloroacetic acid, 114, 115, 122
1, 2, 4-Trichlorodibenzo-*p*-dioxin, 161
1, 2, 4-Trichlorobenzene, 147
1, 1, 2-Trichloroethane, 115
1, 1, 2-Trichlorotrifluoroethane (CFC-113),
    129–130
Trichloroethylene, 17, 115, 271–273
Trichloromethane, see Chloroform
Trichlorophenols, 11, 153

2,4,6-Trichlorophenol, 153
2,4,5-Trichlorophenoxyacetic acid,
    (2,4,5-T), 146
Triclopyr, 211–212
Tricresyl phosphate, 231
Trietazine, 186
1,1,1-Trifluoro-2,2,2-trichloroethane (FC
    113), 129
1,3,5-Trihydroxybenzene, 265
Trihalomethanes (THMs), 113
2,4,6-Trimethylphenol (TMP), 265, 268
Trimethyl phosphate, 228, 236
Trimethyl phosphite, 236
2,4,6-Trinitrotoluene (TNT), 207
Trioctyl phosphate, 231
Triphenyl phosphate, 231
Tris(2,2'-bipyridine)-iron(II), 53
Tris(2-chloroethyl) phosphate, 231
Tris(chloropropyl) phosphate, 231
Tris(dichloropropyl) phospate, 231
Tryptophan, 213
Turnover numbers, 19
Two-electron reduction of oxygen, 42

U

U.S. EPA, drinking water standards, 271
Ultrasonics, 284
Upscaling to industrial processes, 21
Uracil derivatives, 196–198
Uranium, 55–56
Urea, 196

V

Vacuum UV photolysis, 130, 187
Vamidothion, 230

W

Waste water treatment, by excimer
    irradiation, 28–29
Weizmann Institute Central Receiver, 272,
    274
Wood pulp, 95–97
WO_3 as photocatalyst, 4, 54, 55, 132, 156,
    196, 215

X

Xenobiotic organic compounds, 255, 283,
    288
Xenon-excimer light source, 28, 130, 143,
    187

Z

Zeta potentials, 15, 94
ZnO as photocatalyst, 4, 12, 55, 57, 82, 93,
    97, 132, 133, 156, 187, 196, 204,
    206, 215, 254, 255–256, 258
    colloidal, 42, 254
    quantum-sized, 42
ZnO/WO_3 as photocatalyst, 56
ZnS as photocatalyst, 52, 54, 72